Interaction of Mechanics and Mathematics

Henning Struchtrup

Macroscopic Transport Equations for Rarefied Gas Flows

Approximation Methods in Kinetic Theory

With 35 Figures

 Springer

Author

Dr. Henning Struchtrup
University of Victoria
Department of Mechanical Engineering and
Institute for Integrated Energy Systems
PO Box 3055, STN CSC
Victoria, BC, V8W 3P6
Canada

ISSN 1860-6245

ISBN-10 3-540-24542-1 **Springer Berlin Heidelberg New York**
ISBN-13 978-3-540-24542-1 **Springer Berlin Heidelberg New York**

Library of Congress Control Number: 2005925870

Springer is a part of Springer Science+Business Media

springeronline.com

© Springer-Verlag Berlin Heidelberg 2005
Printed in Germany

Typesetting: Data conversion by the author.
Final processing by PTP-Berlin Protago-TEX-Production GmbH, Germany
Cover-Design: Erich Kirchner, Heidelberg
Printed on acid-free paper   89/3141 Yu - 5 4 3 2 1 0

Für Martina

Preface

An important parameter to describe rarefied gases is the Knudsen number Kn, defined as the ratio between the mean free path of the gas, i.e. the average distance traveled by a gas particle between two subsequent collisions, and a macroscopic length describing the flow, e.g. a channel width or the diameter of an object exposed to the flow.

The well known laws of Navier-Stokes and Fourier are applicable only for flows at sufficiently small Knudsen numbers, and fail in the description of flows at Knudsen numbers $Kn \gtrsim 0.05$. These Knudsen numbers are easily reached nowadays, e.g. in microscopic flows or in high altitude flight, and a reliable set of equations that can be solved at low computational cost for the description of these flows is highly desirable.

The basic equation for the description of rarefied gases is the Boltzmann equation which describes the microscopic behavior of the gas from a statistical viewpoint. The Boltzmann equation is valid for all Knudsen numbers, but—due to the detailed microscopic description—its numerical solution is expensive.

This text discusses classical and newer methods to derive macroscopic transport equations for rarefied gases from the Boltzmann equation, for small and moderate Knudsen numbers, i.e. at and above the Navier-Stokes-Fourier level. The resulting equations are compared and tested for a variety of standard problems. The classical methods, due to Chapman and Enskog, and to Grad, yield unsatisfactory equations, which are unstable in case of the Chapman-Enskog expansion (Burnett and super-Burnett equations), and describe unphysical discontinuous shocks in case of the Grad method. Only recently, the author was involved in developing alternative methods, which yield the regularized 13 moment equations (R13 equations) that avoid the shortcomings of the classical equations, but retain their benefits. Naturally, the methods for deriving the R13 equations, and the discussion of the equations are central points in the text.

This book is intended to be accessible not only to experts, but also the novice in kinetic theory, and thus develops the topic starting from the basic description of an ideal gas, and the derivation of the Boltzmann equation, followed by various methods for deriving macroscopic transport equations, and test problems, including shock waves and Couette flow. About forty end-of-chapter problems provide opportunity to deepen the understanding, and numerous references provide directions for further study.

My understanding of the topic benefitted immensely from discussions with many scientists and students, and I wish to thank in particular the following people and institutions: Marcello Anile, Jörg Au, Elvira Barbera, Maurice Bond, Wolfgang Dreyer, Gilberto Kremer, David Levermore, Luc Mieussens, Ingo Müller, Institute for Integrated Energy Systems (Victoria, BC), Institute for Mathematics and its Applications (Minneapolis), Natural Sciences and Engineering Research Council of Canada, Daniel Reitebuch, Tommaso Ruggeri, Adam Schuetze, TMR Project "Asymptotic Methods in Kinetic Theory", Toby Thatcher, University of Victoria, Wolf Weiss, and Yingsong Zheng.

The R13 equations originate from a collaboration with my friend Manuel Torrilhon, to whom I extend special thanks for the successful and exciting cooperation, and also for carefully proof-reading the manuscript.

Special thanks are also due to Marshall Slemrod, for many stimulating discussions over the last years, and for carefully reading and correcting the manuscript on behalf of the editors.

Finally, I thank my lovely wife, Martina Wanders-Struchtrup, and our sweet daughter, Nora, for their continuous understanding, patience, and love.

Victoria, BC   *Henning Struchtrup*
March 2005

Contents

1   Introduction 1
- 1.1   Contents and scope 1
- 1.2   What is an ideal gas? 6
- 1.3   Length scales 7
- 1.4   Pressure and energy 9
- 1.5   Estimates for viscosity and heat conductivity 10
- 1.6   Knudsen number and microscale effects 11
- Problems 12

2   Basic quantities and definitions 15
- 2.1   Phase space and phase density 15
- 2.2   Moments of the phase density 16
  - 2.2.1   Number, mass density, and velocity 16
  - 2.2.2   Energy, pressure, and temperature 17
  - 2.2.3   General moments 19
  - 2.2.4   The relevance of moments 21
- 2.3   Equilibrium and the Maxwellian distribution 21
- 2.4   Boundary conditions for the phase density 22
- Problems 24

3   The Boltzmann equation and its properties 27
- 3.1   The Boltzmann equation 27
  - 3.1.1   Space-time evolution of f 27
  - 3.1.2   Binary collisions 29
  - 3.1.3   Potential and differential cross section 31
  - 3.1.4   The Stosszahlansatz 32
  - 3.1.5   The Boltzmann equation 33
- 3.2   Equilibrium 34
- 3.3   Equation of transfer and conservation laws 35
  - 3.3.1   Equation of transfer 35
  - 3.3.2   Conservation laws 36
  - 3.3.3   Moment equations 38

3.4   Entropy and the H-theorem 38
       3.4.1   H-theorem 38
       3.4.2 $H = k \ln W$—entropy and disorder 39
       3.4.3   Gibbs paradox 40
       3.4.4   Equilibrium and maximum disorder 41
       3.4.5   Irreversibility 42
3.5   Collision frequency 43
3.6   Kinetic Models 45
       3.6.1   The BGK model 46
       3.6.2   The ES-BGK model 47
3.7   The Direct Simulation Monte Carlo method 48
Problems 49

4   The Chapman-Enskog method 53
4.1   Basic principles 53
       4.1.1   The closure problem 53
       4.1.2   Dimensionless and scaled Boltzmann and ES-BGK
               equations 55
       4.1.3   The Chapman-Enskog expansion 56
4.2   The CE expansion for the ES-BGK model 57
       4.2.1   General arguments 57
       4.2.2   Zeroth order: Euler equations 58
       4.2.3   First order: Navier-Stokes-Fourier equations 59
       4.2.4   Second order: Burnett equations 60
4.3   CE expansion for power potentials 62
       4.3.1   First order: Navier-Stokes-Fourier laws 62
       4.3.2   Second order: Burnett equations 67
       4.3.3   Third order: Super-Burnett equations 68
       4.3.4   Augmented Burnett equations 69
       4.3.5   BGK-Burnett equations 70
4.4   The second law 70
       4.4.1   The second law to first order 70
       4.4.2   The H-theorem in higher order expansions 71
Problems 73

5   Moment equations 75
5.1   Moment equations as infinite coupled system 75
5.2   Equations for trace-free central moments 77
       5.2.1   Generic moment equation 77
       5.2.2   Conservation laws 78
       5.2.3   Scalar moments 79
       5.2.4   Vector moments 79
       5.2.5   Rank-2 tensor moments 80
       5.2.6   General equation 80

5.3   Production terms 80
        5.3.1   BGK model 81
        5.3.2   ES-BGK model 81
        5.3.3   Maxwell molecules.................................. 82
        5.3.4   Linear production terms............................ 84
        5.3.5   Eigenfunctions 85

6   Grad's moment method 87
    6.1   Closed moment systems by Grad's method.................. 87
            6.1.1   General outline 87
            6.1.2   Grad's 13 moment equations..................... 89
            6.1.3   Grad's 26 moment equations 91
            6.1.4   Extended moment sets 92
    6.2   Remarks on Grad-type equations 93
            6.2.1   Large moment numbers 93
            6.2.2   Moment systems and its competitors............. 95
            6.2.3   Mathematical properties 95
    6.3   CE expansion of Grad's moment equations 96
            6.3.1   Grad and CE phase density for Maxwell molecules
                    and BGK model.................................. 96
            6.3.2   13 moments 97
            6.3.3   26 moments 99
            6.3.4   Maxwellian iteration 99
    6.4   Reinecke-Kremer-Grad method 101
    6.5   Moments of the collision term 102
    6.6   Entropy maximization/Extended Thermodynamics 104
            6.6.1   Brief outline of entropy maximization 104
            6.6.2   Properties and problems 105
            6.6.3   Linearization and Grad method 106
    Problems 107

7   Regularization of Grad equations 109
    7.1   Grad distributions as pseudo-equilibrium manifolds 109
    7.2   Basic equations 111
    7.3   Expansion around $f_{|13}$ 112
    7.4   Euler and Navier-Stokes-Fourier equations 114
    7.5   Linearized equations 115
    7.6   Discussion 116
    7.7   Jin-Slemrod regularization 118
            7.7.1   Basic equations 118
            7.7.2   Comparison with R13 equations 119

8   Order of magnitude approach 123
   8.1   Introduction 123
   8.2   The order of magnitude of moments 126
      8.2.1   Zeroth and first order expansion.................... 126
      8.2.2   Second order....................................... 127
      8.2.3   Minimal number of moments of order $\mathcal{O}\left(\varepsilon\right)$ 127
   8.3   The transport equations with 2nd order accuracy 129
      8.3.1   The conservation laws and the definition of $\lambda - th$
           order accuracy 129
      8.3.2   Zeroth order accuracy: Euler equations.............. 129
      8.3.3   Collision moments for vectors and tensors 130
      8.3.4   Equations for pressure deviator and heat flux 130
      8.3.5   First order accuracy: Navier-Stokes-Fourier equations .. 131
      8.3.6   Second order accuracy: 13 moment theory 133
      8.3.7   Burnett equations 135
   8.4   Third order accuracy: R13 equations 136
   8.5   Discussion 138
      8.5.1   Higher order accuracy............................. 138
      8.5.2   Comparison with Chapman-Enskog method........... 139
      8.5.3   Comparison with the Grad method 140
      8.5.4   Comparison with the original derivation of the R13
           equations.. 141
      8.5.5   Comparison with consistently ordered extended
           thermodynamics 142
      8.5.6   Comparison with Jin-Slemrod equations.............. 143

9   Macroscopic transport equations for rarefied gas flows 145
   9.1   Relations between the equations........................... 145
   9.2   3-D non-linear equations 146
      9.2.1   Conservation laws 146
      9.2.2   Chapman-Enskog expansion 147
      9.2.3   Moment equations for Maxwell molecules............. 150
      9.2.4   Moment equations for general molecule types 152
   9.3   One-dimensional equations 153
      9.3.1   Conservation laws 154
      9.3.2   Chapman-Enskog expansion 154
      9.3.3   Moment equations for Maxwell molecules............. 155
      9.3.4   Moment equations for general molecule types 157
   9.4   Linear dimensionless equations 157
      9.4.1   Conservation laws 158
      9.4.2   Chapman-Enskog expansion 158
      9.4.3   Moment equations for Maxwell molecules............. 159
      9.4.4   Moment equations for general molecule types 160
   Problems 160

10 Stability and dispersion 161
  10.1 Linear stability 161
    10.1.1 Plane harmonic waves 161
    10.1.2 Linear one-dimensional equations 163
    10.1.3 Euler equations, speed of sound 163
    10.1.4 Linear stability in time 164
    10.1.5 Linear stability in space 166
    10.1.6 Discussion 166
  10.2 Dispersion and Damping 168
  10.3 Stability analysis for the ES-BGK Burnett equations 170
  Problems 172

11 Shock structures 175
  11.1 The 1-D shock structure problem 175
    11.1.1 Basic definitions 175
    11.1.2 Conservation laws and Rankine-Hugoniot relations 176
    11.1.3 Entropy production over the shock 177
    11.1.4 Shock thickness and asymmetry 178
    11.1.5 Shocks and Knudsen number 179
    11.1.6 Solution method 179
  11.2 Comparison with DSMC results 181
    11.2.1 Failure of classical methods 182
    11.2.2 Maxwell molecules 183
    11.2.3 Hard Spheres 185
    11.2.4 Jin-Slemrod equations 186
  11.3 Solution behavior 188
    11.3.1 Transition from Grad's 13 moment equations 188
    11.3.2 Temperature overshoot 189
    11.3.3 Shock thickness 190
    11.3.4 Shock asymmetry 192
    11.3.5 Positivity of the phase density 192
  11.4 Concluding remarks 194

12 Boundary value problems 197
  12.1 Boundary conditions for moments 197
    12.1.1 Basic considerations 197
    12.1.2 Tangential momentum 200
    12.1.3 Energy flux 200
    12.1.4 Maxwell-Smoluchowski boundary conditions 201
    12.1.5 Knudsen layer correction 202
  12.2 Plane Couette flow 203
    12.2.1 Couette geometry and conservation laws 203
    12.2.2 Navier-Stokes-Fourier equations 204
    12.2.3 Grad 13 equations 206
    12.2.4 R13 equations 207

12.3 Bulk equations 208
12.4 Linear Knudsen boundary layers 210
    12.4.1 Scaling and Knudsen layers 210
    12.4.2 Navier-Stokes-Fourier and Grad 13 equations 211
    12.4.3 Burnett equations 211
    12.4.4 Super-Burnett and augmented Burnett equations 212
    12.4.5 Regularized 13 moment equations 214
    12.4.6 The heat flux parallel to the flow 214
    12.4.7 26 and more moments 216
12.5 Superpositions 217
12.6 A boundary condition for normal stress 225
12.7 Further discussion 227
Problems 228

A   Appendix 229
A.1 Tensor index notation 229
A.2 Symmetric and trace-free tensors 231
    A.2.1 Symmetry 231
    A.2.2 Trace-free tensors 231
    A.2.3 Spherical harmonics 233
A.3 Integrals of Gaussians and Maxwellians 234
    A.3.1 Gaussian integrals 234
    A.3.2 Integrals of the Maxwellian 235
    A.3.3 Half-space moments of the Maxwellian 236
    A.3.4 Integrals of the anisotropic Gaussian 238
A.4 The integrals (5.21) 239
A.5 Lagrange multipliers 240
A.6 Equations for the computation of generalized 13 moment
    equations 241
    A.6.1 Moment equations for w_i^a 241
    A.6.2 Moment equations for w_{ij}^a 242
    A.6.3 Coefficients in (8.18, 8.19) 244

References 247

Index 255

1

Introduction

1.1 Contents and scope

A gas at standard conditions (1bar, 25°C) contains ca. 2.43×10^{16} particles per cubic millimeter. Despite this huge number of individual particles, a wide variety of flow and heat transfer problems can be described by a rather low number of partial differential equations, namely the well-known equations of Navier-Stokes and Fourier. Due to the many collisions between particles which effectively distribute disturbances between the particles, the particles behave not as individuals, but as a continuum.

Under standard conditions a particle collides with the others very often, about 10^9 times per second, and travels only very short distances between collisions, about 5×10^{-8}m. Both numbers, known as collision frequency ν and mean free path λ, depend on the number density of the gas. Flow problems in which the typical length scales L are much larger than the mean free path λ, or in which the typical frequencies ω are much smaller than ν, are well described through the laws of Navier-Stokes and Fourier. The Knudsen number Kn = $\lambda/L = \omega/\nu$ is the relevant dimensionless measure to describe these conditions, and the Navier-Stokes-Fourier equations are valid as long as Kn $\ll 1$.

This condition fails to hold when the relevant length scale L becomes comparable to the mean free path λ. This can happen either when the mean free path becomes large, or when the length L becomes small. A typical example of a gas with large mean free path is high altitude flight in the outer atmosphere, where the mean free path must be measured in meters, not nanometers, and the Knudsen number becomes large for, e.g., a spacecraft. Miniaturization, on the other hand, produces smaller and smaller devices, e.g. micro-electro-mechanical systems (MEMS), where the length L approaches the mean free path.

Moreover, the Navier-Stokes-Fourier equations will fail in the description of rapidly changing processes, when the process frequency ω approaches, or exceeds, the collision frequency ν.

The Knudsen number is used to classify flow regimes as follows:

- $Kn \lesssim 0.01$: The hydrodynamic regime, which is very well described by the Navier-Stokes-Fourier equations.
- $0.01 \lesssim Kn \lesssim 0.1$: The slip flow regime, where the Navier-Stokes-Fourier equations can describe the flow well, but must be supplied with boundary conditions that describe velocity slip and temperature jumps at gas-wall interfaces.
- $0.1 \lesssim Kn \lesssim 10$: The transition regime, where the Navier-Stokes-Fourier equations fail, and the gas must be described in greater detail, e.g. by the Boltzmann equation, or by extended macroscopic models.
- $Kn \gtrsim 10$: Free molecular flow, where collisions between the particles do not play an important role, and the flow is dominated by wall/particle interactions.

Rarefied gases are gases outside the hydrodynamic regime, i.e. with $Kn \gtrsim 0.01$. For Knudsen numbers in $0.01 \lesssim Kn \lesssim 1$, the gas still behaves as a continuum, but the Navier-Stokes-Fourier equations loose their validity, and must be replaced by more refined sets of continuum equations which describe the behavior of the gas.

Approximation methods to derive equations that allow to describe processes in rarefied gases, and the evaluation of the resulting equations, are the main topic of this text. Particular emphasis is put on understanding the relations between the different methods, and between the various sets of equations that result from these. Most methods rely on expansions in the Knudsen number Kn, and therefore yield equations that cannot cover the full transition regime, but are restricted to $0.01 \lesssim Kn \lesssim 1$.

A rarefied gas is well described by the Boltzmann equation which describes the gas on the *microscopic* level by accounting for the translation and collisions of the particles. The solution of the Boltzmann equation is the phase density f which is a measure for the likelihood to find atoms at a location $\mathbf{x}$ with microscopic velocities $\mathbf{c}$. The Boltzmann equation is the central equation in the *kinetic theory of gases*.

Macroscopic quantities such as mass density ρ, mean velocity $\mathbf{v}$, temperature T, pressure tensor[1] $\mathbf{p}$, and heat flux vector $\mathbf{q}$ are weighted averages of the phase density, obtained by integration over the microscopic velocity. One way to compute the macroscopic quantities is to first solve the Boltzmann equation, and then integrate over its solution, f. Alternatively, rational methods can be used to deduce *macroscopic transport equations* from the Boltzmann equation, that is transport equations for the macroscopic quantities ρ, $\mathbf{v}$, T, etc. This is particularly suitable for processes at small and moderate Knudsen numbers, which, as it turns out, can be described by a small number of equations.

The phase density and its moments—the macroscopic quantities—are introduced in **Chapter 2**.

---

[1] The pressure tensor has a different sign than the stress tensor $\mathbf{t}$ of fluid dynamics, $\mathbf{p} = -\mathbf{t}$.

The Boltzmann equation and its properties, such as its equilibrium states, the conservation of mass, momentum and energy, and the second law of thermodynamics, are presented and discussed in **Chapter 3**.

The classical method for the derivation of macroscopic equations for rarefied gases is the Chapman-Enskog (CE) method, which relies on an expansion of the phase density around equilibrium in terms of the Knudsen number, $f = f_M + \mathrm{Kn} f^{(1)} + \mathrm{Kn}^2 f^{(2)} + \cdots$, where f_M is the equilibrium phase density, known as Maxwellian distribution. The CE method gives the Euler equations at zeroth order, and the Navier-Stokes-Fourier equations at first order, with explicit expressions for viscosity and heat conductivity. Both sets of equations are cornerstones of gas dynamics in engineering applications.

Unfortunately, the success of the Chapman-Enskog method at zeroth and first order is not continued towards higher order expansions, which yield the Burnett and super-Burnett equations at second and third order, respectively. Both sets of equations suffer from instabilities in transient processes (at high frequencies or small wavelengths) and from unphysical oscillations in steady state processes.

The Chapman-Enskog method is discussed in **Chapter 4**, where it is applied first to the ES-BGK equation, which shares the main features with the Boltzmann equation, but allows an easier, and more transparent, application of the method. Subsequently, the application of the CE method for the full Boltzmann equation is sketched, and the resulting macroscopic equations up to third order are listed.

The Boltzmann equation can be replaced by an infinite set of coupled moment equations, which follows from averaging the Boltzmann equation over a complete set of functions in the microscopic velocity **c**. The infinite set, presented and discussed in **Chapter 5**, is equivalent to the Boltzmann equation and can be used alternatively as a base for finding macroscopic transport equations.

A well-established approach to moment equations is Grad's method, which truncates the infinite system to a finite number of equations and then uses an approximation for the phase density—the Grad distribution $f_{|G}$—to close the system. Best-known is Grad's system of 13 equations with the variables ρ, **v**, T, **p**, **q**, but the method can be applied to arbitrary sets of moments. The method is introduced and performed in **Chapter 6**. There, the relation between the Grad equations and the equations derived from the Chapman-Enskog expansion is extensively discussed. Indeed, the latter can be derived from the Grad equations by a CE expansion of the moments, which implies that the Grad method yields equations at higher orders in the Knudsen number (in a rather unspecific sense).

Other than the higher order equations from the Chapman-Enskog method, i.e. the Burnett and super-Burnett equations, Grad-type equations are stable in transient processes, and thus offer an alternative to higher order equations. However, they form hyperbolic equations, which implies finite transport velocities, and thus discontinuous shocks for velocities that exceed the maximum characteristic velocity of the equations. **Chapter 7** presents a method to regularize Grad-type moment equations by means of a Chapman-Enskog expansion around the Grad distribution $f_{|G}$ instead of expanding around the equilibrium distribution f_M. The method, pioneered by Grad and developed further by M. Torrilhon and the author, is performed on Grad's 13 moment set, and yields the regularized 13 moment, or R13, equations. These are stable, yield smooth shock structures at all velocities, and the Navier-Stokes, Burnett, and super-Burnett equations can be extracted by means of a Chapman-Enskog expansion.

The R13 equations have desirable properties, but a closer look shows that they have two drawbacks: (a) Their derivation takes Grad's moment method for granted, and it would be preferable to have a derivation that is independent of Grad's method, while relating the equations directly to orders in the Knudsen numbers. (b) The R13 equations are derived for a particular model for molecular collisions—interaction with an inverse fifth power potential (Maxwell molecules)—and their generalization to arbitrary collision potentials would be desirable.

Chapter 8 presents an alternative method for deriving macroscopic transport equations that can be performed for any type of molecular interaction, and is based on accounting for the order of magnitude of moments and terms in moment equations through powers in the Knudsen number. For Maxwell molecules, the Euler and NSF equations are obtained in zeroth and first order, Grad's 13 moment equations in second order, and (a variation of) the R13 equations in third order. For non-Maxwellian molecules, the method is developed to second order, where it gives a generalization to Grad's 13 moment equations. The application to third order is discussed, but not performed.

The different sets of transport equations that are derived in Chapters 4-8 are collected in **Chapter 9**, which presents the full non-linear three-dimensional equations along with their one-dimensional form, as well as the linearized dimensionless equations in three space dimensions. Two tables are used to clarify the relations between the different models.

Chapters 10 and 11 present the application of the various equations to standard test problems, namely stability, dispersion and damping (Chapter 10), and one-dimensional shock waves (Chapter 11). Here, the aforementioned strengths and weaknesses of the various equations are proven, e.g. the instability of the Burnett and super-Burnett equations, the occurrence of discontinuities in shock calculations for Grad's equations, and that the R13 equations are stable and yield smooth shock structures.

Boundary value problems are discussed in **Chapter 12**, which begins with the derivation of the standard jump and slip boundary conditions for the Navier-Stokes-Fourier equations. Then the ability of the various sets of macroscopic equations to describe boundary value problems in the transition regime is examined. Particular emphasis is put on linear Knudsen boundary layers, and on non-linear rarefaction effects, e.g. a heat flux not driven by a temperature gradient in Couette flow. Complete boundary conditions are not available for equations above the Navier-Stokes-Fourier equations, and comparisons with solutions of the Boltzmann equation must rely on fitting constants of integration. These comparisons show that higher order models, in particular the R13 equations, can describe boundary value problems in the transition regime very well. The development of boundary conditions for macroscopic models is discussed, but no definitive answers are given.

The equations in the book consider only single-constituent monatomic ideal gases, the standard material in kinetic theory. Of course, kinetic theory is not restricted to monatomic gases, and is applied successfully to a host of materials such as diatomic and polyatomic gases, mixtures, electrons in semiconductors, thermal radiation, phonons, etc. The methods studied here can be used for these materials as well. However, the monatomic ideal gas, and in particular the gas of Maxwell molecules, allows to study the methods most easily. Other applications are not included due to lack of space.

The methods are nontrivial, even for a monatomic ideal gas, and require some knowledge in vector algebra and calculus. The **Appendix** provides some necessary background on tensor index notation and the formalism of trace-free tensors, and on the computation of integrals that appear frequently in the development.

The new methods and ideas presented here are still under development, and there are many presently open problems that must be addressed in the future. These include

- Reliable boundary conditions for all models above the Navier-Stokes-Fourier equations
- Equations at third order for non-Maxwellian molecules.
- Applications to mixtures, diatomic and polyatomic gases.
- Applications to non-classical gases, including photons (thermal radiation), electrons in semiconductors, and phonons.
- Hybrid models, which combine solutions of the Boltzmann equation in regions of large Knudsen numbers, to macroscopic models in regions with lower Knudsen numbers.
- Proper entropy inequalities (H-theorem) for higher order models.

The reminder of this chapter discusses monatomic ideal gases, mean free path, Knudsen number, and rarefaction effects in a rather elementary way. With the inclusion of this and the elementary chapters on phase density (Chapter 2) and Boltzmann equation (Chapter 3), the book should be accessible to the novice in kinetic theory. About forty end-of-chapter problems are intended to help the reader in deepening her understanding of the concepts. Moreover, the reader is encouraged to follow the derivations with his pencil in hand—while the main steps are described, many details had to be left out, and these should be considered as implicit problems, in addition to the explicitly stated end-of-chapter problems.

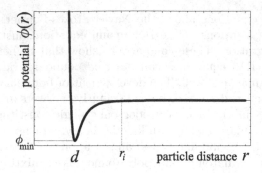

Fig. 1.1. Typical interaction potential $\phi(r)$ between two particles.

The derivation of macroscopic transport equations is only one topic in kinetic theory, and there are many excellent books and handbook articles that cover the topic from a wide variety of viewpoints, see, e.g., [1]-[21].

1.2 What is an ideal gas?

The material presented in this book concerns monatomic ideal gases, and we start by discussing under what circumstances a gas can be considered as ideal.

While in a liquid or a solid the atoms are in constant contact with each other, and exchange energy and momentum at any instant, the molecules in a gas move most of the time in free flight, interrupted by short interactions which we shall call collisions. The particles exchange energy and momentum only during the collisions, and the details of this follow from the interaction potential ϕ. The latter depends on the particle distance r and is typically of the form depicted in Fig. 1.1, where d can be considered as an effective particle diameter. At distances larger then several diameters d the interaction potential is virtually zero, and we can say that the particles do not feel each others presence at large distances.

We introduce the interaction radius r_i as a measure for the maximum distance at which the particles feel each other, and the mean free path λ as the average distance a particle travels in free flight between two collisions.

A gas can be considered as being ideal, if the particles are in free flight most of the time, that is when

$$\frac{r_i}{\lambda} \ll 1 \, .$$

Alternatively, we can consider the mean time for a collision τ_c and the mean time of free flight τ, which leads to the condition

$$\frac{\tau_c}{\tau} \ll 1 \, .$$

The interaction radius r_i must be chosen as that radius where the interaction potential ϕ is sufficiently small, that means small compared to the mean kinetic energy $\bar{e}_p$ of the particle.

An alternative requirement for a gas to behave ideal is that the average interaction potential $\bar{\phi}$ must be small compared to the mean kinetic energy,

$$\frac{\phi}{\bar{e}_p} \ll 1 \, .$$

Obviously, a gas will be an ideal gas if the average particle distance l_d is large, that is at low densities. However, the last condition implies that a hot gas, where the particle energies $\bar{e}_p$ are high, can behave ideally even at larger densities. Then, the potential must be very strong, corresponding to very small distances r, for the particles to feel each other.

1.3 Length scales

In order to gain a better idea of the different lengths introduced, we now have a look at some numbers. For this we consider an ideal gas under standard environmental conditions, that is at the pressure $p_0 = 1$ bar and the temperature $T_0 = 298$K. The ideal gas law relates pressure p, total volume V, total number of particles N, and temperature T as

$$pV = NkT \tag{1.1}$$

where $k = 1.38066 \times 10^{-23}$J/K is Boltzmann's constant. The average volume available for one particle is the inverse of the number density n, i.e.

$$\frac{1}{n} = \frac{V}{N} = \frac{kT}{p} = 4.114 \times 10^{-26} \text{m}^3 \, .$$

This corresponds to a cube with the side length

$$l_d = \sqrt[3]{\frac{1}{n}} = 3.452 \times 10^{-9}\text{m} \, ,$$

which gives a good measure for the average distance of particles at standard conditions.

The effective diameter of gas particles depends on the type of gas and is typically in the range [1]

$$d = 2 \cdots 6 \times 10^{-10}\text{m} \, ,$$

that is one order of magnitude less then the distance l_d.

Next we determine the mean free path λ by means of a simple argument for a gas of hard spheres of diameter d [18]. The mean speed of the particles

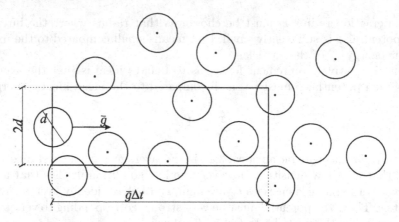

Fig. 1.2. On computing the mean free path λ. The particle with velocity $\bar{g}$ will collide with all particles with center in the cylinder $\pi d^2 \bar{g} \, \Delta t$.

is denoted as $\bar{c}$ and the mean relative speed between two particles is $\bar{g} = \sqrt{2}\bar{c}$ (Problem 2.6). To compute the mean free path, we consider one particle moving through the gas as depicted in Fig. 1.2. During the time Δt the particle would collide with any particle with center in the volume $\pi d^2 \bar{g} \, \Delta t$. The number of collisions during Δt follows through multiplication with the number density n as $n\pi d^2 \bar{g} \, \Delta t$ collisions during Δt. If this expression is set equal to one, $\Delta t = \tau$ is just the average time for one collision,

$$\tau = \frac{1}{n\pi d^2 \bar{g}} \, .$$

This is the mean free time, and its inverse is the mean collision frequency, $\nu = 1/\tau$. The mean free path—the average distance traveled during the time τ—is then given by

$$\lambda = \tau \bar{c} = \frac{1}{\sqrt{2}\pi d^2 n} \, . \tag{1.2}$$

The mean free path at standard conditions is therefore in the range

$$\lambda = 2.275 \times 10^{-7}\mathrm{m} \, \cdots \, 2.53 \times 10^{-8}\mathrm{m} \, ,$$

depending on the value of d. This shows that the mean free path is one to two orders of magnitude larger than the mean particle distance, and two to three orders of magnitude larger than the particle diameter. Thus $d/\lambda \ll 1$, so that a gas at standard conditions can be considered as an ideal gas.

It must be remarked, however, that this simple argument cannot be used for all gases with the obvious example of water vapor: the dipole structure of the molecules introduces long range electrostatic forces so that the interaction potential is large and the relevant interaction radius r_i is considerably larger than the actual molecule size d, so that r_i/λ cannot be ignored anymore.

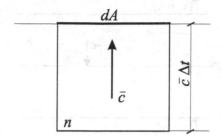

Fig. 1.3. On the number of particles crossing an area dA.

1.4 Pressure and energy

We assume a gas at number density n where all particles have the same speed $\bar{c}$, and 1/6 of the particles moves into each of the 6 directions of space.[2] Then, as can be seen from Fig. 1.3, the particle flux in one direction through an area element dA during the time Δt is given by $\frac{n}{6}\bar{c}\,dA\,\Delta t$.

We use this to relate pressure and particle velocity [16]: A particle colliding with a wall perpendicular to its direction of flight will change its velocity by $2\bar{c}$ and its momentum by $2m\bar{c}$ where m denotes the particle mass. Thus the total momentum exchanged between wall and particles during Δt is $\frac{n}{6}\bar{c}\,dA\,\Delta t\,2m\bar{c}$. Division by $dA\,\Delta t$ yields the force per unit area, that is the pressure, as

$$p = \frac{1}{3}nm\bar{c}^2 \,.$$

The ideal gas law (1.1) can be written as $p = nkT$ and we can identify

$$\bar{c} = \sqrt{3\frac{k}{m}T} \,, \tag{1.3}$$

so that the mean particle velocity is directly related to the temperature. The density of internal energy of the gas is given by the kinetic energy of the particles as $\rho u = n\frac{m}{2}\bar{c}^2$, where $\rho = nm$ is the mass density and u denotes the specific internal energy. Using the above relations we find

$$u = \frac{1}{2}\bar{c}^2 = \frac{3}{2}\frac{k}{m}T = \frac{3}{2}\frac{p}{\rho} \,. \tag{1.4}$$

These relations, although derived under simplifying assumptions, are valid for monatomic gases indeed, as will be seen later.

Finally we note that the particle mass m (with unit kg) is related to the molecular weight (with unit kg / kmol) as $m = M/A$ where $A = 6.0221367 \times 10^{23}\,\text{mol}^{-1}$ is Avogadro's constant, i.e. the number of particles in one mole.

---

[2] Forwards-backwards/up-down/left-right.

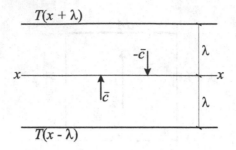

Fig. 1.4. On the computation of the heat flux.

1.5 Estimates for viscosity and heat conductivity

The value of the mean free path plays an important role for the behavior of a gas, and first insight into this can be obtained by developing rough estimates for viscosity and heat conductivity. Later chapters will show how to compute these quantities exactly.

We return to the simple gas model of the last section, for which the mass flow density of particles in one of the six directions is

$$J = \frac{1}{6}mn\bar{c} = \frac{1}{6}\frac{p}{\frac{k}{m}T}\bar{c} \,. \tag{1.5}$$

We consider a gas with a temperature gradient as depicted in Fig. 1.4, and ask for the net energy transport through the layer labeled as $x - x$ in the figure. The crossing particles had their last collision about at a distance λ away from x and carry the temperature of the location where they last collided. The net energy flow through the layer per unit area is therefore

$$q = J\left(u\left(x - \lambda\right) - u\left(x + \lambda\right)\right) = J\left(\frac{3}{2}\frac{k}{m}T\left(x - \lambda\right) - \frac{3}{2}\frac{k}{m}T\left(x + \lambda\right)\right) \,.$$

When the temperature gradient is sufficiently small, the temperatures can be expanded in a Taylor series as $T\left(x \pm \lambda\right) \simeq T \pm \frac{dT}{dx}\lambda$ so that, by means of (1.1–1.5), the energy flux can be written as

$$q = -\frac{1}{2}\frac{\bar{c}p\lambda}{T}\frac{dT}{dx} = -\frac{k}{2\pi d^2}\sqrt{\frac{3}{2}\frac{k}{m}T}\frac{dT}{dx} = -\kappa\frac{dT}{dx} \,. \tag{1.6}$$

This linear relation between heat flux and temperature gradient is known as Fourier's law. κ is called the heat conductivity, and our simple argument gives $\kappa = \frac{k}{2\pi d^2}\sqrt{\frac{3}{2}\frac{k}{m}T}$ which implies that the heat conductivity depends on the properties of the atoms (m, d) and only on temperature, but not on pressure. This will prove to be true in the exact treatment as well.

To compute the viscosity μ, we consider a gas with a gradient in the macroscopic velocity $v(x)$ instead of the temperature gradient. A similar argument as above leads to an expression for the momentum flux through $x - x$ which is equal to the stress σ,

$$\sigma = J\left(v\left(x - \lambda\right) - v\left(x + \lambda\right)\right) = -\frac{m}{3\pi d^2}\sqrt{\frac{3}{2}\frac{k}{m}T}\frac{dv}{dx} = -\mu\frac{dv}{dx} \, .$$

This relation between velocity gradient and stress is known as the Navier-Stokes law. Also the viscosity $\mu = \frac{m}{3\pi d^2}\sqrt{\frac{3}{2}\frac{k}{m}T}$ depends only on temperature, but not on pressure.

The dimensionless ratio between both coefficients is known as the Prandtl number, for which our simplified theory yields

$$\mathrm{Pr} = \frac{5}{2}\frac{k}{m}\frac{\mu}{\kappa} = \frac{5}{3} \tag{1.7}$$

while the measured values for all monatomic gases is very close to $\mathrm{Pr} = \frac{2}{3}$. The difference, of course, is due to the oversimplified argument, which is nevertheless reasonable enough to recover most important features.

1.6 Knudsen number and microscale effects

The laws of Navier-Stokes and Fourier (NSF) are the most important laws to describe fluid flows and heat transfer and are routinely used by engineers. Thus it is not surprising that one of the early goals of kinetic theory was to compute their coefficients, i.e. viscosity μ and heat conductivity κ. The derivation of the NSF laws, which will be discussed in Chapters 4, 7 and 8, is only possible if the mean free path λ is sufficiently small compared to relevant characteristic length scales L. The dimensionless parameter to mathematically describe this condition is the Knudsen number

$$\mathrm{Kn} = \frac{\lambda}{L} \, .$$

As will be seen, the laws of Navier-Stokes and Fourier are only valid if $\mathrm{Kn} \ll 1$. If this condition is violated, non-local effects play a role, so that fluid flow and heat transfer problems can no longer be described by the NSF laws, and a more refined description of the gas is required.

The limitations of the NSF laws can already be seen from our discussion in the previous section: The laws follow when temperature and velocity can be expanded into a Taylor series which breaks off after the second term. The relation to the Knudsen number becomes obvious by introducing a dimensionless length scale $\hat{x} = x/L$, so that, e.g., $T(x + \lambda) \rightarrow T(\hat{x} + \mathrm{Kn}) \simeq T(\hat{x}) + \frac{dT}{d\hat{x}}\mathrm{Kn} + \frac{1}{2}\frac{d^2T}{dx^2}\mathrm{Kn}^2 + \cdots$. If temperature, or velocity, vary rapidly, more terms in the series must be taken into account, which implies that the NSF laws are no longer valid anymore. The same is true when the Knudsen number is not small, that is for Kn $\gtrsim 0.05$ (say). Then microscale effects, or effects of gas rarefaction, become important for the description of a flow [22].

Typical processes where this occurs are

- High altitude flight, where gas pressure and density are very low, and the mean free path becomes so large that the Knudsen number cannot be considered as small even for macroscopic bodies such as space craft.
- Microscopic flows, where the relevant length scales L are very small, so that even at normal pressures and densities the Knudsen number is small. Typical cases are micro channel flows, flow in porous media, etc.
- Propagation of ultrasound, where the frequency ω is so high that the relevant Knudsen number Kn $= \omega\tau$ cannot be considered as small.
- Shock waves, where a flow changes from super-sonic to sub-sonic over few mean free paths, including drastic changes in temperature, pressure and density, so that the gradients are large.
- Boundary value problems at larger Knudsen numbers, where velocity and temperature of the gas differ from those of the wall, i.e. temperature jumps and velocity slip occur. Knudsen boundary layers are observed adjacent to the wall, which have a width of few mean free paths.

The derivation and discussion of equations that—other than the NSF laws—allow the description of microscale effects is at the heart of this book.

Problems

1.1. Length scales
Compute mean particle distance l_d and mean free path λ for helium ($M = 4\,\text{kg}\,/\,\text{kmol}$, $d = 2.3 \times 10^{-10}\,\text{m}$) and argon ($M = 39.95\,\text{kg}\,/\,\text{kmol}$, $d = 4.2 \times 10^{-10}\,\text{m}$) at pressures 0.01 bar, 1 bar, 100 bar, and temperatures 100K, 300K, 1000K. Discuss whether the ideal gas assumption is valid. For what values of pressure/temperature does the assumption break down?

1.2. Time scales
(a) For the same data as in Problem 1.1 estimate the mean free time τ and the collision time τ_c.
(b) Above it was said that the NSF laws cannot be used for Kn > 0.05. Compute the corresponding frequency of a sound wave for a variety of pressure/temperature pairs.

1.3. Particle speed and speed of sound
The speed of sound of a monatomic gas is given by $a = \sqrt{\frac{5}{3}\frac{k}{m}T}$ (see Sec. 10.1.3) and is thus very close to the mean particle speed $\bar{c} = \sqrt{3\frac{k}{m}T}$. Is this surprising? Discuss.

1.4. Viscosity and heat conductivity

(a) The derivation of the relations for heat conductivity and viscosity in this chapter are based on several simplifying assumptions, and not all of these were discussed explicitly. List of as many as you find, and discuss them.

(b) Estimate viscosity and heat conductivity for helium and argon, and compare the values with tabulated values from the literature.

1.5. Mean free path and altitude

The number density in an isothermal atmosphere at temperature T_0 decreases as $n\,(z) = n_0 \exp\left[-\frac{mgz}{kT_0}\right]$, where n_0 is the number density at ground level and $g = 9.81\,\mathrm{m\,/\,s^2}$ is the gravitational acceleration (see Problem 3.5). Using data for air ($M = 29\,\mathrm{kg\,/\,kmol}$) estimate the height at which the mean free path is (a) 1cm, (b) 10cm, (c) 1m, (d) 10m. Discuss the corresponding values of the Knudsen number for a space craft.

1.6. Microscale effects

Think of everyday problems, devices etc. where microscale and rarefaction effects could play a role.

1.7. Measurement of particle diameter

A heated cathode and an anode are mounted inside a glass tube in a distance of 20 cm. The tube is filled with air at a pressure of 1.5×10^{-7} bar and a temperature of 300 K. Measurements show that 90% of the electrons emitted at the cathode reach the anode. Assume that an electron that collides with an air particle will not reach the anode, that electrons and air particles interact as hard spheres, that the diameter of electrons is negligible, and that electron-electron collisions are so unlikely that they can be neglected.

(a) Compute the mean free path of electrons.

(b) The probability that a particle travels the distance s without collision be $p\,(s)$. Accordingly, $p\,(s + ds)$ is the probability that a particle travels the distance $s + ds$ without collision. Give arguments that $p\,(s + ds) = p\,(s)\,p\,(ds)$ and $p\,(ds) = 1 - \frac{ds}{\lambda}$ where λ is the mean free path, and show that $p\,(s) = \exp\left(-\frac{s}{\lambda}\right)$.

(c) Use the results of (b) to compute the diameter of air particles from the experimental data given.

2

Basic quantities and definitions

This chapter introduces the phase density and its moments as the central quantities in kinetic theory. The equilibrium phase density—the Maxwellian distribution—is derived, and its moments are computed. Particle wall interactions are discussed, and Maxwell's boundary conditions for the phase density are introduced.

2.1 Phase space and phase density

A gas consists of a huge number—in the order of 10^{23}—interacting particles α whose physical state is described by their locations $\mathbf{x}^\alpha = \{x_1^\alpha, x_2^\alpha, x_3^\alpha\}$ and their velocities $\mathbf{c}^\alpha = \{c_1^\alpha, c_2^\alpha, c_3^\alpha\}$ at any time t. The (micro-) state of the gas is given by the complete set of the $\{\mathbf{x}^\alpha, \mathbf{c}^\alpha\}$, and each particle can be described through its trajectory in the 6-dimensional space spanned by $\mathbf{x}$ and $\mathbf{c}$, the so-called phase space.

Thus, to describe the gas one could establish the equation of motion for each particle, and then had to solve a set of $\sim 10^{23}$ coupled equations. Clearly this is not feasible, and therefore kinetic theory chooses to describe the state of the gas on the micro-level through the phase density, or distribution function, $f(\mathbf{x}, t, \mathbf{c})$ which is defined such that

$$N_{\mathbf{x},\mathbf{c}} = f(\mathbf{x}, t, \mathbf{c}) \, d\mathbf{x} d\mathbf{c} \qquad (2.1)$$

gives the number of particles that occupy a cell of phase space $d\mathbf{x} d\mathbf{c}$ at time t, see Fig. 2.1. In other words, $N_{\mathbf{x},\mathbf{c}}$ is the number of particles with velocities in $\{\mathbf{c}, \mathbf{c} + d\mathbf{c}\}$ located in the interval $\{\mathbf{x}, \mathbf{x} + d\mathbf{x}\}$ at time t.

With this definition, a certain level of inaccuracy is introduced, since now the state of each particle is only known within an error of $d\mathbf{x} d\mathbf{c}$. It is interesting to note that the phase space concept was developed well before the advent of quantum mechanics which indeed demands a finite size of the phase cell $d\mathbf{x} d\mathbf{c} \propto (h_P/m)^3$, where h_P denotes Planck's constant.

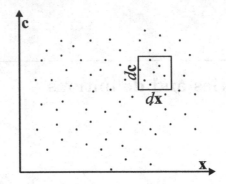

Fig. 2.1. The 6-dimensional phase space, and a cell of phase space. Each dot represents the momentary state of one particle.

The phase density $f(\mathbf{x}, t, \mathbf{c})$ is the central quantity in kinetic theory, since the state of the gas is (almost) completely known when f is known.

2.2 Moments of the phase density

The phase density gives a detailed picture of the state of the gas, quite often more detailed than desired, and is far from our daily experience. More accessible for us are the moments of f, weighted integrals of f that will be discussed now.

2.2.1 Number, mass density, and velocity

From the definition (2.1) follows that the total number of particles in the volume V is given by

$$N = \iiint_V \iiint_{-\infty}^{\infty} f \, d\mathbf{c} \, d\mathbf{x}$$

where the possible values of the location $\mathbf{x}$ are restricted to lie within the volume V, and values for the three components of the velocity vector lie between $-\infty$ and ∞. This implies that relativistic effects of the motion of the particles are ignored, which, however, might play an important role in very hot gases [5][10].

When we integrate only over velocity, and divide by the cell volume $d\mathbf{x}$, we obtain the number density n

$$n = \iiint_{-\infty}^{\infty} f \, d\mathbf{c} \,.$$

We frequently have to integrate over the full velocity space, and in order to condense notation we shall often write one integral sign without limits, so that, e.g.,

$$n = \int f d\mathbf{c} .$$

$F d\mathbf{c} = f/n d\mathbf{c}$ can be interpreted as the probability to find a particle with velocity in $[\mathbf{c}, \mathbf{c} + d\mathbf{c}]$, which is properly normalized since

$$\int F d\mathbf{c} = \int \frac{f}{n} d\mathbf{c} = 1 .$$

Since the mass density is given by $\rho = mn$ we have

$$\rho = m \int f d\mathbf{c} . \tag{2.2}$$

The mean velocity of the particles within the volume $d\mathbf{x}$ is given by

$$\mathbf{v} = \frac{1}{n} \int \mathbf{c} f d\mathbf{c} \quad \text{or} \quad v_i = \frac{1}{n} \int c_i f d\mathbf{c} ,$$

where both equations are identical, and differ only in notation: the first equation is written in symbolic tensor notation, and the second one in index notation. Often we will have to deal with vectors and tensors of higher order for which the index notation is easier to handle, and thus it will be used almost exclusively in the sequel. See Appendix A.1 for a review of the basic elements of index notation.

v_i, or $\mathbf{v}$, is the macroscopic velocity of the gas, also known as the center of mass velocity, or barycentric velocity. The momentum density is given by

$$\rho v_i = m \int c_i f d\mathbf{c} . \tag{2.3}$$

The peculiar velocity C_i of the particles gives the particle speed as measured by an observer moving with the gas at the local velocity v_i, i.e. an observer in the rest-frame,

$$C_i = c_i - v_i . \tag{2.4}$$

Due to this definition, and with (2.3), the first moment of f over C_i vanishes,

$$0 = m \int C_i f d\mathbf{c} . \tag{2.5}$$

Note that $d\mathbf{c} = d\mathbf{C}$ as shown in Problem 2.1.

2.2.2 Energy, pressure, and temperature

The kinetic energy of a particle is given by $\frac{m}{2}c^2$, so that the energy density of the gas is given by

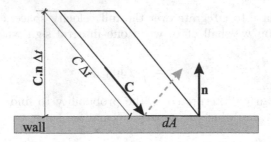

Fig. 2.2. On the calculation of pressure.

$$\rho e = \frac{m}{2} \int c^2 f d\mathbf{c} \,, \tag{2.6}$$

e denotes the specific energy. With (2.4) and (2.5) follows that

$$\rho e = \frac{m}{2} \int \left(C^2 + 2v_k C_k + v^2\right) f d\mathbf{c} = \rho u + \frac{\rho}{2} v^2 \,,$$

where

$$\rho u = \frac{m}{2} \int C^2 f d\mathbf{c} \tag{2.7}$$

is the internal, or thermal, energy of the gas, and $\frac{\rho}{2} v^2$ is the kinetic energy of its macroscopic motion. Thus, the internal energy of an ideal monatomic gas is the kinetic energy of its particles as measured in the rest-frame.

Pressure is the average force on a wall due to the change of momentum of the particles. We consider a wall with normal pointing into the x_1-direction and assume elastic collisions. Then, when the velocity of a particle before collision is $\{C_1, C_2, C_3\}$, its velocity after collision will be $\{-C_1, C_2, C_3\}$, and the corresponding change of momentum is $\{2mC_1, 0, 0\}$. The pressure results from many collisions by particles flying with any speed in all directions that allow to hit the wall, and we use Fig. 2.2 to help with its calculation. In order to collide, a particle must have a velocity component $C_1 < 0$, and all particles in the volume $-\mathbf{C} \cdot \mathbf{n} \, \Delta t \, dA$ will collide during Δt, and contribute $2mC_1$ to momentum change ($\mathbf{C} \cdot \mathbf{n} = C_k n_k$ denotes the scalar product). Accordingly, the total momentum change during Δt is

$$2mC_1 \, C_1 \Delta t \, dA \, f d\mathbf{c} \,.$$

Integration over all admissible velocities gives the total momentum change during Δt, and division by $\Delta t \, dA$ gives the pressure on a wall with normal $\{1, 0, 0\}$ as

$$p = 2m \iiint_{C_1 < 0} C_1^2 f d\mathbf{c} \,.$$

For now we consider isotropic gases only, where the phase density depends solely on the magnitude of the velocity, but not on its direction, so that

$$\iiint_{C_1 < 0} C_1^2 f d\mathbf{c} = \iiint_{C_1 > 0} C_1^2 f d\mathbf{c} \quad \text{and} \quad \int C_1^2 f d\mathbf{c} = \int C_2^2 f d\mathbf{c} = \int C_3^2 f d\mathbf{c} .$$

Combining the last equations allows to write the pressure as

$$p = \frac{1}{3} m \int C^2 f d\mathbf{c} = \frac{2}{3} \rho u , \tag{2.8}$$

where $C^2 = C_k C_k = C_1^2 + C_2^2 + C_3^2$. Here we have found the relation (1.4) by a more accurate argument [18].

From the ideal gas law $p = \rho \frac{k}{m} T$ follows the relation between energy and temperature as

$$u = \frac{3}{2} \frac{k}{m} T , \quad \text{or} \quad T = \frac{1}{3} \frac{m^2}{\rho k} \int C^2 f d\mathbf{c} . \tag{2.9}$$

Temperature in kinetic theory is usually defined by the above relation for all situations, including strong deviations from equilibrium, and we shall follow that tradition. Later, in Sec. 12.1.4, we shall see that the notion of temperature as a meaningful—that is measurable—quantity maybe lost in these extreme cases, since temperature jumps can occur at the thermometer walls.

For notational convenience we shall write the temperature in energy units as

$$\theta = \frac{k}{m} T .$$

The atoms in a monatomic gas have three translatorial degrees of freedom (the three components of velocity), and each degree of freedom contributes $\frac{1}{2} \frac{k}{m} T$ to the specific internal energy, or $\frac{1}{2} \frac{k}{m}$ to the specific heat $c_v = \left(\frac{\partial u}{\partial T} \right)_\rho$. Other than monatomic molecules, e.g. diatomic molecules such as oxygen (O_2) or nitrogen (N_2), three-atomic molecules such as water (H_2O) or carbon dioxide (CO_2), or even larger molecules such as methane (CH_4), have additional degrees of freedom due to rotation and vibration. Each fully activated degree of freedom contributes $\frac{1}{2} \frac{k}{m}$ to the specific heat, but the activation follows quantum mechanical laws which results in temperature dependent specific heats [12][23].

2.2.3 General moments

There are more moments of importance, some of which have a meaningful physical interpretation, such as pressure tensor p_{ij} and heat flux q_i, given as

$$p_{ij} = m \int C_i C_j f d\mathbf{c} \quad \text{and} \quad q_i = \frac{m}{2} \int C^2 C_i f d\mathbf{c} . \tag{2.10}$$

However, additional moments become important when rarefaction effects are considered, although they do not have physical interpretations. It is thus convenient to introduce a general notation for moments, and we define general tensorial moments as

$$\rho^a_{i_1 i_2 \cdots i_n} = m \int C^{2a} C_{i_1} C_{i_2} \cdots C_{i_n} f d\mathbf{c} \, . \tag{2.11}$$

It will prove very useful to use irreducible, or trace-free, moments of the phase density which are defined as

$$u^a_{i_1 \cdots i_n} = \rho^a_{\langle i_1 i_2 \cdots i_n \rangle} = m \int C^{2a} C_{\langle i_1} C_{i_2} \cdots C_{i_n \rangle} f d\mathbf{c} \tag{2.12}$$

where indices in angular brackets denote the symmetric and trace-free part of a tensor. Note that the moments $u^a_{i_1 \cdots i_n}$ used in the following are trace-free by definition, $u^a_{i_1 \cdots i_n} = u^a_{\langle i_1 \cdots i_n \rangle}$, and that this is not made explicit with brackets in order to avoid confusing notation involving several pairs of brackets. Indices in round brackets, as, e.g., in (2.16) below, denote the symmetric part of a tensor. Appendix A.2 gives a short introduction to tensor index notation, and to symmetric and trace-free tensors.

In this notation we identify

$$\rho^0 = u^0 = \rho \;\; , \;\; \rho^0_i = u^0_i = 0 \;\; , \;\; \rho^1 = u^1 = 2\rho e = 3\rho\theta = 3p \;\; , \tag{2.13}$$
$$\rho^0_{ij} = p_{ij} \;\; , \;\; u^0_{ij} = p_{\langle ij \rangle} = \sigma_{ij} \;\; , \;\; \rho^1_i = 2q_i \, .$$

Here, in addition to quantities defined earlier, we have introduced the stress tensor σ_{ij} as the trace-free part of the pressure tensor p_{ij}.

The moments defined above are taken with respect to the peculiar velocity C_k, and are sometimes denoted as *central* moments to distinguish them form the so-called *convective* moments which are defined with respect to the microscopic velocity c_k [24]. The moments

$$F^0_{i_1 i_2 \cdots i_n} = m \int c_{i_1} c_{i_2} \cdots c_{i_n} f d\mathbf{c} \tag{2.14}$$

will be denoted as full convective moments, and we can generalize the definition to

$$F^a_{i_1 i_2 \cdots i_n} = m \int c^{2a} c_{i_1} c_{i_2} \cdots c_{i_n} f d\mathbf{c} = F_{i_1 i_2 \cdots i_n j_1 k_1 \cdots j_m k_m} \delta_{j_1 k_1} \cdots \delta_{j_m k_m} \, . \tag{2.15}$$

Convective and central moments can easily be related by means of (2.4) as

$$F^0_{i_1 i_2 \cdots i_n} = \sum_{k=0}^{n} \binom{n}{k} \rho^0_{(i_1 \cdots i_k} v_{i_{k+1}} \cdots v_{i_n)} \, , \tag{2.16}$$

e.g.,

$$F^0 = \rho \;\; , \;\; F^0_i = \rho v_i \;\; , \;\; F^1 = 2\rho e = 2\rho u + \rho v^2 \;\; , \tag{2.17}$$
$$F^0_{ij} = \rho^0_{ij} + \rho v_i v_j = p_{ij} + \rho v_i v_j \;\; , \;\; \frac{1}{2} F^1_k = \left(\rho u + \frac{1}{2} \rho v^2 \right) v_k + p_{kj} v_j + q_k \; ;$$

F^0_{ij} is the total momentum flux, and $\frac{1}{2} F^1_k$ is the total energy flux (see Sec. 3.3.2).

2.2.4 The relevance of moments

The first reason for our interest in moments is that some, namely mass density (2.2), momentum density (2.3), energy density (2.6) or internal energy density (2.7), pressure tensor (2.10)$_1$, and heat flux (2.10)$_2$ have a meaningful physical interpretation, and therefore are, at least in principle, accessible to measurements.

The second, and maybe more important, reason to consider moments is that under certain conditions, in particular at smaller Knudsen numbers, the knowledge of a small number of moments, e.g. 5, 13, or 26, is sufficient to describe the state of the gas—including mass, momentum and energy transfer—with reasonable accuracy.

Obviously, the complete knowledge of the phase density allows to compute all of its moments, but also the inverse is true: if all moments are known, the distribution can be reconstructed. In those cases where the gas flow is well described through a small, and finite, number of moments, the complete knowledge of the phase density is not required. Then, the description of the gas state through moments is more convenient, since only the relevant moments must be considered, while all others can be ignored.

The phase density $f(\mathbf{x}, t, \mathbf{c})$ must be computed by solving the Boltzmann equation that will be discussed in Chapter 3, and the equations for the moments will be derived from the Boltzmann equation in later chapters.

2.3 Equilibrium and the Maxwellian distribution

A simple argument going back to Maxwell allows us to find the velocity distribution in equilibrium [2][18].

Equilibrium is a state where no changes will occur when the gas is left to itself, and this will imply that the gas is homogeneous, i.e. displays no gradients in any quantity, and the phase density is isotropic, that is independent of the direction $\nu_i = C_i/C$. The following argument considers the gas in the rest-frame, where $v_i = 0$.

An arbitrary atom picked from the gas will have the velocity components $C_k, k = 1, 2, 3$, and the probability to find the component in direction x_k within the interval $[C_k, C_k + dC_k]$ is given by $p(C_k) dC_k$. Note that, due to isotropy, the probability function p is the same for all components. Then, the probability to find a particle with the velocity vector $\{C_1, C_2, C_3\}$ is given by

$$F(C) dC_1 dC_2 dC_3 = p(C_1) p(C_2) p(C_3) dC_1 dC_2 dC_3$$

where $F(C) = f(C)/n$ depends only on the absolute value of velocity, $C = \sqrt{C_1^2 + C_2^2 + C_3^2}$, since the probability must be independent of direction. Thus, F and p are related as

$$F(C) = p(C_1) p(C_2) p(C_3) .$$

Taking the logarithmic derivative of this equation with respect to C_1 we see

$$\frac{\partial}{\partial C_1} \ln F(C) = \frac{\partial}{\partial C_1} \left[\ln p(C_1) + \ln p(C_3) + \ln p(C_3) \right]$$

or

$$\frac{1}{C} \frac{F'(C)}{F(C)} = \frac{1}{C_1} \frac{p'(C_1)}{p(C_1)} = -2\gamma .$$

Since the left and the right side of this equation depend on different variables, γ must be a constant, and integration gives an isotropic Gaussian distribution,

$$F = \frac{f}{n} = A \exp\left[-\gamma C^2\right] \quad \text{and} \quad p(C_k) = \sqrt[3]{A} \exp\left[-\gamma C_k^2\right] ,$$

where A is a constant of integration. The two constants γ and A follow from the conditions that the phase density must reproduce mass and energy density, that is

$$\rho = m \int f d\mathbf{c} \quad \text{and} \quad \rho u = \frac{3}{2}\rho\theta = \frac{m}{2} \int C^2 f d\mathbf{c} .$$

The resulting equilibrium phase density is the Maxwellian distribution

$$f_M = \frac{\rho}{m} \frac{1}{\sqrt{2\pi\theta}^3} \exp\left[-\frac{C^2}{2\theta}\right] . \tag{2.18}$$

The values of the moments (2.12) in equilibrium (E), when the phase density is a Maxwellian, are given by

$$u_{|E}^a = (2a+1)!!\rho\theta^a \ , \quad u_{i_1\cdots i_n|E}^a = 0 \ , \quad n \geq 1 , \tag{2.19}$$

where the double factorial is defined as $(2a+1)!! = \prod_{s=1}^{a}(2s+1)$. Useful equations for integrals of the Maxwellian are given in Appendix A.3.

2.4 Boundary conditions for the phase density

Gas particles that hit a wall will exchange energy and momentum with the wall and be re-emitted into the gas with a new velocity vector. The details of the interaction depend strongly on the microscopic details of wall and gas, and generally it is not possible to describe this interaction accurately. Instead, probabilistic models are used. The most popular, and most simple, model for boundary conditions is due to Maxwell [2][3] and will be used exclusively in this text.

The barycentric velocity v_k of a rarefied gas will differ from the wall velocity v_k^W, so that it has a non-zero slip velocity

$$V_k = v_k - v_k^W \tag{2.20}$$

which is parallel to the wall, since there is no gas flow through the wall,

$$V_k n_k = 0 . \tag{2.21}$$

Moreover, the temperatures of wall, θ_W, and gas directly in front of the wall, θ, differ, so that a temperature jump $\theta - \theta_W$ can be observed. Jump and slip are rarefaction effects, and vanish as Kn $\to 0$.

The boundary conditions are formulated in a frame where the wall is at rest, so that the particle velocity is $C_i^W = c_i - v_i^W$. The normal vector n_i of the wall points inside the gas, so that $C_k^W n_k \leq 0$ for incident particles and $C_k^W n_k \geq 0$ for outgoing particles. The boundary condition must relate the (unknown) distribution f_R of the outgoing particles ($C_k^W n_k \geq 0$) to the (known) distribution of the incoming particles ($C_k^W n_k \leq 0$), denoted as $f_N \left(C_i^W, x_i, t \right)$. It is convenient to write f_N as a function of the tangential velocity $C_i^W - C_k^W n_k n_i$ and the normal velocity $C_k^W n_k$ as $f_N \left(C_i^W - C_k^W n_k n_i, C_k^W n_k, x_i, t \right)$.

The simplest choice for a boundary condition is that of a specularly reflecting wall, where the tangential velocity of a colliding particle remains unchanged, while the normal component of its velocity changes sign. Then, the distribution of the reflected particles is

$$f_R = f_N \left(C_i^W - C_k^W n_k n_i, -C_k^W n_k, x_i, t \right) \quad \text{for } C_k^W n_k \geq 0 .$$

In a specular reflection the particle does not exchange energy with the wall, and exerts only a normal force on the wall. Thus, no energy is transferred between gas and wall—the wall is adiabatic—, and no shear stresses are acting on the wall. Heat transfer and shear are important phenomena, and therefore the model of a specularly reflecting wall is too simple.

Another simple model is that of a perfectly thermalizing wall, where the colliding gas particles interact strongly with the wall, and leave in a Maxwellian distribution which is determined by wall temperature θ_W and velocity v_k^W,

$$f_R = f_W = \frac{\rho_W}{m} \sqrt{\frac{1}{2\pi\theta_W}}^3 \exp\left[-\frac{C_W^2}{2\theta_W} \right] . \tag{2.22}$$

The density ρ_W is the density of the thermalized particles which has to be chosen such that the wall does not accumulate particles.

Also the perfectly thermalizing wall is too simple to describe realistic cases, and therefore Maxwell introduced a combination of both models. He assumes that the fraction $(1 - \chi)$ of the incident particles is reflected elastically, and that the remaining fraction χ is thermalized. χ is called accommodation coefficient, and could, in principle, depend on the particle velocity, but we shall assume it to be a constant.

For simplicity of the notation, we suppress the tangential velocity as well as space and time in the list of arguments of f so that the phase density directly at the wall $\bar{f}$ according to Maxwell's boundary conditions is

$$\bar{f} = \begin{cases} \chi f_W + (1 - \chi) f_N \left(-C_k^W n_k \right) , & C_k^W n_k \geq 0 \\ f_N \left(C_k^W n_k \right) & , C_k^W n_k \leq 0 . \end{cases} \tag{2.23}$$

The velocity normal to the wall vanishes since the wall does not accumulate particles, a condition which can be written as

$$\int C_k^W n_k \bar{f} d\mathbf{c} = 0 \ \text{ or } \int_{n_k C_k^W \geq 0} C_k^W n_k \bar{f} d\mathbf{c} = -\int_{n_k C_k^W \leq 0} C_k^W n_k \bar{f} d\mathbf{c} \ .$$

In the last equation we insert (2.23), substitute $-n_k C_k^W \to n_k C_k^W$, and introduce the peculiar velocity of the particles $C_i = C_i^W - V_k$ with, due to (2.21), $n_k C_k = n_k C_k^W$. The mean value of C_i vanishes by definition, $\int C_k n_k f_N d\mathbf{c} = 0$ or $\int_{n_i C_i \geq 0} C_k n_k f_N d\mathbf{c} = -\int_{n_k C_k \leq 0} C_k n_k f_N d\mathbf{c}$, and thus the condition can be simplified to

$$\int_{n_k C_k^W \geq 0} C_k^W n_k f_W d\mathbf{c} = \int_{n_i C_i \geq 0} C_k n_k f_N d\mathbf{c} \ . \tag{2.24}$$

This equation must be solved for ρ_W.

Problems

2.1. Elements of integration
Show that the elements of integration for microscopic velocity $\mathbf{c}$ and peculiar velocity $\mathbf{C}$ are equal, $d\mathbf{c} = d\mathbf{C}$.

2.2. Gaussian integrals
Perform the proof by induction to confirm (A.4) in Appendix A.3.

2.3. Coefficients in the Maxwell distribution
Confirm (2.18) by computing the coefficients γ and A.

2.4. Properties of the phase density
(a) What percentage of particles in He (Xe) at room temperature (at 1300K) have velocities above 1000m/s (300m/s)?
(b) How many particles have a velocity of exactly 10000m/s?

2.5. Distribution of absolute velocity in equilibrium
f is the distribution of the velocity vector. The distribution of the absolute velocity follows from the integration over the directions. Introduce spherical coordinates so that $C_k = \{\sin \vartheta \sin \varphi, \sin \vartheta \cos \varphi, \cos \vartheta\}_k$, and $d\mathbf{C} = C^2 \sin \vartheta d\vartheta d\varphi dc$. Compute the distribution of absolute velocity in equilibrium as

$$F_M(C) = \iint f_M C^2 \sin \vartheta d\vartheta d\varphi \ .$$

Plot the function and discuss its properties, in particular the number of particles with very large, and very small velocities.

2.6. Mean velocities
(a) For a gas in equilibrium compute the following mean values of velocity:

$$\bar{C} = \frac{\int C f_M dc}{\int f_M dc}, \qquad \sqrt{\overline{C^2}} = \sqrt{\frac{\int C^2 f_M dc}{\int f_M dc}}.$$

Also, compute the most probable velocity C_p, i.e. the maximum of $F_M(C)$, as introduced in Problem 2.5. Compare all three with the speed of sound, and give the velocities at room temperature $(20°C)$ and at $1300K$ for helium, argon and xenon.

(b) Give reasons that the mean relative speed between two particles with velocities c, c_1 is given by

$$\bar{g} = \frac{\int \int g f f_1 dc dc_1}{\int \int f f_1 dc dc_1}$$

where $g = |C - C_1|$. Compute g in equilibrium, where $f = f_M$. For the solution of the integral, introduce the relative velocity and the velocity of the center of mass according to $\mathbf{g} = \mathbf{C} - \mathbf{C}_1$, $\mathbf{c}_s = \frac{1}{2}(\mathbf{C} + \mathbf{C}_1)$. Show that $\bar{g} = \sqrt{2}\bar{C}$.

2.7. Effusion
Consider two containers filled with a monatomic gas at equilibrium that are held at p_1, θ_1 and p_2, θ_2, respectively. The two containers are connected through a small pipe whose diameter is far below the mean free path. Compute the net mass flow between the containers. In particular, consider the case with zero mass flow, and discuss the corresponding values for pressures and temperatures.

2.8. Knudsen heat transfer
A gas is confined between two parallel walls at a distance L which is much smaller than the mean free path $(Kn \gg 1)$. This implies that the gas particles do not collide with each other, but only with the walls. Due to the collisions with the walls, the particles thermalize, and we assume that they leave the walls in Maxwellian distributions of wall temperature. The walls have different temperatures θ_1, θ_2 and therefore heat is transported between the walls. Accordingly, the phase density in the interior must read

$$f = \begin{cases} \frac{\rho_1/m}{\sqrt{2\pi\theta_1}^3} \exp\left[-\frac{C^2}{2\theta_1}\right], & C_x > 0 \\[2ex] \frac{\rho_2/m}{\sqrt{2\pi\theta_2}^3} \exp\left[-\frac{C^2}{2\theta_2}\right], & C_x < 0 \end{cases}$$

(a) The total density of the gas is given as ρ, and the gas is at rest. Compute ρ_1, ρ_2 from this information.

(b) Compute the temperature of the gas in the interior, and discuss the temperature jumps at the walls.

(c) Compute the relation between heat flux $q_x = \frac{m}{2} \int C^2 C_x f d\mathbf{c}$ and temperature difference $\theta_1 - \theta_2$. Note that, since the gas temperature is uniform, heat transfer takes places without a temperature gradient, so that Fourier's law (1.6) is not valid.

2.9. Thermophoresis in Knudsen gas

Consider the Knudsen heat transfer problem above. A small plate is placed between the walls in an angle α measured relative to the plate distance. Gas particles colliding with the plate are elastically reflected. Compute the resulting force on the plate. Generalize to the case of a cylinder of radius R, whose axis is parallel to the walls. (A sphere would be interesting too, but is more cumbersome...) Assume that the object in the flow is larger than a gas particle, but small against the wall distance. With this assumption the object will not distort the distribution function, which is the same as in Problem 2.8. The resulting force on the object will lead to a directed motion of the latter, known as thermophoresis. Discuss this further: does the object move from hot to cold or vice versa? How would size and mass of the object (very large/small, light/heavy) influence that motion?

The Boltzmann equation and its properties

Collisions between particles involve exchange of energy and momentum, and accordingly the phase density f will change in time and space through the collisions. The equation that describes the evolution of f due to collisions and free flight of particles is the Boltzmann equation which was introduced by Boltzmann [25].

In this chapter, the Boltzmann equation is derived, and its most important properties are discussed, including equilibrium conditions, conservation laws, and the H-theorem.

Kinetic models, in particular the BGK and ES-BGK models, are introduced, which share the main properties of the Boltzmann equation, but are mathematically simpler.

3.1 The Boltzmann equation

3.1.1 Space-time evolution of f

In order to develop the evolution equation for the phase density, we consider a fixed volume Ω of phase space that contains

$$N_\Omega = \iint_\Omega f d\mathbf{c} d\mathbf{x}$$

particles, and ask for the change of N_Ω with time. To compute that change, we introduce dA_Ω as a (5-dimensional) surface element of phase space with normal vector $\mathbf{n}$, the phase space vector $\xi_A = \{x_k, c_k\}_A$ ($A = 1, \cdots, 6$, $k = 1, 2, 3$), and the phase space velocity $\dot{\xi}_A = \{\dot{x}_k, \dot{c}_k\}_A$, see Fig. 3.1. All particles that cross dA_Ω during the time dt contribute to the change of N_Ω and by integration over the surface $\partial\Omega$ of Ω we find

$$\frac{dN_\Omega}{dt} = \frac{d}{dt} \iint_\Omega f d\mathbf{c} d\mathbf{x} = -\oint_{\partial\Omega} \dot{\xi}_A n_A f dA_\Omega \, .$$

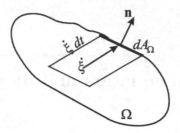

Fig. 3.1. The phase space element Ω.

Gauss's theorem allows to convert the surface integral into a volume integral, and the time derivative can be drawn into the integral since Ω is fixed in time, so that

$$\iint_\Omega \frac{\partial f}{\partial t} d\mathbf{c}d\mathbf{x} = -\iint_\Omega \frac{\partial \dot{\xi}_A f}{\partial x_A} d\mathbf{c}d\mathbf{x} \quad \text{or} \quad \iint_\Omega \left[\frac{\partial f}{\partial t} + \frac{\partial \dot{\xi}_A f}{\partial x_A} \right] d\mathbf{c}d\mathbf{x} = 0 .$$

This equation must hold for any phase space volume Ω, and therefore the expression in the square brackets must vanish. Since $\xi_A = \{x_k, c_k\}_A$ and $\dot{\xi}_A = \{\dot{x}_k, \dot{c}_k\}_A = \{c_k, \dot{c}_k\}_A$, we can expand this into

$$\frac{\partial f}{\partial t} + c_k \frac{\partial f}{\partial x_k} + \frac{\partial \dot{c}_k f}{\partial c_k} = 0 .$$

Here it was also used that $\partial c_k / \partial x_k = 0$ holds, since x_k and c_k are independent variables in phase space. The acceleration $\dot{c}_k = G_k + W_k$ of the particles is due to external forces G_k, e.g. gravity, and to the interaction forces with other particles W_k, where both, G_k and W_k, are forces per unit mass. The external forces considered here are independent of the particle velocity[1] and we finally obtain

$$\frac{\partial f}{\partial t} + c_k \frac{\partial f}{\partial x_k} + G_k \frac{\partial f}{\partial c_k} = \mathcal{S} , \tag{3.1}$$

where $\mathcal{S} = -\frac{\partial W_k f}{\partial c_k}$ describes the change of the phase density due to interaction between particles, i.e. collisions.

In Chapter 1 we learned that the time for a collision is very short. Therefore it is impossible to resolve time on a scale fine enough to follow through the collisions, that means to find an expression for W_k. The idea of the Boltzmann equation is to consider a coarser time scale, where collisions appear as instant changes in velocities, and $\mathcal{S}$ will be determined based on that idea.

---

[1] That excludes the Lorentz force, which, however, could be build in easily.

3.1.2 Binary collisions

For the derivation of the Boltzmann equation detailed information about the interaction between particles is required, in particular about the changes of particle velocities due to the collisions, and we proceed with discussing these.

We consider two particles of equal mass m, with the pre-collision velocities $\mathbf{c}$, $\mathbf{c}^1$, and the corresponding relative speed $\mathbf{g} = \mathbf{c} - \mathbf{c}^1$. After the collision, the particles will have the velocities

$$\mathbf{c}' = \mathbf{c} + \mathbf{a} \ , \quad \mathbf{c}^{1\prime} = \mathbf{c}^1 + \mathbf{b} \ , \quad \mathbf{g}' = \mathbf{c}' - \mathbf{c}^{1\prime} \ ,$$

with vectors $\mathbf{a}$ and $\mathbf{b}$ to be determined. Note that the velocities defined in this section refer to the velocities of the particles before and after they notice the presence of the other particle, that is when their relative distance r is so large that the interaction potential $\phi(r)$ is virtually zero.

The conservation of momentum over the collision requires

$$\mathbf{c} + \mathbf{c}^1 = \mathbf{c}' + \mathbf{c}^{1\prime} \quad \text{so that} \quad \mathbf{b} = -\mathbf{a} \ .$$

The conservation of energy requires that

$$(\mathbf{c})^2 + \left(\mathbf{c}^1\right)^2 = (\mathbf{c}')^2 + \left(\mathbf{c}^{1\prime}\right)^2$$

which implies that

$$\left(\mathbf{c} - \mathbf{c}^1\right) \cdot \mathbf{a} + a^2 = 0 \ .$$

This equation allows us to compute the absolute value $a = \sqrt{\mathbf{a}^2}$ as $a = -\mathbf{k} \cdot \mathbf{g}$ where the unit vector $\mathbf{k} = \mathbf{a}/a$ is the the so-called collision vector. The post-collision velocities can now be expressed as

$$\mathbf{c}' = \mathbf{c} - (\mathbf{k} \cdot \mathbf{g})\,\mathbf{k} \ , \quad \mathbf{c}^{1\prime} = \mathbf{c}^1 + (\mathbf{k} \cdot \mathbf{g})\,\mathbf{k} \ , \quad \mathbf{g}' = \mathbf{g} - 2\,(\mathbf{k} \cdot \mathbf{g})\,\mathbf{k} \ . \qquad (3.2)$$

After multiplying $(3.2)_3$ successively with $\mathbf{k}$, $\mathbf{g}$, and $\mathbf{g}'$, and evaluation of the results, we find

$$\mathbf{k} \cdot \mathbf{g}' = -\mathbf{k} \cdot \mathbf{g} \ , \quad \mathbf{k} = \frac{\mathbf{g} - \mathbf{g}'}{|\mathbf{g} - \mathbf{g}'|} \quad \text{and} \quad g\prime = g \ . \qquad (3.3)$$

Thus, the absolute relative velocity g remains unchanged, and the vector $\mathbf{k}$ halves the angle between $\mathbf{g}$ and $\mathbf{g}'$, see Fig. 3.2. The figure also introduces the collision angle Θ as the angle between $\mathbf{g}$ and $\mathbf{k}$, so that

$$\mathbf{k} \cdot \mathbf{g} = g \cos \Theta \ . \qquad (3.4)$$

Equations (3.2) and (3.3) can be combined to give the pre-collision velocities in terms of the post-collision velocities as

$$\mathbf{c} = \mathbf{c}' - (\mathbf{k} \cdot \mathbf{g}')\,\mathbf{k} \ , \quad \mathbf{c}^1 = \mathbf{c}^{1\prime} + (\mathbf{k} \cdot \mathbf{g}')\,\mathbf{k} \ , \quad \mathbf{g} = \mathbf{g}' - 2\,(\mathbf{k} \cdot \mathbf{g}')\,\mathbf{k} \ . \qquad (3.5)$$

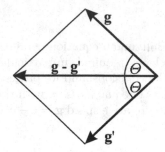

Fig. 3.2. Relative velocities and collision angle Θ. Recall that $\mathbf{k}$ is the unit vector pointing in the direction of $(\mathbf{g} - \mathbf{g}')$.

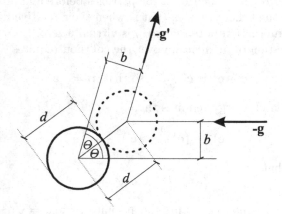

Fig. 3.3. The collision of two hard spheres of diameter d.

The conservation of moment of momentum of the colliding pair of particles requires that the collision takes place in a plane, the collision plane. This condition reads

$$\mathbf{g} \times \mathbf{r} = \mathbf{g}' \times \mathbf{r}' \ \text{ or } \ gb = g'b' \ \text{ so that } \ b\prime = b\,,$$

where $\mathbf{r}$ denotes the relative distance of the particles, and b is the collision parameter, which gives the particle distance perpendicular to the relative velocity $\mathbf{g}$.

Figure 3.3 visualizes collision parameter b, collision angle Θ, and relative velocities $\mathbf{g}$, $\mathbf{g}'$ for the collision of two hard sphere molecules of diameter d, as seen for an observer resting with one particle; the collision vector $\mathbf{k}$ is not shown, in order to not overload the drawing. As can be seen from Fig. 3.3, in the hard sphere case collision angle and parameter are related as

$$\Theta_{HS} = \arcsin \frac{b}{d}\,. \tag{3.6}$$

In general, the relation between Θ, b, and g depends on the interaction potential $\phi(r)$, and it can be shown by integration through the collision that [15]

$$\Theta(b, g) = \int_0^{s^*} \frac{ds}{\sqrt{1 - s^2 - \frac{4\phi\left(\frac{b}{s}\right)}{mg^2}}} \quad \text{with } 1 - (s^*)^2 - \frac{4\phi\left(\frac{b}{s^*}\right)}{mg^2} = 0. \quad (3.7)$$

Here, $s = b/r$ is a dimensionless inverse distance, and $s^* = b/r_{\min}$ is its value at the minimal particle distance $r_{\min}$.

3.1.3 Potential and differential cross section

For most potentials the relation (3.7) is difficult to evaluate. We shall consider only hard spheres where

$$\phi_{HS}(r) = \begin{cases} 0 & , r > d \\ \infty & , r \le d, \end{cases} \quad (3.8)$$

and power potentials where

$$\phi_\gamma(r) = \phi_0 r^{1-\gamma}; \quad (3.9)$$

ϕ_0 and γ are constants. Power potentials are purely repulsive potentials (the interparticle force is given by $-\frac{d\phi}{dr}$), and do not describe particle attraction that leads to condensation at low temperatures. Attractive forces can be ignored as long as the potential trough $\phi_{\min}$ in Fig. 1.1 is small compared to the average kinetic energy of a collision $\frac{m}{2}\bar{g}^2 = \frac{8}{\pi}kT$. This will be the case for temperatures well above the saturation temperature.

For power potentials, (3.7) can be written as

$$\Theta_\gamma(b, g) = \int_0^{s^*} \frac{ds}{\sqrt{1 - s^2 - \frac{4\phi_0}{m}\left(\frac{s}{\hat{s}}\right)^{\gamma-1}}} \quad \text{with } 1 - (s^*)^2 - \frac{4\phi_0}{m}\left(\frac{s^*}{\hat{s}}\right)^{\gamma-1} = 0;$$

that is Θ_γ depends on b and g only through the combination $\hat{s} = bg^{\frac{2}{\gamma-1}}$,

$$\Theta_\nu = \Theta\left(bg^{\frac{2}{\gamma-1}}\right) = \Theta(\hat{s}). \quad (3.10)$$

The differential cross section σ is defined according to

$$\sigma \sin\Theta\, d\Theta = b\, db \quad \text{so that } \sigma = \frac{b}{\sin\Theta \frac{\partial\Theta}{\partial b}}.$$

For hard spheres and power potentials we obtain

$$\sigma_{HS} = d^2 \cos\Theta \quad \text{and} \quad \sigma_\gamma = g^{-\frac{4}{\gamma-1}} \frac{\hat{s}}{\sin\Theta} \frac{d\hat{s}}{d\Theta} = g^{-\frac{4}{\gamma-1}} F(\Theta), \quad (3.11)$$

respectively, where $F(\Theta) = \frac{\hat{s}}{\sin\Theta}\frac{d\hat{s}}{d\Theta}$ is a function of Θ alone.

An important special case are Maxwell molecules, which have a power potential with $\gamma = 5$, where

$$\sigma g = F_M (\Theta) \ . \tag{3.12}$$

As Maxwell noticed, and as will be seen in the sequel, this particular choice allows much easier evaluation of many problems.

3.1.4 The Stosszahlansatz

The knowledge of the collision details allows to compute the change of the phase density $f(\mathbf{x}, t, \mathbf{c})$ through collisions, $\mathcal{S}$, that was introduced in (3.1). The following assumptions are implicitly used in the argument, known as "Stosszahlansatz":

- Only binary collisions take place. Indeed, even in a dense gas collisions between three or more particles are extremely unlikely, and can be safely ignored.
- It needs many collisions to change f significantly. In particular, the phase density does not change during a single collision.
- While f varies in space, it can be considered as constant over the range d of the interatomic forces.
- At every point in space-time $(\mathbf{x}, t)$, the values of f for different velocities are independent (assumption of "molecular chaos").

Each collision $\mathbf{c}, \mathbf{c}^1 \to \mathbf{c}', \mathbf{c}^{1\prime}$ diminishes the phase density $f(\mathbf{x}, t, \mathbf{c})$, while a so-called inverse collision $\mathbf{c}', \mathbf{c}^{1\prime} \to \mathbf{c}, \mathbf{c}^1$ increases $f(\mathbf{x}, t, \mathbf{c})$. To account for loss and gain, we write

$$\mathcal{S} = \mathcal{S}_+ - \mathcal{S}_- \ , \tag{3.13}$$

where $\mathcal{S}_+$ ($\mathcal{S}_-$) is the number density of collisions per unit time that produce (destroy) a phase point with velocity $\mathbf{c}$.

All particles with velocities $\mathbf{c}^1$ in the volume $g \, dt \, b \, db \, d\varepsilon$ depicted in Fig. 3.4 will collide with the particle in the center (velocity $\mathbf{c}$) during dt, so that the single particle experiences

$$f(\mathbf{x}, t, \mathbf{c}^1) \, d\mathbf{c}^1 \, g \, dt \, b \, db \, d\varepsilon$$

such collisions during dt (where the post-collision velocities are $\mathbf{c}', \mathbf{c}^{1\prime}$). The number density of particles with velocity $\mathbf{c}$ is $f(\mathbf{x}, t, \mathbf{c}) \, d\mathbf{c}$ so that the total number density of collisions $\mathbf{c}, \mathbf{c}' \to \mathbf{c}', \mathbf{c}^{1\prime}$ per unit time is

$$\mathcal{S}_- d\mathbf{c} = \int \int_0^{2\pi} \int_0^{\pi/2} f(\mathbf{x}, t, \mathbf{c}) \, f(\mathbf{x}, t, \mathbf{c}^1) \, g \, \sigma \sin \Theta \, d\Theta \, d\varepsilon \, d\mathbf{c}^1 d\mathbf{c} \ . \tag{3.14}$$

Here we have introduced the differential cross section (3.11), and integrated over all possible collision partners, that is over all velocities $\mathbf{c}^1$, all collision

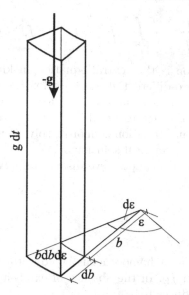

Fig. 3.4. Counting collisions.

angles Θ, and all angles ε which describe the orientation of the collision plane, see Fig. 3.4.

Next we consider inverse collisions $\mathbf{c}', \mathbf{c}^{1'} \to \mathbf{c}, \mathbf{c}^1$. By means of the same argument as above, the number density of these collisions per unit time is

$$f(\mathbf{x}, t, \mathbf{c}')\, f(\mathbf{x}, t, \mathbf{c}^{1'})\, g'\, \sigma'\, \sin\Theta\, d\Theta\, d\varepsilon\, d\mathbf{c}^{1'} d\mathbf{c}' \ ,$$

where the velocities are related according to (3.5). In the previous section we found that $g' = g$, $b' = b$, which implies $\Theta' = \Theta(b', g') = \Theta(b, g) = \Theta$, and therefore $\sigma' = \sigma$, and in Problem 3.3 it is shown that $d\mathbf{c}d\mathbf{c}^1 = d\mathbf{c}'d\mathbf{c}^{1'}$. Summation then gives the number density of collisions that yield a particle with velocity $\mathbf{c}$ as

$$\mathcal{S}_+ d\mathbf{c} = \int\int_0^{2\pi}\int_0^{\pi/2} f(\mathbf{x}, t, \mathbf{c}')\, f(\mathbf{x}, t, \mathbf{c}^{1'})\, g\,\sigma \sin\Theta\, d\Theta\, d\varepsilon\, d\mathbf{c}^1 d\mathbf{c} \ . \quad (3.15)$$

3.1.5 The Boltzmann equation

To simplify the notation, we write

$$f(\mathbf{x}, t, \mathbf{c}) = f \ , \quad f(\mathbf{x}, t, \mathbf{c}^1) = f^1 \ , \quad f(\mathbf{x}, t, \mathbf{c}') = f' \ , \quad f(\mathbf{x}, t, \mathbf{c}^1) = f^{1'} \ ,$$

and combine (3.1) with (3.13–3.15) to finally obtain the Boltzmann equation [25]

$$\frac{\partial f}{\partial t} + c_k \frac{\partial f}{\partial x_k} + G_k \frac{\partial f}{\partial c_k} = \int \int_0^{2\pi} \int_0^{\pi/2} \left(f' f^{1\prime} - f f^1 \right) g \sigma \sin \Theta \, d\Theta \, d\varepsilon \, d\mathbf{c}^1 .$$

$$(3.16)$$

The Boltzmann equation is the central equation in kinetic theory of gases. It is a non-linear integro-differential equation for the phase density f that describes the evolution in space and time of a monatomic ideal gas due to free flight, acceleration through external forces G_k, and collisions. Due to its complexity, the Boltzmann equation cannot be solved easily, and in almost all cases one has to rely on numerical solutions.

In the remainder of this chapter, we discuss important properties of the Boltzmann equation.

3.2 Equilibrium

Equilibrium is defined as a homogeneous steady state, and we ask for the equilibrium distribution f_E. In the absence of external forces ($G_k = 0$), the Boltzmann equation reduces to

$$\int \int_0^{2\pi} \int_0^{\pi/2} \left(f_E' f_E^{1\prime} - f_E f_E^1 \right) g \sigma \sin \Theta \, d\Theta \, d\varepsilon \, d\mathbf{c}^1 = 0 ,$$

and the integral vanishes if

$$f_E' f_E^{1\prime} - f_E f_E^1 = 0 \quad \text{or} \quad \ln f_E' + \ln f_E^{1\prime} = \ln f_E + \ln f_E^1 . \qquad (3.17)$$

For binary collisions of particles of mass m, mass, momentum and energy are conserved,

$$m + m = m + m ,$$
$$m\mathbf{c}' + m\mathbf{c}^{1\prime} = m\mathbf{c} + m\mathbf{c}^1 ,$$
$$\frac{m}{2} \left(\mathbf{c}' \right)^2 + \frac{m}{2} \left(\mathbf{c}^{1\prime} \right)^2 = \frac{m}{2} \left(\mathbf{c} \right)^2 + \frac{m}{2} \left(\mathbf{c}^1 \right)^2 ,$$

and it can be shown that these five quantities[2] are the only collisional invariants [8][18]. According to (3.17), $\ln f_E$ is a collisional invariant, and thus must be a linear combination of the known invariants,

$$\ln f_E = \alpha + \gamma_i c_i - \beta c^2 \quad \text{or, alternatively,} \quad f_E = A \exp \left[-\beta \left(c_i - \Lambda_i \right)^2 \right] \qquad (3.18)$$

(see Problem 3.4). The coefficients are determined by ensuring that this distribution gives the proper values for mass density, momentum density and energy density, i.e. by inserting f_E into

$$\rho = m \int f_E d\mathbf{c} , \quad \rho v_i = m \int c_i f_E d\mathbf{c} , \quad \rho u = \frac{3}{2} \rho \theta = \frac{m}{2} \int C^2 f_E d\mathbf{c} . \qquad (3.19)$$

---

[2] Momentum has three components, while mass and energy are scalar.

This yields

$$A = \frac{\rho}{m} \frac{1}{\sqrt{2\pi\theta}^3} \quad , \quad \beta = \frac{1}{2\theta} \quad , \quad \Lambda_i = v_i \qquad (3.20)$$

that is the Maxwellian (2.18)

$$f_E = f_M = \frac{\rho}{m} \frac{1}{\sqrt{2\pi\theta}^3} \exp\left[-\frac{(c_i - v_i)^2}{2\theta}\right] ,$$

which was found in Sec. 2.3 by means of a different argument.

In general, density $\rho(\mathbf{x}, t)$, temperature $\theta(\mathbf{x}, t)$, and velocity $v_i(\mathbf{x}, t)$ will depend on space and time, and thus a *local Maxwellian*

$$f_M(\mathbf{x}, t, \mathbf{c}) = \frac{\rho(\mathbf{x}, t)}{m} \frac{1}{\sqrt{2\pi\theta(\mathbf{x}, t)}^3} \exp\left[-\frac{(c_i - v_i(\mathbf{x}, t))^2}{2\theta(\mathbf{x}, t)}\right] \qquad (3.21)$$

can be assigned to each space point.

As long as the actual distribution f differs from the local Maxwellian, the effect of collisions is to force the distribution towards the local Maxwellian.

3.3 Equation of transfer and conservation laws

3.3.1 Equation of transfer

As discussed in Sec. 2.2, we shall often be interested in the moments of f rather than f itself. Moments are weighted averages of the distribution function, and evolution equations for moments follow by taking weighted averages of the Boltzmann equation. To simplify notation, we write

$$\rho \langle \psi \rangle = m \int \psi f d\mathbf{c} ,$$

where ψ is any function of $(\mathbf{x}, t, \mathbf{c})$, so that, e.g., $\rho = \rho\langle 1 \rangle$, $\rho v_i = \rho\langle c_i \rangle$, $\rho u = \frac{3}{2}\rho\theta = \rho\langle C^2 \rangle$, etc. The evolution equation for $\langle \psi \rangle$, the equation of transfer, is obtained by multiplying the Boltzmann equation (3.16) with $\psi(\mathbf{x}, t, \mathbf{c})$, and subsequent integration, as[3]

$$\frac{\partial \rho \langle \psi \rangle}{\partial t} + \frac{\partial \rho \langle \psi c_k \rangle}{\partial x_k} = \rho\left\langle \frac{\partial \psi}{\partial t} \right\rangle + \rho\left\langle c_k \frac{\partial \psi}{\partial x_k} \right\rangle + \rho\left\langle G_k \frac{\partial \psi}{\partial c_k} \right\rangle + S_\psi . \qquad (3.22)$$

The production term S_ψ is defined as

---

[3] For the derivation it is used that $\int \frac{\partial}{\partial c_k}(mG_k\psi f)\, d\mathbf{c} = \oint_{c\to\infty} mG_k\psi f n_k dA = 0$. The surface integral has to be taken over the velocity shell for $c \to \infty$, and vanishes since there are no particles with infinite energy, i.e. $f(c \to \infty) = 0$.

$$S_\psi = \int \psi S d\mathbf{c} = m \int\int \int_0^{2\pi} \int_0^{\pi/2} \psi \left(f' f^{1\prime} - f f^1 \right) g\sigma \sin\Theta \, d\Theta \, d\varepsilon \, d\mathbf{c}^1 d\mathbf{c} \ . \tag{3.23}$$

Due to the symmetry of the collision equations (3.2–3.5), the pairs $(\mathbf{c}, \mathbf{c}^1) \leftrightarrows (\mathbf{c}', \mathbf{c}^{1\prime})$ can be exchanged to give the alternative expression

$$S_\psi = m \int\int \int_0^{2\pi} \int_0^{\pi/2} \psi' \left(f f^1 - f' f^{1\prime} \right) g\sigma \sin\Theta \, d\Theta \, d\varepsilon \, d\mathbf{c}^1 d\mathbf{c} \ . \tag{3.24}$$

Exchange the integration variables $\mathbf{c} \leftrightarrows \mathbf{c}^1$ in (3.23), and $\mathbf{c}' \leftrightarrows \mathbf{c}^{1\prime}$ in (3.24) to give

$$S_\psi = m \int\int \int_0^{2\pi} \int_0^{\pi/2} \psi^1 \left(f' f^{1\prime} - f f^1 \right) g\sigma \sin\Theta \, d\Theta \, d\varepsilon \, d\mathbf{c}^1 d\mathbf{c} \ , \tag{3.25}$$

$$S_\psi = m \int\int \int_0^{2\pi} \int_0^{\pi/2} \psi^{1\prime} \left(f f^1 - f' f^{1\prime} \right) g\sigma \sin\Theta \, d\Theta \, d\varepsilon \, d\mathbf{c}^1 d\mathbf{c} \ . \tag{3.26}$$

Summation of (3.23–3.26) yields an expression for S_ψ that shows that the production term will vanish, if ψ is a collisional invariant, viz.

$$S_\psi = \frac{m}{4} \int\int \int_0^{2\pi} \int_0^{\pi/2} \left[\psi + \psi^1 - \psi' - \psi^{1\prime} \right] \left(f' f^{1\prime} - f f^1 \right) g\sigma \sin\Theta d\Theta d\varepsilon d\mathbf{c}^1 d\mathbf{c} \ . \tag{3.27}$$

Finally, by exchanging $(\mathbf{c}', \mathbf{c}^{1\prime}) \leftrightarrows (\mathbf{c}, \mathbf{c}^1)$ in the first term of (3.23), we find another useful formula, where the post-collision velocities $\mathbf{c}', \mathbf{c}^{1\prime}$ appear only in the argument of ψ, but not in f,

$$S_\psi = m \int\int \int_0^{2\pi} \int_0^{\pi/2} \left(\psi' - \psi \right) f f^1 g\sigma \sin\Theta \, d\Theta \, d\varepsilon \, d\mathbf{c}^1 d\mathbf{c} \ . \tag{3.28}$$

3.3.2 Conservation laws

The conservation laws for mass, momentum and energy follow by choosing ψ as a collisional invariant.

For $\psi = 1$, (3.22) with (3.27) reduces to the mass balance (continuity equation)

$$\frac{\partial F^0}{\partial t} + \frac{\partial F^0_k}{\partial x_k} = 0 \quad \text{or} \quad \frac{\partial \rho}{\partial t} + \frac{\partial \rho v_k}{\partial x_k} = 0 \ . \tag{3.29}$$

For $\psi = c_i$, (3.22) with (3.27) reduces to the balance of momentum

$$\frac{\partial F^0_i}{\partial t} + \frac{\partial F^0_{ik}}{\partial x_k} = F^0 G_i \quad \text{or} \quad \frac{\partial \rho v_i}{\partial t} + \frac{\partial \rho v_i v_k + p_{ik}}{\partial x_k} = \rho G_i \ . \tag{3.30}$$

For $\psi = \frac{1}{2} c^2$, (3.22) with (3.27) reduces to the balance of total energy

$$\frac{\partial \frac{1}{2} F^1}{\partial t} + \frac{\partial \frac{1}{2} F_k^1}{\partial x_k} = F^0 G_k v_k$$

or

$$\frac{\partial \rho \left(u + \frac{1}{2} v^2\right)}{\partial t} + \frac{\partial \rho \left(u + \frac{1}{2} v^2\right) v_k + p_{ik} v_i + q_k}{\partial x_k} = \rho G_k v_k \tag{3.31}$$

The moments appearing in these equations were introduced in Sec. 2.2.

The above balance laws are written in divergence form, that is they are of the general form

$$\frac{\partial \rho w_A}{\partial t} + \frac{\partial W_{Ak}}{\partial x_k} = \Pi_A + S_A . \tag{3.32}$$

Here w_A stands for the specific quantities, W_{Ak} is their flux, Π_A is the supply from the outside [15], and S_A is the production due to collisions. Indeed, integration over a fixed volume of space V, and application of Gauss's theorem gives

$$\frac{d}{dt} \int_V \rho w_A dV = - \oint_{\partial V} W_{Ak} n_k dA + \int_V (\Pi_A + S_A) \, dV$$

which shows that w_A in V changes due to the flux W_{Ak} over the surface, and to supply and production. For the conservation laws the production term vanishes, and the supply of energy and momentum is due to the external body force G_k.

Often the flux is split into the convective part $\rho w_A v_k$ and the non-convective part $\overline{W}_{Ak}$ as $W_{Ak} = \rho w_A v_k + \overline{W}_{Ak}$. By means of the mass balance, the general equation (3.32) can then be written as

$$\rho \frac{D w_A}{D t} + \frac{\partial \overline{W}_{Ak}}{\partial x_k} = \Pi_A + S_A . \tag{3.33}$$

Here

$$\frac{D}{Dt} = \frac{\partial}{\partial t} + v_k \frac{\partial}{\partial x_k} \tag{3.34}$$

is the convective time derivative, which describes the change of w_A for an observer resting in the gas, so that he moves with velocity v_k, in contrast to the derivative $\frac{\partial w_A}{\partial t}$ which describes the change at a fixed location.

In this notation, the conservation laws read

$$\frac{D\rho}{Dt} + \rho \frac{\partial v_k}{\partial x_k} = 0 ,$$

$$\rho \frac{Dv_i}{Dt} + \frac{\partial p_{ik}}{\partial x_k} = \rho G_i , \tag{3.35}$$

$$\rho \frac{D\left(u + \frac{1}{2} v^2\right)}{Dt} + \frac{\partial p_{ik} v_i + q_k}{\partial x_k} = \rho G_k v_k .$$

The balance of internal energy follows either by multiplying the momentum balance $(3.35)_2$ with v_i and subtracting the result from the energy balance $(3.35)_3$, or by choosing $\psi = \frac{1}{2} C^2 = \frac{1}{2} (c_i - v_i)^2$, as (see Problem 3.6)

$$\rho \frac{Du}{Dt} + \frac{\partial q_k}{\partial x_k} = -p_{ij} \frac{\partial v_i}{\partial x_j} \, . \tag{3.36}$$

Internal energy is not conserved, and the term on the right hand side describes the production of internal energy due to friction.

The natural variables occurring in the five conservation laws are mass density ρ, velocity v_i, and temperature θ (remember that $u = \frac{3}{2}\theta$), that is five quantities. However, the conservation laws contain also the heat flux vector q_k and the pressure tensor p_{ij}, and thus do not form a closed system of equations. Appropriate expressions for p_{ij} and q_k must be found in order to produce a closed set of equations on the macroscopic level.

3.3.3 Moment equations

The transport equations for the trace free moments $u_{i_1 \cdots i_n}^a$ (2.12) and the convective moments $F_{i_1 \cdots i_n}^\alpha$ (2.15) will be discussed in Chapter 5.

3.4 Entropy and the H-theorem

3.4.1 H-theorem

In the equation of transfer (3.22) we chose, following Boltzmann [25],

$$\psi = -k \ln \frac{f}{y} \tag{3.37}$$

with the Boltzmann constant k and another constant y. We introduce

$$\eta = -k \int f \ln \frac{f}{y} d\mathbf{c} \; , \quad \Phi_\kappa = -k \int c_k f \ln \frac{f}{y} d\mathbf{c} \tag{3.38}$$

and

$$\Sigma = S_{-k \ln \frac{f}{y}} = \frac{mk}{4} \int \int \int_0^{2\pi} \int_0^{\pi/2} \ln \frac{f' f^{1'}}{f f^1} \left(f' f^{1'} - f f^1 \right) g\sigma \sin \Theta \, d\Theta \, d\varepsilon \, d\mathbf{c}^1 d\mathbf{c} \, , \tag{3.39}$$

so that the corresponding transport equation for η reads

$$\frac{\partial \eta}{\partial t} + \frac{\partial \Phi_k}{\partial x_k} = \Sigma \, . \tag{3.40}$$

Closer examination shows that the collision term (3.39) cannot be negative (recall that $g \geq 0$, $\sigma \geq 0$, $\sin \Theta \geq 0$ for Θ in $[0, \pi/2]$). Σ vanishes if $\ln f$ is a collisional invariant, as is the case in equilibrium (Sec. 3.2), so that

$$\Sigma \geq 0 \, .$$

Thus, η always has a positive production which vanishes in equilibrium. Accordingly, η can only grow in an isolated system, where no flux over the surface is allowed ($\oint_{\partial V} \Phi_k n_k dA = 0$), and reaches its maximum in equilibrium, where $\Sigma = 0$. This property of η and in particular the definite sign of Σ are known as the H-theorem.[4]

The H-theorem is equivalent to the second law of thermodynamics, the entropy law, which was introduced on purely phenomenological grounds, see, e.g., [15][18], or standard engineering textbooks [26]–[28]. Therefore η is the entropy density of the gas, and we write

$$\eta = \rho s ,$$

where s denotes the specific entropy. Φ_κ as given in $(3.38)_2$ is the entropy flux, and the non-convective entropy flux $\phi_k = \Phi_k - \rho s v_k$ can be computed according to

$$\phi_\kappa = -k \int C_k f \ln \frac{f}{y} d\mathbf{c} . \tag{3.41}$$

This gives the second law in the form

$$\rho \frac{Ds}{Dt} + \frac{\partial \phi_k}{\partial x_k} = \Sigma \geq 0 ; \tag{3.42}$$

Σ is the entropy production (entropy generation rate).

The specific entropy of the monatomic gas in equilibrium follows by evaluating $(3.38)_1$ with the Maxwellian (3.21) as

$$s = \frac{k}{m} \ln \frac{\theta^{3/2}}{\rho} + s_0 \text{ where } s_0 = \frac{k}{m} \left[\frac{3}{2} + \ln \left(my\sqrt{2\pi}^{-3} \right) \right] . \tag{3.43}$$

This result stands in agreement with classical thermodynamics [15]; s_0 is the entropy constant.

3.4.2 $H = k \ln W$—entropy and disorder

Boltzmann's choice (3.37) allows to relate entropy to microscopic properties, and to interpret entropy as a measure for disorder [15].

We consider a gas of N particles in the volume V, and compute its total entropy as

$$H = \int_V \eta d\mathbf{x} = -k \int_V \int f \ln \frac{f}{y} d\mathbf{c} d\mathbf{x} .$$

The number of particles in a phase cell is, by (2.1),

$$N_{\mathbf{x},\mathbf{c}} = f d\mathbf{x} d\mathbf{c} = fY$$

---

[4] H stands for the capital greek letter "eta". Boltzmann has a different sign for the H-function, which then is decreasing in time. This text follows the tradition to have a positive entropy, which is growing in time.

where $Y = d\mathbf{x}d\mathbf{c}$ is the size of the cell. By writing sums over phase cells instead of integrals, and with $N = \sum_{\mathbf{x},\mathbf{c}} N_{\mathbf{x},\mathbf{c}}$ follows

$$H = -k \sum_{\mathbf{x},\mathbf{c}} N_{\mathbf{x},\mathbf{c}} \ln \frac{N_{\mathbf{x},\mathbf{c}}}{yY} = -k \sum_{\mathbf{x},\mathbf{c}} N_{\mathbf{x},\mathbf{c}} \ln N_{\mathbf{x},\mathbf{c}} + kN \ln yY \ .$$

Use of Stirling's formula, $\ln N! = N \ln N - N$, yields

$$H = -k \sum_{\mathbf{x},\mathbf{c}} (\ln N_{\mathbf{x},\mathbf{c}}! + N_{\mathbf{x},\mathbf{c}}) + kN \ln yY = k \ln \frac{N!}{\prod_{\mathbf{x},\mathbf{c}} N_{\mathbf{x},\mathbf{c}}!} + kN \ln \frac{yY}{N} \ .$$

Choosing $y = N/Y$ finally allows to write

$$H = k \ln W \tag{3.44}$$

where

$$W = \frac{N!}{\prod_{\mathbf{x},\mathbf{c}} N_{\mathbf{x},\mathbf{c}}!} \tag{3.45}$$

is the number of possibilities to distribute N particles into the cells of phase space, so that the cell at $(\mathbf{x}, \mathbf{c})$ contains $N_{\mathbf{x},\mathbf{c}}$ particles.

Equation (3.44) relates the total entropy H to the number of possibilities to realize the state of the gas. The growth of entropy, which is imperative in an isolated system, therefore corresponds to an increasing number of possibilities to realize the state. Since a small number of possibilities refers to an ordered state, and a large number to disorder, we can say that the H-theorem states that disorder must grow in an isolated process. Accordingly, entropy is a measure for disorder.

The relation (3.44) is generally accepted as being valid not only for monatomic ideal gases—where it originated—but for any substance. However, the evaluation for other substances can be quite difficult, or impossible, since it requires a detailed understanding of the microscopic details of phase space, accessible states, etc., in order to determine W properly.

3.4.3 Gibbs paradox

With the choice $y = N/Y$ the entropy constant (3.43) depends on the particle number N, since then

$$s_0 = \frac{k}{m} \left[\frac{3}{2} + \ln \left(\frac{mN}{Y} \sqrt{2\pi}^{-3} \right) \right] \ . \tag{3.46}$$

This dependence on particle number causes entropy to be non-additive, as will be discussed now. Figure 3.5 shows a gas of N particles at temperature θ in two situations, (I) in a container of volume V, and (II) being equally distributed between two containers of volume $V/2$. The equilibrium entropy for the two cases is, by means of (3.43, 3.46),

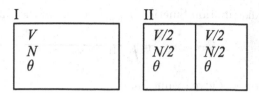

Fig. 3.5. On the Gibbs paradox.

$$H_I = \rho V s_I (V, N, \theta) = \rho V \frac{k}{m} \ln \frac{\theta^{3/2}}{\rho} + \rho V s_{0,I} \,,$$

$$H_{II} = 2 \left(\rho \frac{V}{2} s_{II} \left(\frac{V}{2}, \frac{N}{2}, \theta \right) \right) = \rho V \frac{k}{m} \ln \frac{\theta^{3/2}}{\rho} + \rho V s_{0,II} \,,$$

respectively, and their difference is

$$H_{II} - H_I = \rho V (s_{0,II} - s_{0,I}) = -Nk \ln 2 \,.$$

Therefore, with $y = N/Y$, the entropy of case II is smaller, so that this case appears to have a larger degree of order. This result is referred to as Gibbs paradox, the paradox lying in the fact that adding or removing a wall does not change the state of the gas, so that all its properties, including entropy, should be the same in both cases.

The solution to the paradox lies in reconsidering the counting of states to compute W: W as computed in (3.45) is the proper number of possibilities to distribute N *distinguishable* particles over the space cells. If particles can be distinguished, then situation II is indeed more ordered, since particles are confined to either side of the container, and cannot change to the other side. In situation I, the particles can move freely all over the container, and thus disorder is higher.

In reality, particles cannot be distinguished,[5] and both cases are identical. Proper computation of the number of possibilities W must include that the particles are in fact identical, and if this is done, all results presented above hold, but y—and therefore s_0— turns out to be independent of the particle number N. Problem 3.8 leads through the necessary considerations and computations.

3.4.4 Equilibrium and maximum disorder

If a gas is left to itself, its entropy will increase until it has reached the maximum value it can have in accordance with the values of the conserved quantities mass, momentum and energy. This statement allows for another derivation of the Maxwellian as the equilibrium distribution.

---

[5] It is impossible to "name-tag" an atom.

We ask what distribution function $f_{\max}$ maximizes the entropy

$$\rho s = -k \int f \ln \frac{f}{y} d\mathbf{c}$$

under the constraints of given values of

$$\rho = m \int f d\mathbf{c} \;, \quad 0 = m \int C_i f d\mathbf{c} \;, \quad \rho u = \frac{3}{2}\theta = \frac{m}{2} \int C^2 f d\mathbf{c} \;. \qquad (3.47)$$

The constraints can be incorporated by means of Lagrange multipliers Λ_a (see Appendix A.5) so that we need to maximize

$$\phi = -k \int f \ln \frac{f}{y} d\mathbf{c} + \Lambda_\rho \left(\rho - m \int f d\mathbf{c} \right) + \Lambda_{\rho v_k} \left(0 - m \int C_i f d\mathbf{c} \right)$$

$$+ \Lambda_{\rho u} \left(\frac{3}{2}\theta - \frac{m}{2} \int C^2 f d\mathbf{c} \right) \;. \qquad (3.48)$$

This is a standard problem of variational calculus[6] with the solution

$$f_{\max} = y \exp[-1 - \frac{m}{k}\Lambda_\rho - \frac{m}{k}\Lambda_{\rho v_k} C_k - \frac{m}{2k}\Lambda_{\rho u} C^2] \;.$$

The Lagrange multipliers must be determined so that the constraints (3.47) are fulfilled, and this is just the same problem as in Sec. 3.2, so that $f_{\max} = f_M$.

3.4.5 Irreversibility

We consider the gas on the molecular level, where the initial state $(\mathbf{x}_0^\alpha, \mathbf{c}_0^\alpha)$ is assigned to the particles α at time t_0. As time proceeds, the particles move through space, and interact, and after a time Δt they are in a state $(\mathbf{x}_1^\alpha, \mathbf{c}_1^\alpha)$. Now, the velocity vectors of all particles are reversed, so that the particles are in the state $(\mathbf{x}_1^\alpha, -\mathbf{c}_1^\alpha)$. Time proceeds again, and after the time Δt the particles will be in the state $(\mathbf{x}_0^\alpha, -\mathbf{c}_0^\alpha)$. In other words, if the particle velocity is reversed after some time, the particles will return to their initial locations $\mathbf{x}_0^\alpha$, and the velocity will be the same as initially, only in the opposite direction— the laws of motion on the molecular level are reversible.

Next we consider the same problem on the level of the Boltzmann equation. The initial state is now given by a phase density $f_0(\mathbf{x}, \mathbf{c})$ with the corresponding entropy density η_0. The development in time is described by the Boltzmann equation, so that after some time Δt the phase density $f_1(\mathbf{x}, \mathbf{c})$ is observed. As the phase density evolves, the entropy η of the gas, and therefore

---

[6] The maximum of $J[y] = \int_{x_0}^{x_1} F(x, y, y') dx$, with $y = y(x)$, $y' = \frac{dy}{dx}$, and fixed x_0, x_1, $y(x_0)$, $y(x_1)$, is given by Euler's equation $\frac{\partial F}{\partial y} - \frac{d}{dx}\frac{\partial F}{\partial y'} = 0$. In the present case $y = f$, $x = \mathbf{C}$, and $F = F(\mathbf{C}, f)$ is the integrand in (3.48); note that F does not depend on f'.

disorder, is increasing, $\eta_1 > \eta_0$. Next, particle velocities are reversed, that is the phase density is switched to $f_{1'}(\mathbf{x}, \mathbf{c}) = f_1(\mathbf{x}, -\mathbf{c})$, but, since entropy is a scalar quantity, $\eta_{1'} = \eta_1$. When the phase density evolves again for the time span Δt it cannot return to the state $f_{0'}(\mathbf{x}, \mathbf{c}) = f_0(\mathbf{x}, -\mathbf{c})$ since that state has a smaller entropy, $\eta_{0'} = \eta_0 < \eta_{1'}$. In other words, the evolution of the gas as described by the Boltzmann equation is irreversible.

While the Boltzmann equation is based on the reversible motion of the particles, irreversibility has entered the equation along its derivation (Loschmidt's paradox). Chapman and Cowling locate this in the assumption of molecular chaos in the derivation of the Boltzmann collision term [2]. Molecular chaos requires that the velocities of two colliding particles are uncorrelated, which will be the case when the two particles have not influenced each other some time before the collision. Note that the particles in average have travelled one mean free path before colliding, and thus come from distant regions. Just after the collision, however, the particles's velocities are strongly correlated. When the velocities are reversed, particles that just collided return into the collision, and thus are strongly correlated before the collision—the assumption of molecular chaos does not hold. Cercignani points out that the phase densities in the Boltzmann collision term $\mathcal{S}$ are taken as functions of the pre-collision velocities $(\mathbf{c}, \mathbf{c}^1)$, and this choice introduces a direction in time (from "pre" to "post"), and thus irreversibility [3].

Moreover, recalling Sec. 2.1, we mention that accuracy is lost by describing the state of the gas through the phase density, since it is only known that particles are in a cell of phase space, while the exact phase space location of the particles is not known. This introduces a statistical component into the description of the gas, and thus the Boltzmann equation describes the most likely evolution of a gas. This remark is related to the recurrence argument put forward against Boltzmann by Zermelo: A theorem of Poincaré in classical mechanics shows that every mechanical system will return arbitrarily close to its initial state after some time. Obviously, this cannot happen for solutions of the Boltzmann equation for which entropy must be increasing, while the return to the initial state must involve an eventual decrease of entropy. For a gas, with its large number of particles, the recurrence time is extremely long, and exceeds the age of the universe by far. Thus, an recurrence will not be observed, and must be considered as very unlikely.

3.5 Collision frequency

The number density of collisions involving particles of velocity $\mathbf{c}$ (3.14) can be written as

$$\mathcal{S}_- d\mathbf{c} = \nu f(\mathbf{x}, t, \mathbf{c}) \, d\mathbf{c} \,,$$

where the collision frequency ν is defined as

$$\nu = \int f^1 \, g \, \sigma \sin \Theta \, d\Theta \, d\varepsilon \, d\mathbf{c}^1 \,. \tag{3.49}$$

We evaluate the collision frequency close to equilibrium, by replacing f^1 with the Maxwellian f_M^1. With the dimensionless velocities $\xi_i = C_i^1/\sqrt{2\theta}$, $\eta_i = C_i/\sqrt{2\theta}$ and (3.11) we obtain for hard spheres and power potentials

$$\nu_{HS} = \nu_{HS}^0 L_\infty \text{ and } \nu_\gamma = \nu_\gamma^0 L_\gamma ,$$

respectively, where

$$\nu_{HS}^0 = \frac{\rho}{m\pi^{3/2}}\sqrt{2\theta}d^2 \int \cos\Theta \sin\Theta\, d\Theta \int d\varepsilon = \sqrt{\frac{2}{\pi}}\frac{\rho}{m}\sqrt{\theta}d^2 ,$$

$$\nu_\gamma^0 = \frac{2}{\sqrt{\pi}}\frac{\rho}{m}\sqrt{2\theta}^{\frac{\gamma-5}{\gamma-1}} \int F(\Theta)\sin\Theta d\Theta ,$$

and

$$L_\gamma = \int \exp\left[-\xi^2\right] |\xi_i - \eta_i|^{\frac{\gamma-5}{\gamma-1}} \xi^2 \sin\chi d\chi d\varphi d\xi .$$

The angles are chosen so that $\xi_i = \xi\{\sin\varphi\sin\chi, \cos\varphi\sin\chi, \cos\chi\}$ and $\eta_i = \{0,0,\eta\}$, and with $X(\chi) = \xi^2 + \eta^2 - 2\xi\eta\cos\chi$ we obtain

$$L_\gamma = \int \exp\left[-\xi^2\right] \sqrt{X(\chi)}^{\frac{\gamma-5}{\gamma-1}} \xi^2 \sin\chi d\chi d\zeta d\xi .$$

The relation $dX = 2\xi\eta\sin\chi d\chi$ and integration with respect to X in the proper limits $\left[(\xi-\eta)^2, (\xi+\eta)^2\right]$ yields

$$L_\gamma = 2\pi\frac{\gamma-1}{3\gamma-7}\int_{\xi=0}^\infty \frac{\xi}{\eta}\exp\left[-\xi^2\right]\left[(\xi+\eta)^{\frac{3\gamma-7}{\gamma-1}} - |\xi-\eta|^{\frac{3\gamma-7}{\gamma-1}}\right]d\xi .$$

For Maxwell molecules ($\gamma = 5$) the collision frequency is independent of the particle velocity, and independent of temperature,

$$\nu_{Maxwell} = 4\pi\nu_5^0 \int_{\xi=0}^\infty e^{-\xi^2}\xi^2 d\xi = 2\pi\frac{\rho}{m}\int F(\Theta)\sin\Theta d\Theta = \rho\nu_M^0 .$$

For hard spheres one finds [2]

$$\nu_{HS}(\eta) = \frac{\rho}{m}\sqrt{2\pi\theta}d^2\left\{e^{-\eta^2} + \frac{\sqrt{\pi}}{2}\left(\frac{1}{\eta} + 2\eta\right)\text{erf}\,\eta\right\} .$$

Interestingly, the result for hard spheres corresponds to the power potential $\gamma \to \infty$, when one matches $\nu_{HS}^0 = \nu_\infty^0$.

For other values of γ the collision frequency must be integrated numerically, and Fig. 3.6 shows the normalized dimensionless collision frequency as a function of the dimensionless velocity $\eta = C/\sqrt{2\theta}$ for a variety of values of γ. It can be seen that ν is a monotonously increasing function of speed. This indeed meets the expectations: fast particles should collide more often and should therefore have a larger collision frequency.

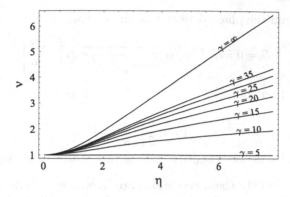

Fig. 3.6. The dimensionless and normalized collision frequency ν as a function of dimensionless microscopic velocity η for some power potentials with exponent γ.

Mean values of the collision frequency can be computed according to

$$\bar{\nu} = \frac{1}{n}\int \nu f d\mathbf{c} = \frac{1}{n}\int\int f f^1 \, g\,\sigma\sin\Theta\,d\Theta\,d\varepsilon\,d\mathbf{c}^1 d\mathbf{c} \;, \tag{3.50}$$

and thus one finds with the Maxwellian (compare to Sec. 1.3 and Problem 2.6)

$$\bar{\nu}_{HS} = \pi d^2 n\bar{g} = \sqrt{16\pi}d^2 n\sqrt{\theta} \text{ and } \bar{\nu}_\gamma = \rho\theta^{\frac{\gamma-5}{2\gamma-2}}\bar{\nu}_\gamma^0 \;, \tag{3.51}$$

where $\bar{\nu}_\gamma^0$ is a constant. Equation (3.51) most notably gives the dependence of the mean collision frequency on density and temperature.

Note that for Maxwell molecules the collision frequency depends only on density, $\bar{\nu}_{Maxwell} = \rho\bar{\nu}_5^0$, and that the power potential $\gamma \to \infty$ agrees with hard spheres when one defines their diameter through $d^2 = \bar{\nu}_\infty^0 m/\sqrt{16\pi}$.

3.6 Kinetic Models

The Boltzmann equation (3.16) is very difficult to handle, mostly due to the complexity of the collision term S. Therefore one is interested in model equations which are simpler than the Boltzmann equation but which should have the same main features of the original collision term.

The most important properties of S are:

1. It guarantees the conservation of mass, momentum and energy, since

$$m\int S d\mathbf{c} = 0 \;, \quad m\int c_i S d\mathbf{c} = 0 \;, \quad \frac{m}{2}\int c^2 S d\mathbf{c} = 0 \;. \tag{3.52}$$

2. The production of entropy is always positive (H-theorem),

$$\Sigma = -k\int \ln f S d\mathbf{c} \geqslant 0 \;.$$

3. In equilibrium the phase density is a Maxwellian, i.e.

$$S = 0 \Longrightarrow f = f_M = \frac{\rho}{m} \frac{1}{\sqrt{2\pi\theta}^3} \exp\left[-\frac{C^2}{2\theta}\right] .$$

4. The Prandtl number (1.7) is close to $\frac{2}{3}$ for all physically meaningful collision factors σ, i.e.

$$\mathrm{Pr} = \frac{5}{2} \frac{k}{m} \frac{\mu}{\kappa} \simeq \frac{2}{3}$$

where μ and κ denote viscosity and thermal conductivity, respectively.

Models that satisfy these conditions are known as kinetic models. Best known among these are the Bhatnager-Gross-Krook model (BGK model) [29]–[31] and the Ellipsoidal-Statistical-BGK model (ES-BGK model) [32], which both will be discussed below, as well as the S-model [33]–[35] and the Liu-model [36]–[38]. All these models employ mean collision frequencies that are independent of the microscopic velocity. Models that incorporate velocity dependent collision frequencies are discussed in [39]–[42].

3.6.1 The BGK model

For motivation of the BGK equation we simplify the Boltzmann collision term in three steps [14].

Step 1: Because of the collisions the phase density will tend to the local Maxwellian f_M. The phase densities $f' f'^1$ in the collision term (3.16) refer to the velocities after the collision—they may be replaced by $f'_M f'^1_M$,

$$S \rightarrow \widehat{S}_M = \int \left(f'_M f'^1_M - f f^1 \right) \sigma g \sin \Theta d\Theta d\varepsilon d\mathbf{c}_1 .$$

Step 2: Since $\ln f_M$ is a linear combination of the collisional invariants we have $f'_M f'^1_M = f_M f^1_M$, and therefore

$$\widehat{S}_M \rightarrow \widetilde{S}_M = f_M \int f^1_M \sigma g \sin \Theta d\Theta d\varepsilon d\mathbf{c}_1 - f \int f^1 \sigma g \sin \Theta d\Theta d\varepsilon d\mathbf{c}_1 .$$

Step 3: The difference between the two integrals may be neglected. This last step leads to the BGK collision term

$$\widetilde{S}_M \rightarrow S_{BGK} = -\nu \left(f - f_M \right)$$

where ν is the collision frequency (3.49). The conservation conditions (3.52) now read

$$m \int \nu \left(f - f_M \right) d\mathbf{c} = 0 , \quad m \int \nu C_i \left(f - f_M \right) d\mathbf{c} = 0 ,$$

$$\frac{m}{2} \int \nu C^2 \left(f - f_M \right) d\mathbf{c} = 0 , \tag{3.53}$$

and are only fulfilled when ν does not depend on the microscopic velocity [40]. Therefore, one either must consider Maxwell molecules, where the collision frequency is a constant, or replace ν by a mean collision frequency $\bar{\nu}$, so that

$$\mathcal{S}_{BGK} = -\bar{\nu}\,(f - f_M) \,. \tag{3.54}$$

In this case, the conservation conditions (3.53) imply that the local Maxwellian f_M and the actual phase density f have the the first five moments in common,

$$\rho = m \int f d\mathbf{c} = m \int f_M d\mathbf{c} \,, \quad 0 = m \int C_i f d\mathbf{c} = m \int C_i f_M d\mathbf{c} \,,$$

$$\rho u = \frac{m}{2} \int C^2 f d\mathbf{c} = \frac{m}{2} \int C^2 f_M d\mathbf{c} \,. \tag{3.55}$$

It must be emphasized that—because of (3.55)—the BGK equation, i.e. the Boltzmann equation with the collision term (3.54), is a non-linear integro-differential equation, just like the Boltzmann equation.

As was just shown, the conservation of mass, momentum and energy is ensured by using the local Maxwellian (property 1), and we check further properties of the BGK model.

Since, with (3.55),

$$k \int \ln f_M \mathcal{S}_M d\mathbf{c} = 0$$

the entropy production can be written as

$$\Sigma = -k \int \mathcal{S}_M \ln f d\mathbf{c} + k \int \mathcal{S}_M \ln f_M d\mathbf{c} = \nu k \int \ln \frac{f}{f_M} (f - f_M) \, d\mathbf{c} \geqslant 0 \,.$$

It follows that the H-theorem is fulfilled (property 2).

In thermodynamic equilibrium (characterized by the subscript E) the BGK collision term vanishes, and this implies the Maxwellian, $f_{|E} = f_M$ (property 3).

The calculation of the Prandtl number (property 4) will be performed in subsequent chapters, where it will be shown that the BGK model leads to $Pr = 1$, and thus fails in the accurate description of the gas. Nevertheless, the BGK model is widely used, since it maintains the properties of the Boltzmann equation reasonably well.

3.6.2 The ES-BGK model

In order to obtain the proper Prandtl number, Holway suggested the ellipsoidal statistical BGK model (ES-BGK model), where the Maxwellian of the standard BGK model is replaced by an anisotropic Gaussian so that the collision term reads [32]

$$\mathcal{S}_{ES} = -\bar{\nu}\,(f - f_{ES}) \tag{3.56}$$

where

$$f_{ES} = \frac{\rho}{m} \frac{1}{\sqrt{\det [2\pi \lambda_{ij}]}} \exp \left[-\frac{1}{2} \lambda_{ij}^{-1} C_i C_j \right] , \qquad (3.57)$$

and the matrix λ_{ij} is given by

$$\lambda_{ij} = \theta \delta_{ij} + b \frac{\sigma_{ij}}{\rho} ; \qquad (3.58)$$

λ_{ij}^{-1} denotes the inverse matrix which must be positive definite, and thus b is restricted to $-\frac{1}{2} \leq b \leq 1$ [32][43]. It is straightforward to show that this model has properties 1 and 3 (Problem 3.9). The parameter b can be adjusted to deliver the proper Prandtl number (property 4.); in Sec. 4.2 it will be shown that $b = \left(1 - \frac{1}{\Pr} \right)$. The proof of property 2 (H-theorem) is non-trivial, and was presented only recently [43].

The ES-BGK model assumes that the collision frequency is independent of the microscopic velocity. The Prandtl number can be adjusted at will. For $b = 0$, the ES-BGK model condenses into the standard BGK model (3.54).

3.7 The Direct Simulation Monte Carlo method

Starting with the next chapter, we shall discuss equations that are derived from the Boltzmann equation under approximations, with the goal to have simpler models that stand in agreement with the Boltzmann equation. The approximative models can be tested by comparing their predictions to experiments, or to predictions of the Boltzmann equation, i.e. numerical solutions of the latter.

Numerical solutions of the Boltzmann equation and of kinetic model equations can be found by discretizing the phase space in a suitable manner by, e.g., finite differences, discrete velocity models, etc.

The most widely used method nowadays is the Direct Simulation Monte Carlo method (DSMC) developed by Bird since the 1960's [1], see, e.g., [44][45] for a wide range of applications.

In short, the DSMC method proceeds as follows [1]: The physical space ($\mathbf{x}$) is divided into cells of finite size $\Delta \mathbf{x}$ which are populated by sample particles. The number of sample particles is much less than the actual number of molecules in the gas that is simulated, and each sample particle describes A_N actual particles. The sample particles move with different microscopic speeds for a short time Δt and then particles that reside in the same space cell are allowed to interact. For the interaction, molecule pairs are picked randomly, and then undergo collisions where the collision parameter is chosen from statistical models. After the collisions are performed, the particles undergo the next period of free flight (with new velocities) for the time Δt and so on. Wall collisions are taken into account, and different molecule models can easily be implemented.

It was shown that the DSMC method is equivalent to solving the Boltzmann equation [46]–[48]. The method is very powerful, and complicated molecular processes can be modelled, including complex molecules with two or more atoms, dense gases, and chemical reactions, for which the proper Boltzmann type transport equations are not even known.

Due to the statistical nature of the method the results contain numerical error in form of stochastic noise. For the computation of steady state processes noise is reduced by running the process for some time and then taking the time average, while for transient processes the process has to be simulated often, and then taking the ensemble average. The error is significantly reduced by increasing the number of sample molecules, and the dramatic progress in computational speed of computers now allows the direct simulation of complicated flows. Of course, the method is well suited for parallel computing.

DSMC simulations will serve as benchmarks for the equations derived later, and for these the Fortran programs that accompany Bird's book [1] were used. Minor adjustments to the programs were made in order to compute additional moments.

Problems

3.1. Collision angle for hard spheres
Compute the collision angle for hard spheres from (3.7) and (3.8).

3.2. Interaction potentials
Draw sketches of the interaction potential for hard spheres and power potentials as function of the interatomic distance r. Discuss the notion of the mean free path for power potentials molecules—does it make sense?

3.3. Velocity space elements
Introduce $\xi_A = \{c, c_1\}$ and $\xi_A' = \{c', c_1'\}$, and show that $\xi_A = A_{AB}\xi_B'$ and $\xi_A' = A_{AB}\xi_B$. Identify the matrix A. The above means that the matrix A is its own inverse. Show that this implies $\det A = 1$ and use this to prove $dc'dc_1' = dcdc_1$.

3.4. Equilibrium distribution
(a) Show that the two equations (3.18) are equivalent, by relating the coefficients $(\alpha, \beta, \gamma_i)$ to the coefficients (A, β, Λ_i).
(b) Insert $(3.18)_2$ into (3.19), and show that (3.20) holds.

3.5. Gas in gravitational field
Assume an isothermal atmosphere, $\theta = \theta_0$, at rest in the gravitational field, where $G_k = \{0, 0, -g\}$ and $g = 9.81 \, \text{m}/\text{s}^2$ is the gravitational acceleration. Show that the equilibrium distribution function is of the form

$$f_E = \exp\left[-\frac{gz}{\theta_0}\right] f_M(\rho_0, \theta_0) .$$

Compute the density ρ as a function of height.

3.6. Energy balances

(a) Chose $\Psi = \frac{1}{2}c^2$ in (3.22), and derive the balance of total energy.

(b) Chose $\Psi = \frac{1}{2}C^2$ in (3.22), and find the balance of internal energy.

(c) The term $\rho G_k v_k$ in the balance of total energy (3.31) describes the power of body forces. Consider the case that the specific body force G_k follows from a time-independent potential Ψ (e.g. $\Psi = gz$) as $G_k = -\frac{\partial \Psi}{\partial x_k}$ and show that the balance can be written as

$$\frac{\partial \rho \left(u + \frac{1}{2}v^2 + \Psi\right)}{\partial t} + \frac{\partial \left[\rho \left(u + \frac{1}{2}v^2 + \Psi\right) v_k + p_{jk}v_j + q_k\right]}{\partial x_k} = 0 .$$

(d) In undergraduate thermodynamics, e.g. [26]–[28], the first law for an open system is commonly written as

$$\frac{dE}{dt} = \sum_{in} \dot{m}_i \left(h + \frac{1}{2}v^2 + gz\right)_{i_i} - \sum_{out} \dot{m}_e \left(h + \frac{1}{2}v^2 + gz\right)_e + \dot{Q} - \dot{W} .$$

This equation follows from the balance given in the previous problem through integration over the control volume. Identify the quantities E, $\dot{m}_e$, $\dot{Q}$, $\dot{W}$.

3.7. Entropy in equilibrium

Using the Maxwell distribution, compute the specific entropy in equilibrium $s_E (\rho, T)$ from (3.38).

3.8. Entropy of fermions and bosons[15]

The Goal is to compute the entropy of fermions and bosons from $S = k \ln W$. For this, compute the number of realizations of a state from the following information: Phase space is divided into cells Ξ_s of volume $d\mathbf{x}d\mathbf{c}$ and in each cell there are $\chi_s = yd\mathbf{x}d\mathbf{c}$ locations ($1/y$ is the phase space volume for one location). The number of particles in Ξ_s is given by $N_s = fd\mathbf{x}d\mathbf{c}$.

(a) Fermions are indistinguishable and cannot be in the same quantum state as another particle (Pauli exclusion principle). Thus, a location can only be occupied by zero or one particle. Show that $W = \prod_s \frac{\chi_s!}{N_s!(\chi_s - N_s)!}$.

(b) Bosons are indistinguishable, and an arbitrary number of bosons can occupy one location. Show that $W = \prod_s \frac{(N_s + \chi_s - 1)!}{N_s!(\chi_s - N_s)!}$.

(c) Replace χ_s and N_s through y and f, turn the sums into integrals, and show that the entropy for fermions (upper sign) and bosons (lower sign) is given by

$$S = -k \int \left(f \ln \frac{f}{y} \pm (y \mp f) \ln \left(1 \mp \frac{f}{y}\right)\right) d\mathbf{x}d\mathbf{c} .$$

Here, you have to assume $\chi_s - 1 \approx \chi_s$ in the boson case.

(d) Determine the equilibrium distribution by maximizing the entropy under the constraints of given particle number and energy according to $N =$

$\int f d\mathbf{x} d\mathbf{c}$ and $E = \frac{m}{2} \int C^2 f d\mathbf{x} d\mathbf{c}$. Show that the equilibrium distribution function assumes the form

$$f_E = \frac{1}{\exp\left(\alpha + \beta \frac{m}{2} C^2\right) \pm 1}$$

where α, β are Lagrange multipliers.

(e) Compute density ρ, specific internal energy u and specific entropy s, and show that they can be written as

$$\rho = y \frac{8\pi\sqrt{2}}{\sqrt{m}} \frac{1}{\sqrt{\beta}} I_2^{\mp}(\alpha) \; , \quad u = \frac{1}{m} \frac{1}{\beta} \frac{I_4^{\mp}(\alpha)}{I_2^{\mp}(\alpha)} \; , \quad s = \frac{k}{m}\alpha + \frac{5}{3} \frac{k}{m} \frac{I_4^{\mp}(\alpha)}{I_2^{\mp}(\alpha)}$$

where

$$I_n^{\mp}(\alpha) = \int_0^\infty \frac{x^n}{\exp\left(\alpha + x^2\right) \pm 1} \; .$$

(f) Show that the Gibbs equation holds as

$$ds = k\beta \left(du + \frac{2}{3}\rho u d\frac{1}{\rho} \right)$$

and identify from this temperature as $T = \frac{1}{k\beta}$.

(g) Classical limit: Assume $\alpha \gg 1$, and show that in this case $I_2^{\mp}(\alpha) = \frac{\sqrt{\pi}}{4}e^{-\alpha}$ and $I_4^{\mp}(\alpha) = \frac{3}{2}\frac{\sqrt{\pi}}{4}e^{-\alpha}$. Show that this results in the Maxwellian, and the well known classical results

$$p = \frac{2}{3}\rho u = \rho\theta \quad \text{and} \quad s = \frac{k}{m}\ln\frac{\theta^{3/2}}{\rho} + \frac{k}{m}\left[\frac{5}{2} + \ln\left(my\sqrt{2\pi}^{-3}\right)\right] \; .$$

Compared with the result we had before (3.43), we have a slightly different entropy constant. Note, that y is a constant, independent of N, and this resolves the Gibbs paradox.

3.9. ES-BGK model

(a) Show that the ES-BGK model conserves mass, momentum, and energy.

(b) Show that for the ES-BGK model the equilibrium phase density is a Maxwellian (property 3). Use that $\mathcal{S}_{ES} = 0$ implies that all moments of $\mathcal{S}_{ES}$ must vanish, and that $\int C_{\langle i}C_{j\rangle}\mathcal{S}_{ES}d\mathbf{c} = 0$ only if $\sigma_{ij} = 0$ (or $b = 0$). Integrals over the anisotropic Gaussian are computed in Appendix A.3.4.

4

The Chapman-Enskog method

The most important success of the Chapman-Enskog method, developed independently by Enskog [49][50] and Chapman [51][52], is that it allows us to compute transport coefficients for macroscopic laws from the microscopic details of the gas. The method yields the laws of Navier-Stokes and Fourier with explicit expressions for viscosity and heat conductivity[1] as functions of density and temperature. This method shows that viscosity and heat conductivity depend only on temperature, and not on density, in accordance with experimental data, and that the temperature dependence is linked to the interaction potential between the particles.

The Chapman-Enskog method relies on an expansion in the Knudsen number, and the Navier-Stokes-Fourier laws are obtained in first order. The method was developed fully to second order, which yields the Burnett equations [53][54], but only partly to third order, which yields the super-Burnett equations [55]. Both sets of equations are not widely used for several reasons, including that no proper boundary conditions for the equations are known (Chapter 12), and that they are unstable in transient problems [56] (Chapter 10).

4.1 Basic principles

4.1.1 The closure problem

For a monatomic gas the conservation laws for mass, momentum, and energy $(3.35)_{1,2}$, (3.36) can be written as

$$\frac{D\rho}{Dt} + \rho\frac{\partial v_k}{\partial x_k} = 0 \,,$$

---

[1] ... and Fick's law of diffusion with explicit expressions for diffusion coefficients, when mixtures are considered. Indeed, the idea of thermal diffusion appeared first in this context.

$$\rho \frac{Dv_i}{Dt} + \frac{\partial p}{\partial x_i} + \frac{\partial \sigma_{ik}}{\partial x_k} = \rho G_i \,, \tag{4.1}$$

$$\frac{3}{2}\rho \frac{D\theta}{Dt} + \frac{\partial q_k}{\partial x_k} = -\left(p\delta_{ij} + \sigma_{ij}\right)\frac{\partial v_i}{\partial x_j} \,,$$

where we have split the pressure tensor in its trace and trace-free parts, i.e. pressure and stress tensor, $p_{ij} = p\delta_{ij} + \sigma_{ij}$ with $\sigma_{kk} = 0$ (2.13). Moreover, we used $u = \frac{3}{2}\theta$ and we recall the ideal gas law, $p = \rho\theta$.

A macroscopic theory for a gas flow aims at solving the flow problem at hand on the level of macroscopic quantities, i.e. the moments of the phase density, including mass density ρ, velocity v_i, and temperature θ. The five conservation laws (4.1) are the starting point for any macroscopic approach.

Besides the five variables ρ, v_i, θ, which appear under the time derivatives, the conservation laws contain 8 additional quantities, i.e. the components of heat flux q_k and stress[2] σ_{ij}. Thus, additional equations are required in order to obtain a closed system, and different approaches will lead to different solutions of this so-called closure problem. The main two roads to this end are moment methods, where additional partial differential equations for stress and heat flux, and other moments, are considered, and the Chapman-Enskog method that will be discussed now.

While moment methods extend the number of variables by adding more moments, the Chapman-Enskog (CE) method considers only the five variables

$$U_A = \{\rho, v_i, \theta\}_A \,, \tag{4.2}$$

and finds equations for stress and heat flux in terms of the variables, and their space derivatives. This is achieved by deriving an approximation

$$f_{|CE} = f_{|CE}\left(U_A, \frac{\partial U_A}{\partial x_k}, \frac{\partial^2 U_A}{\partial x_{k_1}\partial x_{k_2}}, \cdots ; C_i\right)$$

to the phase density from the Boltzmann equation in the limit of small Knudsen numbers, and then computing stress and heat flux from the approximation as

$$\sigma_{ij|CE} = m\int C_{\langle i}C_{j\rangle}f_{|CE}d\mathbf{c} \quad \text{and} \quad q_{i|CE} = \frac{m}{2}\int C^2 C_i f_{|CE}d\mathbf{c} \,. \tag{4.3}$$

The approximation $f_{|CE}$ is a series in the Knudsen number Kn, and yields the Euler equations at zeroth order, the Navier-Stokes-Fourier equations at second order, the Burnett equations at second order, and the super-Burnett equations at third order.

Excellent treatments of the Chapman-Enskog procedure can be found in the classical literature on kinetic theory, e.g. [2][3][6][11][13].

---

[2] The stress σ_{ij} (4.3)$_1$ is symmetric by definition. A symmetric tensor has 6 independent components, but a symmetric and trace-free tensor of second order has only 5 independent elements, since $\sigma_{kk} = \sigma_{11} + \sigma_{22} + \sigma_{33} = 0$.

We shall go through the method for the ES-BGK model first, where we will consider the expansion up to second order [57]. Since the collision term of the ES-BGK model is linear in f, as opposed to the quadratic integral expression in the Boltzmann equation, the method is considerably easier to apply to the ES-BGK equation, and thus becomes more transparent. Then, we briefly discuss the application of the CE method to the Boltzmann equation, and report on results from literature. The results obtained for the ES-BGK model will be compared to those for the full Boltzmann equation.

Finally, the H-theorem (second law) will be discussed for the Navier-Stokes-Fourier and Burnett equations.

4.1.2 Dimensionless and scaled Boltzmann and ES-BGK equations

The Knudsen number enters the Boltzmann equation (3.16) when it is made dimensionless by introducing dimensionless variables as

$$f = \frac{n_0}{\bar{C}^3}\hat{f}\ , \ \mathbf{x} = L\hat{\mathbf{x}}\ , \ g = \sqrt{2}\bar{C}\hat{g}\ , \ \mathbf{c} = \bar{C}\hat{\mathbf{c}}\ , \ t = \frac{L}{\bar{C}}\hat{t}\ , \ \sigma = \pi d^2\hat{\sigma}\ , \ G_k = \frac{\bar{C}^2}{L}\hat{G}_k$$

where n_0 is a reference number density, $\bar{C} = \sqrt{\frac{8}{\pi}\theta_0}$ is the mean velocity of a particle at reference temperature θ_0, and d is an effective molecule diameter. Written in terms of the dimensionless quantities, the Boltzmann equation reads

$$\frac{\partial \hat{f}}{\partial \hat{t}} + \hat{c}_k \frac{\partial \hat{f}}{\partial \hat{x}_k} + \hat{G}_k \frac{\partial \hat{f}}{\partial \hat{c}_k} = \frac{1}{\mathrm{Kn}}\hat{S} \qquad (4.4)$$

with the dimensionless collision term

$$\hat{S} = \int \int_0^{2\pi} \int_0^{\pi/2} \left(\hat{f}'\hat{f}^{1\prime} - \hat{f}\hat{f}^1\right) \hat{g}\,\hat{\sigma}\sin\Theta\,d\Theta\,d\varepsilon\,d\hat{c}^1\ ,$$

and the Knudsen number

$$\mathrm{Kn} = \frac{1}{L\sqrt{2}\pi d^2 n_0} = \frac{\lambda}{L}\ .$$

λ denotes the mean free path (1.2).

The transport equation for the ES-BGK model is obtained by replacing the Boltzmann collision term S with the ES-BGK collision term (3.56). In dimensionless form the equation agrees with (4.4), if one chooses

$$\hat{S} \to \hat{S}_{ES} = -\left(\hat{f} - \hat{f}_{ES}\right) \ \text{ and } \ \mathrm{Kn} = \frac{\bar{C}}{L\bar{\nu}} = \frac{\lambda}{L}\ .$$

Thus, the Knudsen number Kn scales the collision terms of Boltzmann and ES-BGK equations (4.4). The CE method is an expansion method for small Knudsen numbers, and can be performed on the dimensionless equation (4.4).

However, to avoid dimensionless variables, the method is often performed on the dimensional Boltzmann equation in which a scaling parameter ε is introduced so that

$$\frac{\partial f}{\partial t} + c_k \frac{\partial f}{\partial x_k} + G_k \frac{\partial f}{\partial c_k} = \frac{1}{\varepsilon} \mathcal{S} . \tag{4.5}$$

The parameter ε is a formal smallness parameter which plays the role of the Knudsen number for monitoring the order of terms and quantities appearing in the equations. At the end of all calculations ε will be set to unity which is equivalent to re-inserting the dimensions into the corresponding dimensionless equations to remove the Knudsen number.

Since the right hand side of the scaled Boltzmann equation (4.5) must remain finite, it follows that $\mathcal{S} \to 0$ in the limit $\varepsilon \to 0$ (or Kn $\to 0$) which implies that

$$f = f_M \text{ as } \varepsilon \to 0 ; \tag{4.6}$$

f_M is the local Maxwellian (3.21).

4.1.3 The Chapman-Enskog expansion

The basic idea of the Chapman-Enskog method is to expand the phase density f into a series in the (formal) smallness parameter ε as

$$f_{|CE} = f^{(0)} + \varepsilon f^{(1)} + \varepsilon^2 f^{(2)} + \varepsilon^3 f^{(3)} + \cdots . \tag{4.7}$$

An important condition on the expansion (4.7) is that the main variables $U_A = \{\rho, v_i, \theta\}_A$ are the same at any level of expansion, which implies that

$$\rho = m \int f^{(0)} d\mathbf{c} , \quad \rho v_i = m \int c_i f^{(0)} d\mathbf{c} , \quad \frac{3}{2}\rho\theta = \frac{m}{2} \int C^2 f^{(0)} d\mathbf{c} , \tag{4.8}$$

and

$$0 = m \int f^{(\alpha)} d\mathbf{c} , \quad 0 = m \int c_i f^{(\alpha)} d\mathbf{c} , \quad 0 = \frac{m}{2} \int C^2 f^{(\alpha)} d\mathbf{c} \text{ for } \alpha \geq 1 , \tag{4.9}$$

for all values of ε. The conditions (4.9) are known as the compatibility conditions. From (4.6)-(4.9) follows that $f^{(0)}$ must be the local Maxwellian,

$$f^{(0)} = f_M = \frac{\rho}{m} \frac{1}{\sqrt{2\pi\theta}^3} \exp\left[-\frac{C^2}{2\theta}\right] . \tag{4.10}$$

Stress and heat flux, however, change with the order of the expansion as

$$\sigma_{ij} = \varepsilon \sigma_{ij}^{(1)} + \varepsilon^2 \sigma_{ij}^{(2)} + \varepsilon^3 \sigma_{ij}^{(3)} + \cdots , \quad q_i = \varepsilon q_i^{(1)} + \varepsilon^2 q_i^{(2)} + \varepsilon^3 q_i^{(3)} + \cdots \tag{4.11}$$

where

$$\sigma_{ij}^{(\alpha)} = m \int C_{\langle i} C_{j\rangle} f^{(\alpha)} d\mathbf{c} , \quad q_i^{(\alpha)} = \frac{m}{2} \int C^2 C_i f^{(\alpha)} d\mathbf{c} . \tag{4.12}$$

Here we used the fact that stress and heat flux vanish for the Maxwellian, $\sigma_{ij}^{(0)} = q_i^{(0)} = 0$.

Inserting the expansion (4.11) into the conservation laws (4.1), and solving for the time derivatives, gives

$$\frac{D\rho}{Dt} = -\rho\frac{\partial v_k}{\partial x_k} \ , \quad \frac{Dv_i}{Dt} = G_i - \frac{1}{\rho}\frac{\partial p}{\partial x_i} - \frac{1}{\rho}\sum_\alpha \varepsilon^\alpha \frac{\partial \sigma_{ik}^{(\alpha)}}{\partial x_k} \ ,$$

$$\frac{D\theta}{Dt} = -\frac{2}{3}\theta\frac{\partial v_j}{\partial x_j} - \frac{2}{3}\frac{1}{\rho}\sum_\alpha \varepsilon^\alpha \frac{\partial q_k^{(\alpha)}}{\partial x_k} - \frac{2}{3}\sum_\alpha \varepsilon^\alpha \frac{\sigma_{ij}^{(\alpha)}}{\rho}\frac{\partial v_i}{\partial x_j} \ .$$

This allows us to write the time derivatives formally as a series in ε,

$$\frac{D\rho}{Dt} = \sum_\alpha \varepsilon^\alpha \frac{D_\alpha \rho}{Dt} \ , \quad \frac{Dv_i}{Dt} = \sum_\alpha \varepsilon^\alpha \frac{D_\alpha v_i}{Dt} \ , \quad \frac{D\theta}{Dt} = \sum_\alpha \varepsilon^\alpha \frac{D_\alpha \theta}{Dt}$$

with

$$\frac{D_0\rho}{Dt} = -\rho\frac{\partial v_k}{\partial x_k} \ , \quad \frac{D_0 v_i}{Dt} = G_i - \frac{1}{\rho}\frac{\partial \rho\theta}{\partial x_i} \ , \quad \frac{D_0\theta}{Dt} = -\frac{2}{3}\theta\frac{\partial v_j}{\partial x_j} \tag{4.13}$$

and, for $\alpha \geq 1$,

$$\frac{D_\alpha\rho}{Dt} = 0 \ , \quad \frac{D_\alpha v_i}{Dt} = -\frac{1}{\rho}\frac{\partial \sigma_{ik}^{(\alpha)}}{\partial x_k} \ , \quad \frac{D_\alpha\theta}{Dt} = -\frac{2}{3}\frac{\sigma_{ij}^{(\alpha)}}{\rho}\frac{\partial v_i}{\partial x_j} - \frac{2}{3}\frac{1}{\rho}\frac{\partial q_k^{(\alpha)}}{\partial x_k} \ . \tag{4.14}$$

The Maxwellian f_M depends on the variables U_A and the above expansion can be used to expand its time derivative according to

$$\frac{Df_M}{Dt} = \frac{D_0 f_M}{Dt} + \varepsilon\frac{D_1 f_M}{Dt} + \cdots \quad \text{where} \quad \frac{D_\alpha f_M}{Dt} = \frac{\partial f_M}{\partial U_A}\frac{D_\alpha U_A}{Dt} \ . \tag{4.15}$$

The time derivatives of $f^{(\alpha)}$ can be expanded similarly, as

$$\frac{Df^{(\alpha)}}{Dt} = \frac{D_0 f^{(\alpha)}}{Dt} + \varepsilon\frac{D_1 f^{(\alpha)}}{Dt} + \cdots \ ; \tag{4.16}$$

the meaning of $D_\beta f^{(\alpha)}/Dt$ will be discussed later.

The above relations will be needed as the CE method proceeds step by step.

4.2 The CE expansion for the ES-BGK model

4.2.1 General arguments

Written with the convective time derivative, the scaled ES-BGK transport equation reads

$$\frac{Df}{Dt} + C_k \frac{\partial f}{\partial x_k} + G_k \frac{\partial f}{\partial c_k} = -\frac{1}{\varepsilon} \bar{\nu} \left(f - f_{ES} \right) . \tag{4.17}$$

For the application of the CE method, the anisotropic Gaussian f_{ES} (3.57) must be expanded in ε as well,

$$f_{ES} = f_{ES}^{(0)} + \varepsilon f_{ES}^{(1)} + \varepsilon^2 f_{ES}^{(2)} + \cdots \tag{4.18}$$

The next step is to insert the expansions (4.7, 4.18) into the ES-BGK equation and to use (4.15, 4.16). We shall be interested in a second order theory, and for this we need to include terms up to first order in ε, that is

$$\frac{D_0 f_M}{Dt} + C_k \frac{\partial f_M}{\partial x_k} + G_k \frac{\partial f_M}{\partial c_k} + \bar{\nu} \left(f^{(1)} - f_{ES}^{(1)} \right)$$
$$+ \varepsilon \left[\frac{D_1 f_M}{Dt} + \frac{D_0 f^{(1)}}{Dt} + C_k \frac{\partial f^{(1)}}{\partial x_k} + G_k \frac{\partial f^{(1)}}{\partial c_k} + \bar{\nu} \left(f^{(2)} - f_{ES}^{(2)} \right) \right] = \mathcal{O} \left(\varepsilon^2 \right) . \tag{4.19}$$

This equation allows us to compute $f^{(1)}$ and $f^{(2)}$. For a higher order theory, i.e. to compute $f^{(\alpha)}$ for $\alpha \geq 3$, more terms in the above equation would have to be considered.

The $f_{ES}^{(\alpha)}$ in (4.18) are computed from f_{ES} as follows: f_{ES} contains the inverse of the matrix (3.58), for which insertion of (4.11) yields

$$\lambda_{ij} = \theta \delta_{ij} + \frac{b}{\rho} \left(\varepsilon \sigma_{ij}^{(1)} + \varepsilon^2 \sigma_{ij}^{(2)} + \cdots \right) .$$

Taylor expansion within terms of second order gives

$$\det \left[\lambda_{ij} \right] = \theta^3 \left[1 - \frac{\varepsilon^2}{2} \left(\frac{b}{p} \right)^2 \sigma_{kn}^{(1)} \sigma_{kn}^{(1)} \right] ,$$

$$\lambda_{ij}^{-1} = \frac{\delta_{ij}}{\theta} - \varepsilon \frac{b}{p\theta} \sigma_{ij}^{(1)} + \varepsilon^2 \frac{b}{p\theta} \left[\frac{b}{p} \sigma_{ik}^{(1)} \sigma_{kj}^{(1)} - \sigma_{ij}^{(2)} \right] ,$$

and therefore

$$f_{ES}^{(0)} = f_M , \quad f_{ES}^{(1)} = f_M \frac{b}{2\rho\theta^2} \sigma_{ij}^{(1)} C_{\langle i} C_{j \rangle} , \tag{4.20}$$

$$f_{ES}^{(2)} = f_M \frac{b^2}{4p^2} \left[\sigma_{kn}^{(1)} \sigma_{kn}^{(1)} - \frac{2}{\theta} \left(\sigma_{ik}^{(1)} \sigma_{kj}^{(1)} - \frac{p}{b} \sigma_{ij}^{(2)} \right) C_i C_j + \frac{\sigma_{ij}^{(1)} \sigma_{kl}^{(1)}}{2\theta^2} C_i C_j C_k C_l \right] .$$

4.2.2 Zeroth order: Euler equations

The solution at zeroth order was obtained already in (4.10) where we found

$$f^{(0)} = f_{ES}^{(0)} = f_M \quad \text{and} \quad \sigma_{ij}^{(0)} = q_i^{(0)} = 0 .$$

In this case, the conservation laws (4.1) reduce to the well-known Euler equations of gas dynamics,

$$\frac{D\rho}{Dt} + \rho\frac{\partial v_k}{\partial x_k} = 0 \ , \quad \rho\frac{Dv_i}{Dt} + \frac{\partial \rho\theta}{\partial x_i} = \rho G_i \ , \quad \frac{3}{2}\rho\frac{D\theta}{Dt} + \rho\theta\frac{\partial v_i}{\partial x_i} = 0 \ . \quad (4.21)$$

4.2.3 First order: Navier-Stokes-Fourier equations

To obtain the first order expression for the phase density, we need to consider the terms of order ε^0 in (4.19), which we write for convenience as

$$f^{(1)} = f^{(1)}_{ES} - \frac{1}{\bar{\nu}} f_M \left[\frac{D_0 \ln f_M}{Dt} + C_k \frac{\partial \ln f_M}{\partial x_k} + G_k \frac{\partial \ln f_M}{\partial c_k} \right] .$$

The differential $\mathrm{d}f_M$ of the Maxwellian is given by

$$\mathrm{d}f_M = f_M \, \mathrm{d}\left(\ln f_M\right) = f_M \left[\frac{\mathrm{d}\rho}{\rho} + \left(\frac{C^2}{2\theta} - \frac{3}{2} \right) \frac{\mathrm{d}\theta}{\theta} - \frac{C_i}{\theta} \mathrm{d}\left(c_i - v_i\right) \right]$$

and when the necessary derivatives are computed, we obtain

$$f^{(1)} = f_M \frac{b}{2\rho\theta^2} \sigma^{(1)}_{kl} C_{\langle k} C_{l\rangle}$$
$$- f_M \frac{1}{\bar{\nu}} \left[\frac{C_{\langle k} C_{l\rangle}}{\theta} \frac{\partial v_{\langle k}}{\partial x_{l\rangle}} + C_i \left(\frac{C^2}{2\theta} - \frac{5}{2} \right) \frac{1}{\theta} \frac{\partial \theta}{\partial x_k} \right.$$
$$+ \frac{1}{\rho} \left(\frac{D_0\rho}{Dt} + \rho\frac{\partial v_n}{\partial x_n} \right) + \left(\frac{C^2}{2\theta} - \frac{3}{2} \right) \left(\frac{1}{\theta}\frac{D_0\theta}{Dt} + \frac{2}{3}\frac{\partial v_n}{\partial x_n} \right)$$
$$\left. + \frac{C_k}{\theta} \left(\frac{D_0 v_k}{Dt} + \frac{1}{\rho}\frac{\partial \rho\theta}{\partial x_k} - G_k \right) \right] .$$

The last two lines vanish due to the definition for $D_0 U_A/Dt$ (4.13), and the result for the first order contribution to the phase density is

$$f^{(1)} = f_M \frac{b}{2\rho\theta^2} \sigma^{(1)}_{kl} C_{\langle k} C_{l\rangle} - f_M \frac{1}{\bar{\nu}} \left[\frac{C_{\langle k} C_{l\rangle}}{\theta} \frac{\partial v_{\langle k}}{\partial x_{l\rangle}} + C_k \left(\frac{C^2}{2\theta} - \frac{5}{2} \right) \frac{1}{\theta} \frac{\partial \theta}{\partial x_k} \right] .$$

It is straightforward to show that $f^{(1)}$ satisfies the compatibility relations (4.9). Stress and heat flux at first order follow from (4.12) as[3]

$$\sigma^{(1)}_{ij} = -\frac{2}{1 - b} \frac{p}{\bar{\nu}} \frac{\partial v_{\langle i}}{\partial x_{k\rangle}} = -2\mu\frac{\partial v_{\langle i}}{\partial x_{k\rangle}} \quad \text{and} \quad q^{(1)}_i = -\frac{5}{2}\frac{p}{\bar{\nu}}\frac{\partial \theta}{\partial x_i} = -\kappa\frac{\partial \theta}{\partial x_i} \ . \quad (4.22)$$

The constitutive relations (4.22) are the laws of Navier-Stokes and Fourier where

---

[3] Recall that we employ energy units for temperature. Thus, the heat conductivity κ is related to the usual heat conductivity $\hat{\kappa}$ as $\hat{\kappa} = \frac{k}{m}\kappa$.

$$\mu = \frac{1}{1-b}\frac{p}{\bar{\nu}} \quad \text{and} \quad \kappa = \frac{5}{2}\frac{p}{\bar{\nu}} \tag{4.23}$$

are viscosity and thermal conductivity, respectively. The Prandtl number (1.7) is related to the coefficient b of the ES-BGK model by

$$\Pr = \frac{5}{2}\frac{\mu}{\kappa} = \frac{1}{1-b} .$$

Its measured value, $\Pr = 2/3$, is obtained for $b = -1/2$.

The classical BGK model (3.54) is obtained for $b = 0$ which yields $\Pr = 1$, i.e. the classical BGK models fails in producing the proper Prandtl number.

For power potentials and hard spheres the collision frequency is of the form $\bar{\nu} = \bar{\nu}^0 p \theta^{-\omega}$ where $\omega = \frac{\gamma+3}{2\gamma-2}$ (3.51). It follows that the viscosity is a function of temperature alone, given by

$$\mu = \mu_0 \left(\frac{\theta}{\theta_0}\right)^\omega ; \tag{4.24}$$

$\mu_0 = \frac{\Pr}{\bar{\nu}^0}\theta_0^\omega$ is the viscosity at a reference temperature θ_0. ω will be denoted as viscosity exponent. Measurements confirm that μ depends on temperature, but not on density, and give $\omega \simeq 0.8$, corresponding to a power potential with $\gamma \simeq 23/3$. For hard sphere molecules ($\gamma \to \infty$) and Maxwell molecules ($\gamma = 5$) one finds $\omega_{HS} = 1/2$ and $\omega_5 = 1$, respectively.

By means of the Navier-Stokes and Fourier laws (4.22), the phase density can be rewritten as

$$f^{(1)} = -\frac{\theta^\omega}{\bar{\nu}^0 p}f_M \left[\Pr \frac{C_i C_k}{\theta}\frac{\partial v_{\langle i}}{\partial x_{k\rangle}} + C_k \left(\frac{C^2}{2\theta} - \frac{5}{2}\right)\frac{1}{\theta}\frac{\partial\theta}{\partial x_k}\right] . \tag{4.25}$$

This form is most convenient for the next step in the expansion.

4.2.4 Second order: Burnett equations

The second order correction to the phase density is obtained from (4.19) as

$$f^{(2)} = f_{ES}^{(2)} - \frac{1}{\bar{\nu}}\left(\frac{D_1 f_M}{Dt} + \frac{D_0 f^{(1)}}{Dt} + C_k \frac{\partial f^{(1)}}{\partial x_k} + G_k \frac{\partial f^{(1)}}{\partial c_k}\right) .$$

To obtain an explicit expression for $f^{(2)}$ in terms of the variables $U_A = (\rho, v_i, \theta)$ and their gradients, requires to take the derivatives of $f^{(1)}$ and to use (4.15). Moreover, from the previous section we know that $f^{(1)} = f^{(1)}(U_A, \partial U_A/\partial x_k)$, so that

$$\frac{D_0 f^{(1)}}{Dt} = \frac{\partial f^{(1)}}{\partial U_A}\frac{D_0 U_A}{Dt} + \frac{\partial f^{(1)}}{\partial\left(\frac{\partial U_A}{\partial x_k}\right)}\frac{D_0}{Dt}\left(\frac{\partial U_A}{\partial x_k}\right) ,$$

where

$$\frac{D_0}{Dt}\left(\frac{\partial U_A}{\partial x_k}\right) = \frac{\partial}{\partial x_k}\frac{D_0 U_A}{Dt} - \frac{\partial U_A}{\partial x_r}\frac{\partial v_r}{\partial x_k} \ .$$

The computation of $f^{(2)}$ is straightforward, but tedious, see [57] for details including the explicit expression of $f^{(2)}$. The resulting expressions for stress and heat flux are the Burnett equations for the ES-BGK model [53][2],

$$\sigma_{ij}^{(2)} = \frac{\mu^2}{p}\left[\varpi_1\frac{\partial v_k}{\partial x_k}S_{ij} - \varpi_2\left(\frac{\partial}{\partial x_{\langle i}}\left(\frac{1}{\rho}\frac{\partial p}{\partial x_{j\rangle}}\right) + \frac{\partial v_k}{\partial x_{\langle i}}\frac{\partial v_{j\rangle}}{\partial x_k} + 2\frac{\partial v_k}{\partial x_{\langle i}}S_{j\rangle k}\right)\right.$$
$$\left.+\varpi_3\frac{\partial^2\theta}{\partial x_{\langle i}\partial x_{j\rangle}} + \varpi_4\frac{\partial\theta}{\partial x_{\langle i}}\frac{\partial\ln p}{\partial x_{j\rangle}} + \varpi_5\frac{1}{\theta}\frac{\partial\theta}{\partial x_{\langle i}}\frac{\partial\theta}{\partial x_{j\rangle}} + \varpi_6 S_{k\langle i}S_{j\rangle k}\right] \ , \quad (4.26)$$

$$q_i^{(2)} = \frac{\mu^2}{\rho}\left[\theta_1\frac{\partial v_k}{\partial x_k}\frac{\partial\ln\theta}{\partial x_i} - \theta_2\left(\frac{2}{3}\frac{\partial^2 v_k}{\partial x_k\partial x_i} + \frac{2}{3}\frac{\partial v_k}{\partial x_k}\frac{\partial\ln\theta}{\partial x_i} + 2\frac{\partial v_k}{\partial x_i}\frac{\partial\ln\theta}{\partial x_k}\right)\right.$$
$$\left.+\theta_3 S_{ik}\frac{\partial\ln p}{\partial x_k} + \theta_4\frac{\partial S_{ik}}{\partial x_k} + 3\theta_5 S_{ik}\frac{\partial\ln\theta}{\partial x_k}\right] \ , \quad (4.27)$$

with the abbreviation

$$S_{ij} = \frac{\partial v_{\langle i}}{\partial x_{j\rangle}} = \frac{1}{2}\frac{\partial v_i}{\partial x_j} + \frac{1}{2}\frac{\partial v_j}{\partial x_i} - \frac{1}{3}\frac{\partial v_k}{\partial x_k}\delta_{ij}$$

for the trace free velocity gradient, and the Burnett coefficients [53][57]

$$\varpi_1 = \frac{4}{3}\left(\frac{7}{2} - \omega\right), \ \varpi_2 = 2, \ \varpi_3 = \frac{2}{\mathrm{Pr}}, \ \varpi_4 = 0, \ \varpi_5 = \frac{2\omega}{\mathrm{Pr}}, \ \varpi_6 = 8,$$

$$\theta_1 = \frac{5}{3}(1-b)^2\left(\frac{7}{2} - \omega\right), \ \theta_2 = \frac{5}{2\,\mathrm{Pr}^2}, \ \theta_3 = -\frac{2}{\mathrm{Pr}}, \ \theta_4 = \frac{2}{\mathrm{Pr}} \ ,$$

$$\theta_5 = \frac{2}{3\,\mathrm{Pr}}\left(\frac{7}{2}\left(1 + \frac{1}{\mathrm{Pr}}\right) + \omega\right) \ . \quad (4.28)$$

The Burnett coefficients (4.28) depend on the two parameters ω and b, which define the collision term. The viscosity exponent ω enters the equations through the temperature dependence of the collision frequency that appears in $f^{(1)}$, see (4.25). For $\mathrm{Pr} = 2/3$ they reduce to

$$\varpi_1 = \frac{4}{3}\left(\frac{7}{2} - \omega\right) \ , \ \varpi_2 = 2 \ , \ \varpi_3 = 3 \ , \ \varpi_4 = 0 \ , \ \varpi_5 = 3\omega \ , \ \varpi_6 = 8 \ ,$$

$$\theta_1 = \frac{15}{4}\left(\frac{7}{2} - \omega\right) \ , \ \theta_2 = \frac{45}{8} \ , \ \theta_3 = -3 \ , \ \theta_4 = 3 \ , \ \theta_5 = \frac{35}{4} + \omega \ . \quad (4.29)$$

4.3 CE expansion for power potentials

4.3.1 First order: Navier-Stokes-Fourier laws

Expansion

We briefly discuss the CE expansion for the full Boltzmann equation to first order, which gives detailed expressions for viscosity and heat conductivity in dependence of the interaction potential.

At first order, the phase density is conveniently written as

$$f = f^{(0)} + \varepsilon f^{(1)} = f_M \left(1 + \varepsilon \varphi\right) , \qquad (4.30)$$

and due to the compatibility conditions (4.9) φ must obey

$$0 = m \int \varphi f_M d\mathbf{c} , \quad 0 = m \int c_i \varphi f_M d\mathbf{c} , \quad 0 = \frac{m}{2} \int C^2 \varphi f_M d\mathbf{c} . \qquad (4.31)$$

The Boltzmann equation with all terms of order ε^1 and higher omitted assumes the form (linearized Boltzmann equation[4])

$$\frac{D_0 f_M}{Dt} + C_k \frac{\partial f_M}{\partial x_k} + G_k \frac{\partial f_M}{\partial c_k} = f_M \int f_M^1 \left(\varphi' + \varphi^{1'} - \varphi - \varphi^1\right) \sigma g \sin \Theta d\Theta d\varepsilon d\mathbf{c}^1 . \qquad (4.32)$$

For the derivation it was used that $f_M' f_M^{1'} = f_M f_M^1$. The left hand side is readily evaluated, and the resulting integral equation for φ reads, with $C_i = \sqrt{2\theta}\xi_i$,

$$\xi_{\langle i}\xi_{j\rangle} 2 \frac{\partial v_{\langle i}}{\partial x_{j\rangle}} + \left(\xi^2 - \frac{5}{2}\right) \xi_i \sqrt{2\theta} \frac{\partial \ln \theta}{\partial x_i} = \mathcal{I}\left[\varphi\right] \qquad (4.33)$$

with the integral operator

$$\mathcal{I}\left[\varphi\right] = \int f_M^1 \left(\varphi' + \varphi^{1'} - \varphi - \varphi^1\right) \sigma g \sin \Theta d\Theta d\varepsilon d\mathbf{c}^1 . \qquad (4.34)$$

Thus, the Chapman-Enskog expansion for the Boltzmann equation involves solving an integral equation, which adds additional difficulty as compared to applying the method to the ES-BGK model. Only for Maxwell molecules can an exact solution of the integral equation be found, while for other molecule models approximations must be used.

---

[4] The linear Boltzmann equation appears for determining the first order contribution $f^{(1)} = f_M \varphi$. Finding higher order terms requires to consider the non-linear Boltzmann equation.

Decomposition of problem

Any solution of the integral equation (4.33) can be split into a homogeneous and an inhomogeneous part, $\varphi^{(h)}$ and $\varphi^{(i)}$, respectively. The solution $\varphi^{(h)}$ of the homogeneous integral equation $\mathcal{I}[\varphi] = 0$ must be a linear combination of the collisional invariants,

$$\varphi^{(h)} = \alpha + \beta_i \xi_i + \gamma \xi^2 \; ;$$

however, the compatibility conditions (4.31) imply $\alpha = \gamma = 0$. The inhomogeneous solution $\varphi^{(i)}$ must be linear in $\frac{\partial \ln \theta}{\partial x_i}$ and $\frac{\partial v_{\langle i}}{\partial x_{j\rangle}}$, since (4.33) is linear in φ, and therefore $\varphi^{(i)}$ must read

$$\varphi^{(i)} = -A(\xi)\,\xi_i \sqrt{2\theta}\frac{\partial \ln \theta}{\partial x_i} - B(\xi)\,\xi_{\langle i}\xi_{j\rangle}\frac{\partial v_{\langle i}}{\partial x_{j\rangle}} \; .$$

The gradients of velocity and temperature can be controlled independently, and thus the unknown functions $A(\xi)$ and $B(\xi)$ solve

$$\mathcal{I}[A\xi_i] = -\left(\xi^2 - \frac{5}{2}\right)\xi_i \quad \text{and} \quad \mathcal{I}\left[B(\xi)\,\xi_{\langle i}\xi_{j\rangle}\right] = -2\xi_{\langle i}\xi_{j\rangle} \; . \tag{4.35}$$

The compatibility relation (4.31)$_2$ assumes the form

$$0 = m \int \xi_i \xi_k \left(\beta_k - A(\xi)\sqrt{2\theta}\frac{\partial \ln \theta}{\partial x_k}\right) f_M d\mathbf{c} \; ,$$

so that β_k must point in the direction of the temperature gradient, $\beta_k = \beta\sqrt{2\theta}\frac{\partial \ln \theta}{\partial x_k}$. Without loss of generality we can set $\beta = 0$ (that is we absorb β into A) and have the new compatibility condition

$$0 = m \int \xi^2 A(\xi) f_M d\mathbf{c} \; . \tag{4.36}$$

Viscosity and heat conductivity

We use the results obtained so far to express viscosity and heat conductivity through the functions $A(\xi)$ and $B(\xi)$. The distribution function is (4.30)

$$f = f_M \left(1 - A(\xi)\,\xi_i \sqrt{2\theta}\frac{\partial \ln \theta}{\partial x_i} - B(\xi)\,\xi_{\langle i}\xi_{j\rangle}\frac{\partial v_{\langle i}}{\partial x_{j\rangle}}\right) , \tag{4.37}$$

and computation of stress and heat flux gives the laws of Navier-Stokes and Fourier as

$$\sigma_{ij} = -2\mu\frac{\partial v_{\langle i}}{\partial x_{j\rangle}} \quad \text{and} \quad q_i = -\kappa\frac{\partial \theta}{\partial x_i}$$

with

$$\mu = \frac{2}{15\pi^{3/2}} p \int \exp\left[-\xi^2\right] B\xi^4 d\boldsymbol{\xi} \; ,$$

$$\kappa = \frac{2}{3\pi^{3/2}} p \int \exp\left[-\xi^2\right] A\xi^4 d\boldsymbol{\xi} = \frac{2}{3\pi^{3/2}} p \int \exp\left[-\xi^2\right] A\xi_i \left(\xi^2 - \frac{5}{2}\right) \xi_i d\boldsymbol{\xi} \; .$$

We define

$$[\psi, \varphi] = \frac{m\pi^{3/2}}{\rho} \int \exp\left[-\xi^2\right] \psi \mathcal{I}\left[\varphi\right] d\boldsymbol{\xi} \tag{4.38}$$

so that, by means of the integral equations,

$$\mu = \frac{1}{10\pi^3} \frac{\rho p}{m} \left[B\left(\xi\right) \xi_{\langle i} \xi_{j\rangle}, B\left(\xi\right) \xi_{\langle i} \xi_{j\rangle} \right] \quad \text{and} \quad \kappa = \frac{2}{3\pi^3} \frac{\rho p}{m} \left[A\left(\xi\right) \xi_i, A\left(\xi\right) \xi_i \right] \; . \tag{4.39}$$

In order to obtain the full solution of the problem, we still need to determine $A\left(\xi\right)$ and $B\left(\xi\right)$.

Expansion in Sonine polynomials

The integral equations are usually solved by means of an expansion into Sonine polynomials, which are defined as [2][13]

$$S_{l+\frac{1}{2}}^{(r)} \left(\xi^2\right) = \sum_{p=0}^{r} \frac{\left(-\xi^2\right)^p \Gamma\left(l + \frac{3}{2} + r\right)}{p! \, (r-p)! \Gamma\left(l + \frac{3}{2} + p\right)} \tag{4.40}$$

and obey the orthonormality condition

$$\int_0^{\infty} \exp\left[-\xi^2\right] S_{l+\frac{1}{2}}^{(r)} \left(\xi^2\right) S_{l+\frac{1}{2}}^{(s)} \left(\xi^2\right) \xi^{2l+2} d\xi = \frac{\Gamma\left(l + \frac{3}{2} + r\right)}{2r!} \delta_{rs} \; .$$

In particular we have $S_{3/2}^{(1)} \left(\xi^2\right) = \left(\frac{5}{2} - \xi^2\right)$ and $S_{5/2}^{(0)} \left(\xi^2\right) = 1$ so that the integral equations (4.35) can be written as

$$\mathcal{I}\left[A\left(\xi\right) \xi_i\right] = S_{\frac{3}{2}}^{(1)} \left(\xi^2\right) \xi_i \quad \text{and} \quad \mathcal{I}\left[B\left(\xi\right) \xi_{\langle i} \xi_{j\rangle}\right] = -2S_{\frac{5}{2}}^{(0)} \left(\xi^2\right) \xi_{\langle i} \xi_{j\rangle} \; . \tag{4.41}$$

For their solution, the functions A and B are expanded as

$$A\left(\xi\right) = -\sum_{r=1}^{n_a} a_r S_{\frac{3}{2}}^{(r)} \left(\xi^2\right) \quad \text{and} \quad B\left(\xi\right) = -\sum_{r=0}^{n_b} b_r S_{\frac{5}{2}}^{(r)} \left(\xi^2\right) \; . \tag{4.42}$$

Note that the summation in the first equation starts at $r = 1$, since $a_0 = 0$ to ensure (4.36). An exact solution would require an infinite number of terms, that is $n_a, n_b \to \infty$. The solution presented below is only possible, however, if n_a and n_b remain finite, and that is assumed from now on. By taking advantage of the orthogonality we see that the integral equations (4.41) reduce to two linear systems for the coefficients a_r and b_r, viz.

$$\sum_{r=1}^{n_a} \alpha_{sr} a_r = \frac{15}{4}\pi^3\frac{m}{\rho}\delta_{s1} \;\Rightarrow\; a_r = \frac{15}{4}\pi^3\frac{m}{\rho}\alpha_{r1}^{-1}$$

$$\sum_{r=0}^{n_b} \beta_{sr} b_r = 5\pi^3\frac{m}{\rho}\delta_{s0} \;\Rightarrow\; b_r = 5\pi^3\frac{m}{\rho}\beta_{r0}^{-1}$$

with

$$\alpha_{sr} = \left[S_{\frac{3}{2}}^{(s)}\left(\xi^2\right)\xi_i, S_{\frac{3}{2}}^{(r)}\left(\xi^2\right)\xi_i \right] \text{ and } \beta_{sr} = \left[S_{\frac{5}{2}}^{(s)}\left(\xi^2\right)\xi_{\langle i}\xi_{j\rangle}, S_{\frac{5}{2}}^{(r)}\left(\xi^2\right)\xi_{\langle i}\xi_{j\rangle} \right].$$
(4.43)

Insertion of the expansion (4.42) into the equations for viscosity and heat conductivity (4.39) and use of the above results yields after a short calculation

$$\mu = \frac{p}{2}b_0 = \frac{5\pi^3}{2}kT\beta_{00}^{-1} \text{ and } \kappa = \frac{5}{2}pa_1 = \frac{75\pi^3}{8}kT\alpha_{11}^{-1}.$$
(4.44)

In order to obtain the desired explicit expressions for for μ and κ we need to compute the matrices α_{rs} and β_{rs}, and their inverses, from which we only require the elements α_{11}^{-1}, β_{00}^{-1}. Their values will depend on the numbers n_a, n_b of coefficients in the Sonine polynomial expansion (4.42).

Obviously, the above arguments would not change much, if another set of polynomials would be used instead of Sonine polynomials. The use of Sonine polynomials is most effective, since for Maxwell molecules the matrices (4.43) turn out to be purely diagonal [21], and the solution is exact, since then $\alpha_{11}^{-1} = 1/\alpha_{11}$ and $\beta_{00}^{-1} = 1/\beta_{00}$. For other molecule types, the results are not too different, and a small number of terms is sufficient in the expansion. We also note that Sonine polynomials are closely related to the eigenfunctions of the linearized Boltzmann collision term for Maxwell molecules [21][13].

The simplest approximation

The simplest approximation assumes just one element in each expansion (4.42), $n_a = 1$ and $n_b = 0$, in which case we only need to compute

$$\alpha_{11} = \left[S_{\frac{3}{2}}^{(1)}\left(\xi^2\right)\xi_i, S_{\frac{3}{2}}^{(1)}\left(\xi^2\right)\xi_i \right] \text{ and } \beta_{00} = \left[S_{\frac{5}{2}}^{(0)}\left(\xi^2\right)\xi_{\langle i}\xi_{j\rangle}, S_{\frac{5}{2}}^{(0)}\left(\xi^2\right)\xi_{\langle i}\xi_{j\rangle} \right].$$

With the result of Problem 4.3 and (3.11), follows for power potentials

$$\alpha_{11} = \frac{1}{4}\int \exp\left[-\xi^2-\xi_1^2\right]\mathcal{D}\left[\xi^2\xi_i\right]\mathcal{D}\left[\xi^2\xi_i\right]g^{\frac{\nu-5}{\gamma-1}}F\left(\Theta\right)\sin\Theta d\Theta d\varepsilon d\boldsymbol{\xi} d\boldsymbol{\xi}^1,$$

$$\beta_{00} = \frac{1}{4}\int \exp\left[-\xi^2-\xi_1^2\right]\mathcal{D}\left[\xi_{\langle i}\xi_{j\rangle}\right]\mathcal{D}\left[\xi_{\langle i}\xi_{j\rangle}\right]g^{\frac{\nu-5}{\gamma-1}}F\left(\Theta\right)\sin\Theta d\Theta d\varepsilon d\boldsymbol{\xi} d\boldsymbol{\xi}^1.$$

where $\mathcal{D}\left[\psi\right] = \psi+\psi^1-\psi'-\psi^{1'}$, and it was used that $\mathcal{D}\left[\xi_i\right] = 0$. For the further treatment it is convenient to introduce dimensionless relative and center of mass velocities defined as $\eta_i = \xi_i - \xi_i'$, $\zeta_i = \frac{1}{2}\left(\xi_i + \xi_i^1\right)$, $g = \sqrt{2}\theta\eta$ so that

$\xi^2 + \xi_1^2 = 2\zeta^2 + \frac{1}{2}\eta^2$, $d\xi d\xi^1 = d\eta d\zeta$, $\mathcal{D}\left(\xi^2 \xi_i\right) = \zeta_i \left(\eta_i \eta_j - \eta_i' \eta_j'\right)$, $\mathcal{D}\left(\xi_{\langle i}\xi_{j\rangle}\right) = \frac{1}{2}\left(\eta_i \eta_j - \eta_i' \eta_j'\right)$. After integration over ζ it can be seen that both coefficients are equal,

$$\alpha_{11} = \beta_{00} = \frac{1}{8}\sqrt{\frac{\pi}{2}}^{-3} \sqrt{2\theta}^{\frac{\gamma-5}{\gamma-1}} \int e^{-\frac{\eta^2}{2}} \left(\eta^4 - \eta_i \eta_j \eta_i' \eta_j'\right) \eta^{\frac{\gamma-5}{\gamma-1}} F\left(\Theta\right) \sin\Theta d\Theta d\varepsilon d\eta .$$

With $(3.2)_3$ follows that $\left(\eta^4 - \eta_i \eta_j \eta_i' \eta_j'\right) = 4\eta^4 \cos^2 \Theta \sin^2 \Theta$ and thus

$$\alpha_{11} = \beta_{00} = \pi^{5/2} \sqrt{\theta}^{\frac{\gamma-5}{\gamma-1}} 2^{\frac{3\gamma-7}{\gamma-1}} \Gamma\left(\frac{4\gamma-6}{\gamma-1}\right) \chi^{(2,3)} .$$

where

$$\chi^{(2,3)} = 2\pi \int_0^{\pi/2} \cos^2 \Theta \sin^3 \Theta F\left(\Theta\right) d\Theta .$$

We introduce

$$\omega = \frac{\gamma+3}{2\gamma-2} \quad \text{and} \quad \mu_0 = \theta_0^\omega 2^{\frac{5-\nu}{\nu-1}} \frac{\Gamma\left(\frac{7}{2}\right)}{\Gamma\left(\frac{4\nu-6}{\nu-1}\right)} \frac{m}{3\chi^{(2,3)}}$$

and then obtain from (4.44) the first approximations for viscosity and heat conductivity for power potentials as

$$\mu^{[1]} = \mu_0 \left(\frac{\theta}{\theta_0}\right)^\omega \quad \text{and} \quad \kappa^{[1]} = \frac{15}{4}\mu \quad \Rightarrow \quad \text{Pr} = \frac{2}{3} . \tag{4.45}$$

Comparison with the viscosity for the ES-BGK model (4.24) shows agreement between both results for viscosity and heat conductivity as long as the Prandtl number in the ES-BGK model is chosen as Pr = 2/3. In particular it is confirmed that both coefficients depend only on temperature, and not on the gas density, where the dependence on temperature depends on the exponent γ in the power potential (hard spheres: $\gamma \to \infty$, $\omega_{HS} = 1/2$; Maxwell molecules: $\gamma = 5$, $\omega_5 = 1$). For Maxwell molecules we obtain explicitly

$$\mu = \frac{p}{\frac{\rho}{m} 3\chi^{(2,3)}} = \frac{\theta}{\frac{3}{m}\chi^{(2,3)}} . \tag{4.46}$$

Consideration of larger values for n_a and n_b yields corrected values, e.g. one obtains for hard spheres ($\gamma \to \infty$) and $n_a = n_b = 4$ [51][2][54],

$$\mu^{[4]} = 1.016\mu^{[1]} \quad \text{and} \quad \kappa^{[4]} = 1.025\kappa^{[1]} .$$

The corresponding Prandtl number is Pr $= \frac{5}{2}\frac{\mu}{\kappa} = \frac{2}{3}\frac{1.016}{1.025} = 0.661$.

A further increase of n_a, n_b leads to no considerable change. As said before, the first approximation is exact for Maxwell molecules. For power potentials with exponents $5 < \gamma < \infty$ the correction factors are between those for Maxwell molecules (1.0), and hard spheres (1.016, 1.025).

4.3.2 Second order: Burnett equations

From the preceding treatment of the first order expansion it is clear that the CE method applied to the full Boltzmann equation is disproportionately more difficult than for the ES-BGK model, although the results of both agree within the approximations made.

The same can be said for the higher order expansions. The computation of the Burnett equations for the ES-BGK model is cumbersome already, and is even more involved for the full Boltzmann equation with power potentials. The resulting equations are just of the same form as for the ES-BGK model (4.26, 4.27), and differ only in the Burnett coefficients, which depend on the interaction model used, and the number of coefficients considered in the expansion into Sonine polynomials [2][6][13].

If only one term in the Sonine expansion is considered, the reported Burnett coefficients agree exactly with the coefficients obtained for the ES-BGK model with $Pr = 2/3$, as shown in (4.29).

With a sufficient number of expansion coefficients one finds the Burnett coefficients for hard spheres as [2][13]

$$\varpi_1 = 1.014 \times \frac{4}{3}\left(\frac{7}{2} - \omega\right) \; , \; \varpi_2 = 1.014 \times 2 \; , \; \varpi_3 = 0.806 \times 3 \; , \; \varpi_4 = 0.681 \; ,$$

$$\varpi_5 = 0.806 \times 3\omega - 0.990 \; , \; \varpi_6 = 0.928 \times 8 \; ,$$

$$\theta_1 = 1.035 \times \frac{15}{4}\left(\frac{7}{2} - \omega\right) \; , \; \theta_2 = 1.035 \times \frac{45}{8} \; , \; \theta_3 = -1.030 \times 3 \; ,$$

$$\theta_4 = 0.806 \times 3 \; , \; \theta_5 = 0.918 \times \frac{35}{4} + 0.806 \times \omega - 0.150 \; .$$

Most of the coefficients are not too different from their counterparts in the simplest approximation (4.29), but some coefficients, most notably ϖ_4 and ϖ_5, differ substantially.

Slemrod showed that for all interaction potentials with spherical symmetry the coefficients are related such that $\varpi_3 + \varpi_4 + \theta_3 = 0$ [58].

One also finds the Burnett equations written with time derivatives in the factors on ϖ_2 and θ_2, in which case they read

$$\sigma_{ij}^{(2)} = \frac{\mu^2}{p}\left[\varpi_1\frac{\partial v_k}{\partial x_k}S_{ij} + \varpi_2\left(\frac{DS_{ij}}{Dt} - 2\frac{\partial v_k}{\partial x_{\langle i}}S_{j\rangle k}\right)\right.$$

$$\left. + \varpi_3\frac{\partial^2\theta}{\partial x_{\langle i}\partial x_{j\rangle}} + \varpi_4\frac{\partial\theta}{\partial x_{\langle i}}\frac{\partial\ln p}{\partial x_{j\rangle}} + \varpi_5\frac{1}{\theta}\frac{\partial\theta}{\partial x_{\langle i}}\frac{\partial\theta}{\partial x_{j\rangle}} + \varpi_6 S_{k\langle i}S_{j\rangle k}\right] \; , \quad (4.47)$$

$$q_i^{(2)} = \frac{\mu^2}{\rho}\left[\theta_1\frac{\partial v_k}{\partial x_k}\frac{\partial\ln\theta}{\partial x_i} + \theta_2\frac{1}{\theta}\left(\frac{D}{Dt}\frac{\partial\theta}{\partial x_i} - \frac{\partial\theta}{\partial x_k}\frac{\partial v_k}{\partial x_i}\right)\right.$$

$$\left. + \theta_3 S_{ik}\frac{\partial\ln p}{\partial x_k} + \theta_4\frac{\partial S_{ik}}{\partial x_k} + 3\theta_5 S_{ik}\frac{\partial\ln\theta}{\partial x_k}\right] \; . \quad (4.48)$$

The strict CE expansion requires the time derivatives $\frac{D_0 S_{ij}}{Dt}$ and $\frac{D_0}{Dt}\frac{\partial \theta}{\partial x_i}$, and in this case $\sigma_{ij}^{(2)}$ and $q_i^{(2)}$ are purely of second order. The use of the full time derivatives $\frac{DS_{ij}}{Dt}$ and $\frac{D}{Dt}\frac{\partial \theta}{\partial x_i}$ introduces higher order terms, which, however, are allowed in a second order theory, since their influence should be small.

4.3.3 Third order: Super-Burnett equations

The Chapman-Enskog expansion to third order yields the super-Burnett equations. Their computation is extremely cumbersome, and it seems that the full three-dimensional non-linear super-Burnett equations were never derived.

The linearized super-Burnett equations, where all products of gradients are ignored, read [55][59]

$$\sigma_{ij}^{(3)} = \frac{\mu^3}{p^2}\left(\frac{5}{3}\theta\frac{\partial^2}{\partial x_{\langle i}\partial x_{j\rangle}}\frac{\partial v_k}{\partial x_k} - \frac{4}{3}\theta\frac{\partial^2}{\partial x_k \partial x_k}\frac{\partial v_{\langle i}}{\partial x_{j\rangle}}\right),$$

(4.49)

$$q_i^{(3)} = \frac{\mu^3}{p^2}\left(-\frac{157}{16}\theta\frac{\partial^3 \theta}{\partial x_i \partial x_k \partial x_k} - \frac{5}{8}\frac{\theta^2}{\rho}\frac{\partial^3 \rho}{\partial x_i \partial x_k \partial x_k}\right).$$

The non-linear super-Burnett equations were only derived for purely one-dimensional geometry, and the results differ somewhat between authors. We shall explain their derivation from moment systems in Sec. 6.3.3 [60], but give the results already here. The third order contribution to the stress is

$$\sigma_{\langle 11\rangle}^{(3)} = \frac{\mu^3}{p^2}\left[\frac{47}{3}\frac{1}{\rho}\frac{\partial\rho}{\partial x}\frac{\partial\theta}{\partial x}\frac{\partial v}{\partial x} - \frac{64}{9}\frac{\theta}{p^2}\left(\frac{\partial\rho}{\partial x}\right)^2\frac{\partial v}{\partial x} + \frac{40}{9}\frac{\theta}{\rho}\frac{\partial^2\rho}{\partial x^2}\frac{\partial v}{\partial x} - \frac{2}{3}\frac{\theta}{\rho}\frac{\partial\rho}{\partial x}\frac{\partial^2 v}{\partial x^2}\right.$$
$$\left. -7\frac{1}{\theta}\left(\frac{\partial\theta}{\partial x}\right)^2\frac{\partial v}{\partial x} - \frac{47}{9}\frac{\partial\theta}{\partial x}\frac{\partial^2 v}{\partial x^2} - \frac{31}{9}\frac{\partial^2\theta}{\partial x^2}\frac{\partial v}{\partial x} + \frac{2}{9}\theta\frac{\partial^3 v}{\partial x^3} + \frac{16}{27}\left(\frac{\partial v}{\partial x}\right)^3\right],$$

(4.50)

and the third order contribution to the heat flux is

$$q_1^{(3)} = \frac{\mu^3}{p^2}\left[-\frac{8035}{336}\frac{\partial\theta}{\partial x}\left(\frac{\partial v}{\partial x}\right)^2 + \frac{166}{21}\frac{\theta}{\rho}\frac{\partial\rho}{\partial x}\left(\frac{\partial v}{\partial x}\right)^2 + \frac{949}{168}\theta\frac{\partial^2 v}{\partial x^2}\frac{\partial v}{\partial x}\right.$$
$$+\frac{917}{8}\frac{1}{\rho}\frac{\partial\rho}{\partial x}\left(\frac{\partial\theta}{\partial x}\right)^2 - \frac{1137}{16}\frac{\theta}{\rho^2}\left(\frac{\partial\rho}{\partial x}\right)^2\frac{\partial\theta}{\partial x} + \frac{397}{16}\frac{\theta}{\rho}\frac{\partial\rho}{\partial x}\frac{\partial^2\theta}{\partial x^2}$$
$$+\frac{701}{16}\frac{\theta}{\rho}\frac{\partial^2\rho}{\partial x^2}\frac{\partial\theta}{\partial x} - \frac{813}{16}\frac{1}{\theta}\left(\frac{\partial\theta}{\partial x}\right)^3 - \frac{1451}{16}\frac{\partial\theta}{\partial x}\frac{\partial^2\theta}{\partial x}$$
$$\left.-\frac{157}{16}\theta\frac{\partial^3\theta}{\partial x^3} - \frac{41}{8}\frac{\theta^2}{\rho^2}\frac{\partial^2\rho}{\partial x^2}\frac{\partial\rho}{\partial x} - \frac{5}{8}\frac{\theta^2}{\rho}\frac{\partial^3\rho}{\partial x^3} + \frac{23}{4}\frac{\theta^2}{\rho^3}\left(\frac{\partial\rho}{\partial x}\right)^3\right]. \quad (4.51)$$

All coefficients in these equations agree with those of Fiscko and Chapman [61]. Shavaliyev agrees with most coefficients, but found slightly different values for the underlined coefficients [55]. In the equation (4.50) for stress, he has at the underlined terms

$$\left[-\frac{40}{3},\frac{32}{3}\right] \quad \text{instead of} \quad \left[-\frac{64}{9},\frac{40}{9}\right],$$

and in the equation for the heat flux (4.51), he has

$$\left[-\frac{9005}{168},\frac{271}{21},\frac{421}{42}\right] \quad \text{instead of} \quad \left[-\frac{8035}{336},\frac{166}{21},\frac{949}{168}\right].$$

In [60] non-linear contributions in the collision term were ignored, which gives for these three coefficients

$$\left[-\frac{2913}{112},\frac{188}{21},\frac{199}{56}\right]$$

while all other coefficients agree. Note that the signs of the coefficients agree in all three cases.

4.3.4 Augmented Burnett equations

The instability of the original Burnett equations lead Zhong et al. to a modification where they added some terms of super-Burnett order to the Burnett equations with the goal to stabilize the equations, and they termed their new set of equations as "augmented Burnett equations" [62][63]. For the one-dimensional case they added the following linear terms to the Burnett equations

$$\sigma_{\langle 11\rangle}^{(A)} = \frac{\mu^3}{p^2}\left(\frac{2}{9}\theta\frac{\partial^3 v}{\partial x^3}\right) \quad , \quad q_1^{(A)} = \frac{\mu^3}{p^2}\left(\frac{11}{16}\theta\frac{\partial^3\theta}{\partial x^3} - \frac{5}{8}\frac{\theta^2}{\rho}\frac{\partial^3\rho}{\partial x^3}\right). \qquad (4.52)$$

Comparison with the true super-Burnett terms (4.49) shows that Zhong et al. have in the heat flux a coefficient of $(11/16)$ instead of $(-157/16)$, citing Wang-Chang [64] as the source of the coefficient. This coefficient contradicts all other results on the super-Burnett equations as cited in the previous section, and must be considered as not being based on the Chapman-Enskog expansion of the Boltzmann equation. The important difference to the proper super-Burnett terms is the difference in sign, which yields stability in time.

Furthermore, Zhong et al. generalize their augmented equations in an ad hoc manner to three dimensions as

$$\sigma_{ij}^{(A)} = \frac{\mu^3}{p^2}\left(\frac{1}{3}\theta\frac{\partial^2}{\partial x_k\partial x_k}\frac{\partial v_{\langle i}}{\partial x_{j\rangle}}\right),$$

$$q_i^{(A)} = \frac{\mu^3}{p^2}\left(\frac{11}{16}\theta\frac{\partial^3\theta}{\partial x_i\partial x_k\partial x_k} - \frac{5}{8}\frac{\theta^2}{\rho}\frac{\partial^3\rho}{\partial x_i\partial x_k\partial x_k}\right). \qquad (4.53)$$

Comparison with (4.49) shows that there is not only the wrong factor (11/16) instead of ($-157/16$) in the heat flux, but also that the three dimensional ad hoc term $\sigma_{ij}^{(A)}$ is not proper in its tensorial structure, since it differs from the super-Burnett relation (4.49)$_1$.

4.3.5 BGK-Burnett equations

While the Burnett equations with the proper Prandtl number are unstable, the BGK-Burnett equations, that is (4.26–4.28) with $b = 0$ turn out to be stable, see Sec. 10.3, and [57][65][66][67]. This, somewhat surprising, result allows stable numerical simulations of the BGK-Burnett equations [67], but it must be noted that the BGK model, and thus its Burnett equations, cannot describe actual gases with sufficient accuracy, as can be seen from (a) the wrong Prandtl number $\mathrm{Pr}_{BGK} = 1$, and (b), the differences between the actual Burnett coefficients (4.29), and the BGK-Burnett coefficients, which follow from (4.28) for Pr = 1 as

$$\varpi_1 = \frac{4}{3}\left(\frac{7}{2} - \omega\right) , \ \varpi_2 = 2, \ \varpi_3 = 2, \ \varpi_4 = 0, \ \varpi_5 = 2\omega, \ \varpi_6 = 8 ,$$

$$\theta_1 = \frac{5}{3}\left(\frac{7}{2} - \omega\right) , \ \theta_2 = \frac{5}{2}, \ \theta_3 = -2, \ \theta_4 = 2 , \theta_5 = \frac{2}{3}(7+\omega) .$$

Thus, numerical simulations based on the BGK-Burnett equations can at most describe a trend, but certainly not the details of a rarefied gas flow.

4.4 The second law

4.4.1 The second law to first order

In Sec. (3.4) we saw that the H-theorem is an essential property of the Boltzmann equation. According to the H-theorem there exits a balance law for entropy $\eta = \rho s$ with a positive production. We can write this as

$$\rho\frac{Ds}{Dt} + \frac{\partial\phi_k}{\partial x_k} = \Sigma \geq 0 \tag{4.54}$$

with the definitions of entropy density and non-convective entropy flux (3.38, 3.41)

$$\rho s = -k\int f\ln\frac{f}{y}d\mathbf{c} \ , \ \ \phi_\kappa = -k\int C_k f\ln\frac{f}{y}d\mathbf{c} . \tag{4.55}$$

According to the CE expansion, the first order solution for the phase density is given as

$$f = f_M\left(1 + \varepsilon\varphi\right) ,$$

and the corresponding expressions for entropy and its flux follow from an expansion of the logarithm, $\ln(1 + \varepsilon\varphi) \simeq \varepsilon\varphi$. For the entropy we obtain by means of (4.31) to first order

$$\rho s = -k \int f_M (1 + \varepsilon\varphi) \left(\ln \frac{f_M}{y} + \varepsilon\varphi \right) d\mathbf{c} = -k \int f_M \ln \frac{f_M}{y} d\mathbf{c} .$$

This is just the entropy in equilibrium (3.43)

$$s = \frac{k}{m} \ln \frac{\theta^{3/2}}{\rho} + s_0 ,$$

and we conclude that to first order in the Knudsen number the gas can be described locally as if it were in equilibrium. This result gives justification to the hypothesis of local equilibrium that forms the basis of the classical theory of thermodynamics of irreversible processes [68][69].

The corresponding first order entropy flux is obtained as

$$\phi_k = -k \int C_k f_M \varepsilon\varphi \left(\ln \frac{f_M}{y} + 1 \right) d\mathbf{c} .$$

Since $\ln f_M + 1 = -C^2/2\theta + F(\rho, T)$, and $\int C_k f_M \varphi d\mathbf{c} = 0$ we obtain, after setting the formal parameter $\varepsilon = 1$,

$$\phi_k = \frac{1}{T} \frac{m}{2} \int C_k C^2 f_M \varphi d\mathbf{c} = \frac{q_k}{T} . \tag{4.56}$$

Also this relation between heat flux and entropy flux is well-known in classical thermodynamics (Problem 4.4). In the context of the CE expansion it becomes clear that (4.56) does not have general validity, but is restricted to processes with sufficiently small Knudsen numbers.

The corresponding entropy production is best computed by inserting the expressions for entropy and entropy flux into (4.54), and eliminating the time derivatives of density and temperature by means of the balance laws (4.1). This yields with the laws of Navier-Stokes and Fourier a non-negative production of entropy

$$\Sigma = -\frac{1}{T} \sigma_{ij} \frac{\partial v_{\langle i}}{\partial x_{j\rangle}} - \frac{q_k}{T^2} \frac{\partial T}{\partial x_k} = \frac{2\mu}{T} \frac{\partial v_{\langle i}}{\partial x_{j\rangle}} \frac{\partial v_{\langle i}}{\partial x_{j\rangle}} + \frac{\kappa}{T^2} \frac{\partial T}{\partial x_k} \frac{\partial T}{\partial x_k} \geq 0 . \tag{4.57}$$

Thus we can summarize the first order CE expansion gives results in accordance with the Boltzmann equation such that the H-theorem is fulfilled.

4.4.2 The H-theorem in higher order expansions

Next we discuss the second law for higher order expansions [70]. Note that Boltzmann's H-theorem states that the entropy production is positive for any distribution function f. From the discussion in this chapter it is clear that the phase density as obtained from the Chapman-Enskog method can be written as

$$f = f_M (1 + \Phi) \quad \text{where} \quad \Phi = \varepsilon \phi^{(1)} + \varepsilon^2 \phi^{(2)} + \ldots \tag{4.58}$$

The $\phi^{(n)}$ are products between polynomials in the peculiar velocity and derivatives in space and time of density, temperature and velocity. The simplest example is given by the the first Chapman-Enskog expansion for the ES-BGK model, or Maxwell molecules (4.25, 4.37); see [57] for the phase density of the Burnett equations.

Due to its definition as a number density in phase space the phase density ought to be positive, and, moreover, the definition (4.55) makes only sense for positive f. The Chapman-Enskog expression (4.58), however, will become negative since Φ will fall below (-1) for large C.

In the proper range of applicability, the gradients of temperature and velocity are rather small, and Φ falls below (-1) only for very large C. However, Φ is always multiplied by the Maxwellian which suppresses the negative contributions for the large velocities in question.

In order to fix our arguments, we introduce a velocity bound $C_{\max}$ defined such that $f \simeq 0$ for $C > C_{\max}$ and demand that

$$\Phi > -1 \quad \text{for} \quad C < C_{\max} . \tag{4.59}$$

With this in mind, we turn the attention to the entropy production. Due to conservation of mass, momentum and energy, $\int S \ln f_M d\mathbf{c} = 0$ and we find the entropy production as

$$\sigma = -k \int S \ln (1 + \Phi) \, d\mathbf{c} .$$

This equation makes sense under the assumption that $C_{\max}$ is a suitable bound for the collision term S as well, so that S is negligible above $C_{\max}$ where the argument of the logarithm becomes negative.

Condition (4.59) is the prerequisite for expanding the logarithm into a Taylor series, viz.

$$\sigma = -k \int S \left[\Phi - \frac{\Phi^2}{2} + \cdots \right] d\mathbf{c} \tag{4.60}$$

$$= -k \int S \left[\varepsilon \phi^{(1)} + \varepsilon^2 \left(-\frac{1}{2} \left(\phi^{(1)} \right)^2 + \phi^{(2)} \right) + \cdots \right] d\mathbf{c} .$$

It must be emphasized that the series does not converge if $\Phi < -1$. In other words: the series expansion is wrong, if the requirement (4.59) is violated.

Thus, if (4.59) holds, everything turns out well. In particular, we are allowed to compute the entropy production by means of the expansion (4.60). On the other hand, if (4.59) is not fulfilled, the expansion is not allowed and will, if used, give incorrect results, e.g. a negative value for the entropy production.

Clearly, (4.59) is not easy to evaluate, in particular because one has to provide the bound C_{max} first. As example we consider the first Chapman-Enskog expansion for Maxwell molecules (4.25). In this case (4.59) may be written as

$$\frac{\mu}{p}\left|\xi_i\xi_j\frac{\partial v_{\langle i}}{\partial x_{j\rangle}}\right| \leq 1 \quad , \quad \frac{15}{4}\frac{\mu}{p\sqrt{\theta}}\left|\xi_k\left(1-\frac{\xi^2}{5}\right)\frac{\partial\theta}{\partial x_k}\right| \leq 1 , \tag{4.61}$$

where ξ_i is a suitable dimensionless velocity with absolute value of the order of $C_{max}/\sqrt{\theta}$.

For the first order expansion, however, only the first term in the series (4.60) is required for the entropy production,

$$\sigma = -k\int S\phi^{(1)}d\mathbf{c} , \tag{4.62}$$

which turns out to be positive for *all* gradients of temperature and velocity as was shown above. Since the expansion is not allowed if the conditions (4.59) or (4.61) are violated, this result must be seen as a mere coincidence.

In [71] it was shown that the entropy production according to the Burnett equations may become negative, but the finite radius of convergence was left out of consideration. Those authors found, for the Burnett equations, conditions similar to (4.61), which limit the values of the gradients so that the H-theorem remains valid. The violation of the H-theorem most probably is related to the instability of the Burnett and super-Burnett equations [72]. See also [73], where the authors show that the realizability of some moments of the first order Chapman-Enskog phase density is not assured if the gradients become too steep.

Problems

4.1. First order equations
(a) Recalculate *all* equations of Sec. 4.2.3.
(b) Show that (4.25) indeed fulfills the compatibility conditions (4.9).

4.2. A property of $[\psi, \varphi]$
Show that (4.38) can be written as

$$[\psi, \varphi] = \frac{1}{4}\int \exp\left[-\xi^2 - \xi_1^2\right] \mathcal{D}\left[\psi\right]\mathcal{D}\left[\varphi\right]\sigma g\sin\Theta d\Theta d\varepsilon d\boldsymbol{\xi} d\boldsymbol{\xi}^1 .$$

where $\mathcal{D}\left[\psi\right] = \psi + \psi^1 - \psi' - \psi^{1\prime}$.

4.3. Super-Burnett equations
Linearize (4.50, 4.51) and simplify (4.49) for one-dimensional processes, and show that the results agree.

4.4. Classical Form of the second law

The classical form of the second law for a closed volume (no convective flux through the boundary) reads

$$\frac{dS}{dt} \geq \sum_{\alpha} \frac{\dot{Q}_{\alpha}}{T_{\alpha}}$$

where $\dot{Q}_{\alpha} = -\int_{A_{\alpha}} q_k n_k dA$ is the heat flow over the boundary A_{α} which has the temperature T_{α}. Show that this equation follows from (4.54) with the approximations (3.43, 4.56).

4.5. Thermophoresis in Chapman-Enskog gas

Consider a spherical disk in a heat conducting gas at rest, where the phase density is given as ($\xi_k = C_k/\sqrt{2\theta}$)

$$f d\mathbf{c} = \frac{\rho}{m} \frac{\exp\left[-\xi^2\right]}{\sqrt{\pi}^3} \left(1 + \frac{4}{5} \frac{q_k}{p\sqrt{2\theta}} \left(\xi^2 - \frac{5}{2}\right) \xi_k\right) d\boldsymbol{\xi} \ .$$

Under the assumption that the gas particles colliding with the disk are specularly reflected, compute the resulting force on the disk. Compute also the force on a cylinder and a sphere. Compare with the result from Problem 2.9. Assume reasonable data for pressure, temperature and heat flux, and compute the force: discuss its effect on actual particles. You might also think about what happens when the gas particles are thermalized in collision.

5

Moment equations

Moment equations derived from the Boltzmann equation offer an alternative description of the ideal gas. When a complete set of moments is considered, that is if the moments are based on a complete functional space, the description of the gas through moments is equivalent to the description through the phase density. This chapter presents equations which form the base for the discussion in later parts of this text.

5.1 Moment equations as infinite coupled system

In Sec. 2.2 we introduced the trace-free moments as (2.12)

$$u_{i_1 \cdots i_n}^{\alpha} = \rho_{\langle i_1 i_2 \cdots i_n \rangle}^{\alpha} = m \int C^{2\alpha} C_{\langle i_1} C_{i_2} \cdots C_{i_n \rangle} f d\mathbf{c} . \qquad (5.1)$$

These are based on the monomials $C^{2\alpha+n}$ and the spherical harmonics $\nu_{\langle i_1} \cdots \nu_{i_n \rangle}$ $(\nu_i = C_i/C)$ and form a complete set of functions [21], see also Appendix A.2.3. The resulting moment equations, presented below, form an infinite coupled set of partial differential equations which is equivalent to the Boltzmann equation.

The approximative methods discussed later allow drastic reductions in the number of moments required, and thus yield systems of partial differential equations that can be handled. Note that the integration over the velocity space removes the velocity vector C_i as an independent variable, and replaces the continuous variable C_i by a finite, and small, number of moments.

The moments (2.14),

$$F_{i_1 i_2 \cdots i_n} = F_{i_1 i_2 \cdots i_n}^0 = m \int c_{i_1} c_{i_2} \cdots c_{i_n} f d\mathbf{c} , \qquad (5.2)$$

form a complete set as well, since they can be split into the trace-free moments (5.1) by means of (2.16) and (A.1). While the full moments are not convenient for detailed calculations, their moment equations have an instructive structure, and that is why we begin our discussion with them.

The moment equations are obtained by multiplying the Boltzmann equation

$$\frac{\partial f}{\partial t} + c_k \frac{\partial f}{\partial x_k} + G_k \frac{\partial f}{\partial c_k} = \mathcal{S}$$

with $mc_{i_1}c_{i_2}\cdots c_{i_n}$, and subsequent integration over velocity space, see Sec. 3.3. Since $\mathbf{x}$, t and $\mathbf{c}$ are independent variables this gives

$$\frac{\partial F_{i_1 i_2 \cdots i_n}}{\partial t} + \frac{\partial F_{i_1 i_2 \cdots i_n k}}{\partial x_k} = \Pi_{i_1 i_2 \cdots i_n} + n F_{(i_1 i_2 \cdots i_{n-1}} G_{i_n)} , \qquad (5.3)$$

where the indices in round brackets denote a symmetric tensor, see Appendix A.2. Here,

$$\Pi_{i_1 i_2 \cdots i_n} = m \int c_{i_1} c_{i_2} \cdots c_{i_n} \mathcal{S} d\mathbf{c} \qquad (5.4)$$

is the production of the moment $F_{i_1 i_2 \cdots i_n}$ due to collisions, which vanishes for conserved quantities. Thus, the equations for the full moments are balance laws in divergence form (3.32). Taking conservation of mass ($\rho = F$), momentum ($\rho v_i = F_i$) and energy ($\rho e = \frac{1}{2} F_{kk}$) into account, the first few equations read

$$\frac{\partial F}{\partial t} + \frac{\partial F_k}{\partial x_k} = 0 ,$$

$$\frac{\partial F_i}{\partial t} + \frac{\partial F_{ik}}{\partial x_k} = FG_i ,$$

$$\frac{\partial F_{ij}}{\partial t} + \frac{\partial F_{ijk}}{\partial x_k} = \Pi_{\langle ij \rangle} + 2F_{(i}G_{j)} ,$$

$$\frac{\partial F_{ijk}}{\partial t} + \frac{\partial F_{ijkl}}{\partial x_l} = \Pi_{ijk} + 3F_{(ij}G_{k)} .$$

The structure of the moment equations is such that the equation for the moment $F_{i_1 i_2 \cdots i_n}$ (where that moment is under the time derivative) contains the next higher moment $F_{i_1 i_2 \cdots i_n k}$ as its flux (in the divergence) and the next lower moment $F_{i_1 i_2 \cdots i_{n-1}}$ in the supply term (as a factor on the external force G_k). Since the equation at order n contains the moments at orders $n \pm 1$, the equations are coupled, and form an infinite set of equations.

Moreover, the equations contain the production terms $\Pi_{i_1 i_2 \cdots i_n}$ (5.4) which are related to the distribution function f through the collision term $\mathcal{S}$. In order to have a complete set of equations for the moments, it is necessary to express the production terms in terms of the moments. Generally this can only be done, if a relation between the moments and the distribution function is known. For select cases, however, the production terms can be evaluated without knowing the phase density, and these are Maxwell molecules and BGK models, including the ES-BGK model. The discussion in this chapter will consider only production terms for the afore mentioned "good" cases. What needs to be done for other interaction models is discussed in Sec. 6.5.

The equations for the full moments (5.3) appear in divergence form (3.32), and this is also so for the convective moments $F_{i_1 i_2 \cdots i_n}^a$ (2.15), and their trace-free parts $F_{\langle i_1 i_2 \cdots i_n \rangle}^a$ where the corresponding balance laws read

$$\frac{\partial F_{i_1 i_2 \cdots i_n}^a}{\partial t} + \frac{\partial F_{i_1 i_2 \cdots i_n k}^a}{\partial x_k} = \Pi_{i_1 i_2 \cdots i_n}^a + n F_{(i_1 i_2 \cdots i_{n-1}}^a G_{i_n)} + 2a F_{i_1 i_2 \cdots i_n k}^{a-1} G_k \, ,$$

(5.5)

$$\frac{\partial F_{\langle i_1 i_2 \cdots i_n \rangle}^a}{\partial t} + \frac{\partial F_{\langle i_1 \cdots i_n k \rangle}^a}{\partial x_k} + \frac{n}{2n+1} \frac{\partial F_{\langle i_1 \cdots i_{n-1} \rangle}^{a+1}}{\partial x_{i_n \rangle}}$$

$$= \Pi_{\langle i_1 i_2 \cdots i_n \rangle}^a + \frac{n(2n+2a+1)}{2n+1} F_{\langle i_1 i_2 \cdots i_{n-1}}^a G_{i_n \rangle} + 2a F_{\langle i_1 \cdots i_n k \rangle}^{a-1} G_k \, . \quad (5.6)$$

5.2 Equations for trace-free central moments

The trace-free central moments (5.1) are convenient variables to consider, and are well-suited for the approximative methods to be discussed later. While equivalent to the equations for convective moments (5.3, 5.5, 5.6), they appear to be considerably more complex. This is merely a consequence of the compact notation for the convective moments, which is expanded here, see (2.16). At first glance, the equations presented below might appear forbiddingly complicated, but they are not, as will become clear as we proceed.

5.2.1 Generic moment equation

Multiplication of the Boltzmann equation with $mC^{2a} C_{\langle i_1} \cdots C_{i_n \rangle}$ and subsequent integration over velocity space yields, after some rearrangement, the general equation for the trace-free moments (5.1) [74],

$$\frac{Du_{i_1 \cdots i_n}^a}{Dt} + 2a u_{i_1 \cdots i_n k}^{a-1} \left[\frac{Dv_k}{Dt} - G_k \right]$$

$$+ \frac{n(2a+2n+1)}{2n+1} u_{\langle i_1 \cdots i_{n-1}}^a \left[\frac{Dv_{i_n \rangle}}{Dt} - G_{i_n \rangle} \right] + \frac{\partial u_{i_1 \cdots i_n k}^a}{\partial x_k}$$

$$+ \frac{n}{2n+1} \frac{\partial u_{\langle i_1 \cdots i_{n-1} \rangle}^{a+1}}{\partial x_{i_n \rangle}} + 2a u_{i_1 \cdots i_n kl}^{a-1} \frac{\partial v_k}{\partial x_l} + 2a \frac{n+1}{2n+3} u_{\langle i_1 \cdots i_n}^a \frac{\partial v_{k \rangle}}{\partial x_k}$$

$$+ 2a \frac{n}{2n+1} u_{k \langle i_1 \cdots i_{n-1}}^a \frac{\partial v_k}{\partial x_{i_n \rangle}} + n u_{k \langle i_1 \cdots i_{n-1}}^a \frac{\partial v_{i_n \rangle}}{\partial x_k} + u_{i_1 \cdots i_n}^a \frac{\partial v_k}{\partial x_k}$$

$$+ \frac{n(n-1)}{4n^2-1} (2a+2n+1) u_{\langle i_1 \cdots i_{n-2}}^{a+1} \frac{\partial v_{i_{n-1}}}{\partial x_{i_n \rangle}} = \mathcal{P}_{i_1 \cdots i_n}^a \, . \quad (5.7)$$

Note that all moments are trace-free, and additional trace-free tensors are made explicit by means of angular brackets. The derivation of the equation above requires multiple use of (A.3). The set of infinitely many moment equations $(a \rightarrow \infty, n \rightarrow \infty)$ is equivalent to the Boltzmann equation.

The production term is given as

$$P^a_{i_1 \cdots i_n} = m \int C^{2\alpha} C_{\langle i_1} \cdots C_{i_n \rangle} S d\mathbf{c} \tag{5.8}$$

and causes the same difficulty as (5.4): It is possible only for few collision models—including Maxwell molecules, BGK model, and ES-BGK model—to explicitly express the production term through the moments. The production terms for these models will be computed in Sec. 5.3 below. For other models, the production term can only be computed if a relation between moments and phase density is known (Sec. 6.5). Due to conservation of mass, momentum and energy, we have in any case

$$\mathcal{P}^0 = \mathcal{P}^0_i = \mathcal{P}^1 = 0 . \tag{5.9}$$

5.2.2 Conservation laws

First, we consider the conservation laws, that is those equations which, by (5.9), have no production. For $a = 0, n = 0$ we obtain the mass balance

$$\frac{Du^0}{Dt} + u^0 \frac{\partial v_k}{\partial x_k} = 0 ,$$

and for $a = 1, n = 0$ we find the balance of internal energy as

$$\frac{Du^1}{Dt} + \frac{\partial u^1_k}{\partial x_k} + 2u^0_{kl} \frac{\partial v_k}{\partial x_l} + \frac{5}{3} u^0 \frac{\partial v_k}{\partial x_k} = 0 .$$

Note that by means of (2.13) these two equations can be brought into their usual textbook form

$$\frac{D\rho}{Dt} + \rho \frac{\partial v_k}{\partial x_k} = 0 , \tag{5.10}$$

$$\frac{3}{2} \rho \frac{D\theta}{Dt} + \rho \theta \frac{\partial v_k}{\partial x_k} + \frac{\partial q_k}{\partial x_k} + \sigma_{kl} \frac{\partial v_k}{\partial x_l} = 0 . \tag{5.11}$$

For the choice $a = 1, n = 1$ we obtain the balance of momentum

$$\rho \frac{Dv_i}{Dt} + \theta \frac{\partial \rho}{\partial x_i} + \rho \frac{\partial \theta}{\partial x_i} + \frac{\partial \sigma_{ik}}{\partial x_k} = \rho G_i . \tag{5.12}$$

There are no further moment equations with vanishing production terms.

5.2.3 Scalar moments

For scalar moments ($n = 0$), the general equation (5.7) reduces to

$$\frac{Du^a}{Dt} + 2au_k^{a-1}\left[\frac{Dv_k}{Dt} - G_k\right] + \frac{\partial u_k^a}{\partial x_k} + 2au_{kl}^{a-1}\frac{\partial v_k}{\partial x_l} + \frac{2a+3}{3}u^a\frac{\partial v_k}{\partial x_k} = \mathcal{P}^a \;.$$

Next, the difference between the scalar variables and their equilibrium values (2.19) is introduced as

$$w^a = u^a - u^a_{|E} \;, \tag{5.13}$$

note that $w^0 = w^1 = 0$. The variables w^a are convenient, since they vanish in equilibrium, as do the $u^a_{i_1\cdots i_n}$ for $n \geq 1$, see (2.19).

Thus, almost all variables vanish in equilibrium, only mass density ρ, velocity v_i and temperature θ remain different from zero. These five quantities correspond to the conserved quantities mass density ρ, momentum density ρv_i and energy density $\rho\left(\frac{3}{2}\theta + \frac{1}{2}v^2\right)$.

The scalar equations for the new variables, where all time derivatives of ρ, θ, v_i are replaced by means of the conservation laws (5.10–5.12), are

$$\frac{Dw^a}{Dt} - \frac{2a}{3}(2a+1)!!\theta^{a-1}\frac{\partial q_k}{\partial x_k} - \frac{2a}{3}(2a+1)!!\theta^{a-1}\sigma_{kl}\frac{\partial v_k}{\partial x_l} + 2au_{kl}^{a-1}\frac{\partial v_k}{\partial x_l}$$
$$- 2au_k^{a-1}\frac{\partial\theta}{\partial x_k} - 2au_k^{a-1}\theta\frac{\partial\ln\rho}{\partial x_k} - 2a\frac{u_k^{a-1}}{\rho}\frac{\partial\sigma_{kl}}{\partial x_l} + \frac{\partial u_k^a}{\partial x_k} + \frac{2a+3}{3}w^a\frac{\partial v_k}{\partial x_k} = \mathcal{P}^a. \tag{5.14}$$

Of course, for $a = 0$ and $a = 1$ the equations are identically fulfilled, so that the above equation makes sense only for $a \geq 2$. The double factorial $a!!$ enters the equation through (2.19).

5.2.4 Vector moments

For vectors ($n = 1$), the general equation (5.7) reduces to

$$\frac{Du_i^a}{Dt} + 2au_{ik}^{a-1}\left[\frac{Dv_k}{Dt} - G_k\right] + \frac{2a+3}{3}u^a\left[\frac{Dv_i}{Dt} - G_i\right] + \frac{\partial u_{ik}^a}{\partial x_k} + \frac{1}{3}\frac{\partial u^{a+1}}{\partial x_i}$$
$$+ 2au_{ikl}^{a-1}\frac{\partial v_k}{\partial x_l} + \frac{4a}{5}u_{\langle i}^a\frac{\partial v_{k\rangle}}{\partial x_k} + \frac{2a}{3}u_k^a\frac{\partial v_k}{\partial x_i} + u_k^a\frac{\partial v_i}{\partial x_k} + u_i^a\frac{\partial v_k}{\partial x_k} = \mathcal{P}_i^a \;.$$

Introducing the non-equilibrium moments w^a in the vector equation, and replacing the time derivatives of the velocity by means of the momentum balance (5.12), we find

$$\frac{Du_i^a}{Dt} + \frac{a\,(2a+3)!!}{3}\rho\theta^a\frac{\partial\theta}{\partial x_i} - \frac{2a+3}{3}w^a\frac{\partial\theta}{\partial x_i} - 2au_{ik}^{a-1}\frac{\partial\theta}{\partial x_k}$$

$$- 2au_{ik}^{a-1}\theta\frac{\partial\ln\rho}{\partial x_k} - \frac{2a+3}{3}w^a\theta\frac{\partial\ln\rho}{\partial x_i} - 2a\frac{u_{ik}^{a-1}}{\rho}\frac{\partial\sigma_{kl}}{\partial x_l}$$

$$- \frac{2a+3}{3}\frac{w^a}{\rho}\frac{\partial\sigma_{ik}}{\partial x_k} - \frac{(2a+3)!!}{3}\theta^a\frac{\partial\sigma_{ik}}{\partial x_k} + \frac{\partial u_{ik}^a}{\partial x_k} + \frac{1}{3}\frac{\partial w^{a+1}}{\partial x_i}$$

$$+ 2au_{ikl}^{a-1}\frac{\partial v_k}{\partial x_l} + \frac{2a+5}{5}u_i^a\frac{\partial v_k}{\partial x_k} + \frac{2a+5}{5}u_k^a\frac{\partial v_i}{\partial x_k} + \frac{2a}{5}u_k^a\frac{\partial v_k}{\partial x_i} = \mathcal{P}_i^a\,. \quad (5.15)$$

This equation is relevant for $a \geq 1$, and identically fulfilled for $a = 0$ ($w^0 = w^1 = u_i^0 = 0$).

5.2.5 Rank-2 tensor moments

After replacing the time derivatives of velocity by means of (5.12), the equations for tensors of rank 2 read

$$\frac{Du_{ij}^a}{Dt} - 2a\frac{u_{ijk}^{a-1}}{\rho}\left(\frac{\partial\sigma_{kl}}{\partial x_l} + \frac{\partial\theta\rho}{\partial x_k}\right) - \frac{2}{5}(2a+5)\frac{u_{\langle i}^a}{\rho}\left(\frac{\partial\sigma_{j\rangle k}}{\partial x_k} + \frac{\partial\theta\rho}{\partial x_{j\rangle}}\right) + \frac{\partial u_{ijk}^a}{\partial x_k}$$

$$+ \frac{2}{5}\frac{\partial u_{\langle i}^{a+1}}{\partial x_{j\rangle}} + 2au_{ijkl}^{a-1}\frac{\partial v_k}{\partial x_l} + \frac{6a}{7}u_{\langle ij}^a\frac{\partial v_{k\rangle}}{\partial x_k} + \frac{4a}{5}u_{k\langle i}^a\frac{\partial v_k}{\partial x_{j\rangle}} + 2u_{k\langle i}^a\frac{\partial v_{j\rangle}}{\partial x_k}$$

$$+ u_{ij}^a\frac{\partial v_k}{\partial x_k} + \frac{2}{15}(2a+5)w^{a+1}\frac{\partial v_{\langle i}}{\partial x_{j\rangle}} + \frac{2}{15}(2a+5)!!\rho\theta^{a+1}\frac{\partial v_{\langle i}}{\partial x_{j\rangle}} = \mathcal{P}_{ij}^a\,. \quad (5.16)$$

5.2.6 General equation

For moments of order higher than two, the general equation reads after the time derivatives of velocity are eliminated

$$\frac{Du_{i_1\cdots i_n}^a}{Dt} - \frac{n\,(2a+2n+1)}{2n+1}\frac{u_{\langle i_1\cdots i_{n-1}}^a}{\rho}\left(\frac{\partial\sigma_{i_n\rangle k}}{\partial x_k} + \frac{\partial\theta\rho}{\partial x_{i_n\rangle}}\right)$$

$$- 2a\frac{u_{i_1\cdots i_n k}^{a-1}}{\rho}\left(\frac{\partial\sigma_{kl}}{\partial x_l} + \frac{\partial\theta\rho}{\partial x_k}\right) + \frac{\partial u_{i_1\cdots i_n k}^a}{\partial x_k} + \frac{n}{2n+1}\frac{\partial u_{\langle i_1\cdots i_{n-1}}^{a+1}}{\partial x_{i_n\rangle}} + 2au_{i_1\cdots i_n kl}^{a-1}\frac{\partial v_k}{\partial x_l}$$

$$+ 2a\frac{n+1}{2n+3}u_{\langle i_1\cdots i_n}^a\frac{\partial v_{k\rangle}}{\partial x_k} + 2a\frac{n}{2n+1}u_{k\langle i_1\cdots i_{n-1}}^a\frac{\partial v_k}{\partial x_{i_n\rangle}} + nu_{k\langle i_1\cdots i_{n-1}}^a\frac{\partial v_{i_n\rangle}}{\partial x_k}$$

$$+ u_{i_1\cdots i_n}^a\frac{\partial v_k}{\partial x_k} + \frac{n\,(n-1)}{4n^2-1}(2a+2n+1)u_{\langle i_1\cdots i_{n-2}}^{a+1}\frac{\partial v_{i_{n-1}}}{\partial x_{i_n\rangle}} = \mathcal{P}_{i_1\cdots i_n}^a\,. \quad (5.17)$$

5.3 Production terms

In order to have the moment equations presented above explicit in the moments, the production terms (5.8),

$$\mathcal{P}^a_{i_1\cdots i_n} = m \int C^{2\alpha} C_{\langle i_1} \cdots C_{i_n \rangle} S d\mathbf{c} \,,$$

must be expressed in terms of moments.

5.3.1 BGK model

We consider the BGK model first, since it is the simplest of our models. The collision term is given by (3.54),

$$\mathcal{S}_{BGK} = -\bar{\nu} \left(f - f_M \right) \,,$$

and the corresponding productions are given by

$$\mathcal{P}^a_{i_1\cdots i_n} = -\bar{\nu} m \int C^{2\alpha} C_{\langle i_1} \cdots C_{i_n \rangle} \left(f - f_M \right) d\mathbf{c} = -\bar{\nu} \left(u^a_{i_1\cdots i_n} - u^a_{i_1\cdots i_n | E} \right) \,,$$

where $u^a_{i_1\cdots i_n | E}$ are the moments of the Maxwellian. With (2.19) and (5.13) it follows that

$$\mathcal{P}^a = -\bar{\nu} w^a \,, \quad \mathcal{P}^a_{i_1\cdots i_n} = -\bar{\nu} u^a_{i_1\cdots i_n} \,.$$

Note that $w^0 = w^1 = u^0_i = 0$, which correspond to the conservation laws $\mathcal{P}^0 = \mathcal{P}^1 = \mathcal{P}^0_1 = 0$. Also recall that the mean collision frequency is a function of density and temperature, see (3.51).

5.3.2 ES-BGK model

Next, we consider the ES-BGK model (3.56), where

$$\mathcal{S}_{ES} = -\bar{\nu} \left(f - f_{ES} \right)$$

so that the corresponding production terms are given by

$$\mathcal{P}^a_{i_1\cdots i_n} = -\bar{\nu} m \int C^{2\alpha} C_{\langle i_1} \cdots C_{i_n \rangle} \left(f - f_{ES} \right) d\mathbf{c} = -\bar{\nu} \left(u^a_{i_1\cdots i_n} - u^a_{i_1\cdots i_n | ES} \right) \,.$$

$u^a_{i_1\cdots i_n | ES}$ denotes the moments of the anisotropic Gaussian (3.57) and their computation is outlined in Appendix A.3.4. The Gaussian integrals of the first 26 moments are,

$$u^0_{|ES} = \rho \,,$$

$$u^1_{|ES} = \rho \lambda_{kk} = 3\rho\theta \,,$$

$$u^2_{|ES} = 2\rho \lambda_{ij} \lambda_{ij} + \rho \left(\lambda_{kk} \right)^2 = 15\rho\theta^2 + 2b^2 \frac{\sigma_{kj}\sigma_{jk}}{\rho} \,,$$

$$u^0_{i|ES} = u^1_{i|G} = 0 \,,$$

$$u^0_{ij|ES} = \rho \lambda_{\langle ij \rangle} = b\sigma_{ij} \,,$$

$$u^1_{ij|ES} = 2\rho \lambda_{k\langle i} \lambda_{j \rangle k} + \rho \lambda_{kk} \lambda_{\langle ij \rangle} = 7b\theta\sigma_{ij} + 2\frac{b^2}{\rho} \sigma_{k\langle i} \sigma_{j \rangle k} \,,$$

$$u^0_{ijk|ES} = 0 \,.$$

Accordingly, the conservation laws for mass, momentum and energy are fulfilled, $\mathcal{P}^0 = \mathcal{P}^1 = \mathcal{P}^0_1 = 0$, which was not proven in Sec. 3.6.2. The nonvanishing production terms can be written as

$$\mathcal{P}^2 = -\frac{2}{3}\frac{p}{\mu}\left(w^2 - \frac{1}{2\rho}\sigma_{kj}\sigma_{jk}\right),$$

$$\mathcal{P}^1_i = -\frac{4}{3}\frac{p}{\mu}q_i,$$

$$\mathcal{P}^0_{ij} = -\frac{p}{\mu}\sigma_{ij},$$

$$\mathcal{P}^1_{ij} = -\frac{2}{3}\frac{p}{\mu}\left(u^1_{ij} + \frac{7}{2}\theta\sigma_{ij} - \frac{1}{2\rho}\sigma_{k\langle i}\sigma_{j\rangle k}\right),$$

$$\mathcal{P}^0_{ijk} = -\frac{2}{3}\frac{p}{\mu}u^0_{ijk}.$$

(5.18)

where the relation between viscosity and collision frequency (4.23) was used, that is $\bar{\nu} = \mathrm{Pr}\frac{p}{\mu}$ with $\mathrm{Pr} = 2/3$ ($b = -1/2$).

5.3.3 Maxwell molecules

To compute the trace-free production terms $\mathcal{P}^a_{i_1\cdots i_n}$ for Maxwell molecules, we use the representation (3.28) together with the expression for the differential cross section (3.12), to compute the full production terms

$$\mathcal{P}^a_{i_1\cdots i_n} = m \int\int\int_0^{2\pi}\int_0^{\pi/2}\left((c')^{2a}C'_{i_1}\cdots C'_{i_n} - C^{2a}C_{i_1}\cdots C_{i_n}\right)$$
$$\times ff^1 F_M(\Theta)\sin\Theta\,d\Theta\,d\varepsilon\,d\mathbf{c}^1 d\mathbf{c}. \quad (5.19)$$

Equations (3.2, 3.4) must be used to express the post-collision velocities as $C'_i = (C_i - gk_i\cos\Theta)$. Again, we consider only the production terms for the first 26 moments, for which we have the conservation laws $\mathcal{P}^0 = \mathcal{P}^1 = \mathcal{P}^0_1 = 0$ and find at first

$$\mathcal{P}^2 = m\int\left(-4gC^2C_kI_k + 2g^2C^2I_{kk} + 4g^2C_kC_jI_{jk}\right.$$
$$\left. -4g^3C_jI_{jkk} + g^4I_{jjkk}\right)ff^1d\mathbf{c}^1 d\mathbf{c},$$

$$\mathcal{P}^1_i = m\int\left(-2gC_iC_kI_k - gC^2I_i + g^2C_iI_{kk} + 2g^2C_kI_{ki} - g^3I_{ikk}\right)ff^1\,d\mathbf{c}^1 d\mathbf{c},$$

$$\mathcal{P}^0_{ij} = m\int\left(g^2I_{\langle ij\rangle} - 2gC_{\langle i}I_{j\rangle}\right)ff^1\,d\mathbf{c}^1 d\mathbf{c},$$

(5.20)

$$\mathcal{P}^1_{ij} = m\int\left(-2gC_{\langle i}C_{j\rangle}C_kI_k - 2gC^2C_{\langle i}I_{j\rangle} + g^2C_{\langle i}C_{j\rangle}I_{kk} + 4g^2C_kC_{\langle i}I_{j\rangle k}\right.$$
$$\left. +g^2C^2I_{\langle ij\rangle} - 2g^3C_{\langle i}I_{j\rangle kk} - 2g^3C_kI_{kij} + g^4I_{\langle ij\rangle kk}\right)ff^1 d\mathbf{c}^1 d\mathbf{c},$$

$$\mathcal{P}^0_{ijk} = m\int\left(-3gC_{\langle i}C_jI_{k\rangle} + 3g^2C_{\langle i}I_{jk\rangle} - g^3I_{\langle ijk\rangle}\right)ff^1\,d\mathbf{c}^1 d\mathbf{c}.$$

The integrals

$$I_{i_1 \cdots i_n} = \int_0^{2\pi} \int_0^{\pi/2} k_{i_1} \cdots k_{i_n} \cos^n \Theta \sin \Theta F_M (\Theta) \, d\Theta \, d\varepsilon \qquad (5.21)$$

are computed in Appendix A.4 as

$$I_i = \chi^{(2,1)} \frac{g_i}{g} \, ,$$

$$I_{ij} = \left(\chi^{(2,1)} - \frac{3}{2} \chi^{(2,3)} \right) \frac{g_i g_j}{g^2} + \frac{1}{2} \chi^{(2,3)} \delta_{ij} \, , \qquad (5.22)$$

$$I_{ijk} = = \left(\chi^{(2,1)} - \chi^{(2,3)} - \frac{5}{2} \chi^{(4,3)} \right) \frac{g_i g_j g_k}{g^3} + \frac{3}{2} \chi^{(4,3)} \delta_{(ij} \frac{g_k)}{g} \, ,$$

$$I_{ijkk} = \left(\chi^{(2,1)} - \chi^{(2,3)} - \frac{3}{2} \chi^{(4,3)} \right) \frac{g_i g_j}{g^2} + \frac{1}{2} \chi^{(4,3)} \delta_{ij} \, ,$$

with

$$\chi^{(r,s)} = 2\pi \int_0^{\pi/2} \cos^r \Theta \sin^s \Theta F_M (\Theta) \, d\Theta \, .$$

Combining (5.20), (5.22), $g_i = \left(C_i - C_i^1 \right)$, (4.46), and the definitions of moments finally we find [20]

$$\mathcal{P}^2 = -\frac{2}{3} \frac{p}{\mu} \left(w^2 + \frac{\sigma_{ij} \sigma_{ij}}{\rho} \right) \, ,$$

$$\mathcal{P}_i^1 = -\frac{4}{3} \frac{p}{\mu} q_i \, ,$$

$$\mathcal{P}_{ij}^0 = -\frac{p}{\mu} \sigma_{ij} \, , \qquad (5.23)$$

$$\mathcal{P}_{ij}^1 = -\frac{7}{6} \frac{p}{\mu} \left(u_{ij}^1 - \theta \sigma_{ij} + \frac{4}{7} \frac{1}{\rho} \sigma_{k \langle i} \sigma_{j \rangle k} \right) \, ,$$

$$\mathcal{P}_{ijk}^0 = -\frac{3}{2} \frac{p}{\mu} u_{ijk}^0 \, .$$

More production terms for Maxwell molecules can be found in [20].

Comparison of (5.23) with the corresponding expressions for the ES-BGK model (5.18) shows that the production terms for stress and heat flux, $\mathcal{P}_{ij}^0$ and $\mathcal{P}_i^1$, agree between the two collision models. Indeed, the ES-BGK model was constructed specifically to yield this agreement, which implies that both models agree up to second order terms in the Knudsen number, as was seen in the previous chapter where we showed that both models give the same viscosity and heat conductivity, and the same Burnett coefficients. Other production terms, however, do not agree.

5.3.4 Linear production terms

To simplify calculations, we shall sometimes consider linearized production terms, which can be computed from the linearized Boltzmann collision term [13][24]. Obviously, linearization simplifies the treatment, and this is why the linearized Boltzmann equation is widely used.

Linearization of the production terms on the moment level is simply performed by ignoring all products of the non-equilibrium quantities $\left(w^a, u^a_{i_1\cdots i_n}\right)$, and keeping only terms which are linear in these. The resulting production terms can be written in compact form as

$$\mathcal{P}^a = -\frac{p}{\mu}\sum_b C^{(0)}_{ab}\theta^{a-b}w^b \quad (a, b \geq 2) ,$$

$$\mathcal{P}^a_i = -\frac{p}{\mu}\sum_b C^{(1)}_{ab}\theta^{a-b}u^b_i \quad (a, b \geq 1) , \qquad (5.24)$$

$$\mathcal{P}^a_{i_1\cdots i_n} = -\frac{p}{\mu}\sum_b C^{(n)}_{ab}\theta^{a-b}u^b_{i_1\cdots i_n} \quad (a, b \geq 0) .$$

The dimensionless matrices $C^{(n)}_{ab}$ depend on the interaction model used. For the BGK model, they are simply unit matrices,

$$C^{(0)}_{ab} = \delta_{ab}, \ (a, b \geq 2) \ ; \ C^{(1)}_{ab} = \delta_{ab}, \ (a, b \geq 1) \ ; \ C^{(n)}_{ab} = \delta_{ab}, \ (n \geq 2; \ a, b \geq 0) \ .$$

For Maxwell molecules, the matrices $C^{(n)}_{ab}$ are of lower triangular form, see [21, 20]. Here we just give those entries that we shall need later, viz.[1]

$$C^{(0)}_{ab} = \begin{bmatrix} \frac{2}{3} & 0 & \cdots & 0 \\ \vdots & \ddots & \ddots & \vdots \\ & & \ddots & 0 \\ & & & \ddots \end{bmatrix} \quad (a, b \geq 2) ,$$

$$C^{(1)}_{ab} = \begin{bmatrix} \frac{2}{3} & 0 & \cdots & 0 \\ -\frac{14}{3} & 1 & \ddots & \vdots \\ \vdots & & \ddots & 0 \\ & & & \ddots \end{bmatrix} \quad (a, b \geq 1) ,$$

$$C^{(2)}_{ab} = \begin{bmatrix} 1 & 0 & \cdots & 0 \\ -\frac{7}{6} & \frac{7}{6} & \ddots & \vdots \\ \vdots & & \ddots & 0 \\ & & & \ddots \end{bmatrix} \quad (a, b \geq 0) ,$$

---

[1] The dots stand for coefficients that are not needed, and thus their numerical values are not given.

$$
\mathcal{C}_{ab}^{(3)} = \begin{bmatrix} \frac{3}{2} & 0 & \cdots & 0 \\ \vdots & \ddots & \ddots & \vdots \\ & & & 0 \\ & & & & \ddots \end{bmatrix} \quad (a, b \geq 0) \ .
$$

All matrices $\mathcal{C}_{ab}^{(n)}$ must have only non-zero eigenvalues, which implies that the matrices are invertible. If this were not the case, that is if an eigenvalue was zero, we could combine the moment equations for the non-conserved quantities $\left(w^a, u_{i_1 \cdots i_n}^b \right)$ to construct another conservation law for a new quantity $\lambda \left(w^a, u_{i_1 \cdots i_n}^b \right)$. However, only mass, momentum and energy are conserved, and all other moments are non-equilibrium variables, that is they vanish in equilibrium. This is obviously so for the triangular matrices of the Maxwell molecules, and the unit matrices of the BGK model.

5.3.5 Eigenfunctions

An alternative set of moments are the eigenfunctions of the linearized collision term for Maxwell molecules,

$$
g_{\langle i_1 \cdots i_n \rangle}^r = m \int \psi_{\langle i_1 \cdots i_n \rangle}^r f d\mathbf{c}
$$

The eigenfunctions are combinations of trace-free tensors and Sonine polynomials,

$$
\psi_{\langle i_1 \cdots i_n \rangle}^r = \sqrt{\frac{(2n+1)!!\sqrt{\pi}}{2n!}} S_{n+\frac{1}{2}}^r \left(\frac{C^2}{2\theta} \right) \frac{1}{\sqrt{2\theta}^n} C_{\langle i_1} \cdots C_{i_n \rangle} \ ,
$$

and are orthonormal with respect to integration over the Maxwellian in the sense that

$$
m A_{\langle j_1 \cdots j_m \rangle} \int \psi_{\langle i_1 \cdots i_n \rangle}^r \psi_{\langle j_1 \cdots j_m \rangle}^s f_M d\mathbf{c} = \rho A_{\langle i_1 \cdots i_n \rangle} \delta_{rs} \delta_{nm} \ .
$$

The corresponding production terms for the linearized collision term (4.32) are given by

$$
\mathcal{G}_{\langle i_1 \cdots i_n \rangle}^r = m \int \psi_{\langle i_1 \cdots i_n \rangle}^r \mathcal{S}_{lin} d\mathbf{c} = -\varsigma_n^r g_{\langle i_1 \cdots i_n \rangle}^r \ .
$$

See, e.g., [21][24] for the computation of the eigenvalues ς_n^r.

We shall mention eigenfunctions occasionally, but shall not use them.

6

Grad's moment method

Grad's moment method is the best known alternative to the Chapman-Enskog method [8][9]. We first discuss its principles and then consider the well-known 13 moment case as well as the 26 moment case in some detail. The method will be discussed, and some emphasis will be put on the relation between the Grad method and its phase density on one hand, and the Chapman-Enskog method on the other, as well as on the discussion of related methods, in particular extended thermodynamics [17].

6.1 Closed moment systems by Grad's method

6.1.1 General outline

Following Grad [8][9], we assume that the state of the gas can be described accurately by an extended set of moments

$$u_A = \int \Psi_A\left(c_k\right) f \, d\mathbf{c} \, ,$$

where $\Psi_A\left(c_k\right)$ is a vector of polynomials of the (peculiar) velocity. For instance in Grad's 13 moment theory one has $\Psi_A^{[13]} = m\left\{1, c_i, \frac{1}{2}C^2, C_{\langle i}C_{j\rangle}, \frac{1}{2}C^2C_i\right\}$, corresponding to the moments $u_A^{[13]} = \{\rho, \rho v_i, \rho u, \sigma_{ij}, q_i\}$, but the number of moments considered might be extended to large numbers. Which moments, and how many, one has to take into account generally depends on the process under consideration, as described, e.g., by the Knudsen number.

Multiplication of the Boltzmann equation (3.16) by Ψ_A and subsequent integration over velocity space yields the moment equations

$$\frac{\partial u_A}{\partial t} + \frac{\partial F_{Ak}}{\partial x_k} = P_A \quad \text{with} \quad F_{Ak} = \int \Psi_A c_k f \, d\mathbf{c} \quad , \quad P_A = \int \Psi_A \mathcal{S} f \, d\mathbf{c} \, , \quad (6.1)$$

with the fluxes of the moments F_{Ak} and their production terms P_A. Note that the production terms of mass, momentum and energy vanish, and that we ignore external forces for this argument.

The moment equations $(6.1)_1$ do not form a closed system of partial differential equations for the moments, since they contain the fluxes and the production terms which are not a priori related to the variables u_A. To solve this closure problem, Grad constructed a phase density by an expansion of the Maxwellian into Hermite polynomials as

$$f_{|G} = \left(a + a_i \frac{\partial}{\partial C_i} + a_{ij} \frac{\partial^2}{\partial C_i \partial C_j} + a_{ijk} \frac{\partial^3}{\partial C_i \partial C_j \partial C_k} + \cdots \right) f_M .$$

The expansion coefficients must be determined such that the Grad phase density reproduces the basic variables u_A as its moments. Obviously, one needs to consider as many coefficients $a_{i_1 \cdots i_n}$ as one has variables u_A. Thus, the phase density assumes the form

$$f_{|G} = f_{|G} \left(u_A \left(\mathbf{x}, t \right), C_i \right) ,$$

that is it depends on space and time only through the moments. This phase density can now be used to compute the fluxes F_{Ak} and production terms P_A as functions of the moments,

$$F_{Ak} = F_{Ak} \left(u_B \right) , \quad P_A = P_A \left(u_B \right) ,$$

and the system of moment equations is closed.

Some important differences to the Chapman-Enskog method can be seen already: (a) The set of variables is extended. (b) The phase density depends only on the variables u_A and *not* on their gradients. (c) The phase density is developed independently from the Boltzmann equation, which comes, however, into play through the moment equations. The relations between both methods will be discussed further in Sec. 6.3.

After performing the derivatives of the Maxwellian (2.18), the Grad phase density assumes the form

$$f_{|G} = f_M \left(a - \frac{1}{\theta} a_i C_i + \frac{1}{\theta} a_{ij} \left(\frac{1}{\theta} C_i C_j - \delta_{ij} \right) \right.$$
$$\left. + \frac{1}{\theta^2} a_{ijk} \left(3 C_{(i} \delta_{jk)} - \frac{1}{\theta} C_i C_j C_k \right) + \cdots \right) .$$

Since we prefer to work with trace-free moments, it is a matter of convenience to introduce new expansion coefficients $\lambda^a_{\langle i_1 \cdots i_n \rangle}$ and rewrite the expansion as

$$f_{|G} = f_M \sum_{n=0} \sum_{a=0} \lambda^a_{\langle i_1 \cdots i_n \rangle} C^{2a} C_{\langle i_1} \cdots C_{i_n \rangle} . \tag{6.2}$$

The set of 13 moments $u_A^{[13]} = \{ \rho, \rho v_i, \rho u, \sigma_{ij}, q_i \}$ contains only those moments which have an intuitive physical meaning. With no physical meaning for the other moments, it arises the question which moments one should consider for extended moment systems. A priori, there are no restrictions on the choice of moments, but in order to have a somewhat systematic approach the sets of moments can be chosen by the following argument:

The basic quantities are the conserved quantities, mass, momentum and energy, given by

$$u_A^{[5]} = \int \Psi_A^{[5]} f d\mathbf{c} \quad \text{with} \quad \Psi_A^{[5]} = m\left\{1, c_i, C^2\right\} .$$

In order to close the system, we have either to find constitutive equations for stress and heat flux,[1] or we can add these to the list of variables, i.e. choose

$$u_A^{[13]} = \int \Psi_A^{[13]} f d\mathbf{c} \quad \text{with} \quad \left\{1, c_i, \frac{1}{2}C^2, C_i C_j, \frac{1}{2}C^2 C_i\right\} .$$

A closer look shows that in this case we need additional equations for the fluxes of heat flux and pressure tensor, i.e. for u_{ijk}^0, u_{ij}^1 and u^2. Again, we have either to find constitutive equations for these fluxes, or we can add them to the list of variables. If we proceed in this manner, we come to the following choices of moments

$$\Psi_A^{(\alpha)} = m\left\{1, c_i, C_i C_j, \ldots, C_{i_1} C_{i_2} \cdots C_{i_\alpha}, C^2 C_{i_1} C_{i_2} \cdots C_{i_{\alpha-1}}\right\} , \tag{6.3}$$

$\alpha = 1, 2, 3, \ldots$, corresponding to $5, 13, 26, 45, 71, 105, 148, 201, 265, 341, 430, \cdots$ moments.

Other authors chose their set of moments differently. A popular choice is to consider complete moments so that [17]

$$\Psi_A^{(\alpha)} = m\left\{1, c_i, C_i C_j, \ldots, C_{i_1} C_{i_2} \cdots C_{i_\alpha}\right\} ,$$

$\alpha = 1, 2, 3, \ldots$, corresponding to $4, 10, 20, 35, 56, 84, 120, 165, 220, 286, 364, \cdots$ moments.

The 5-moment case is trivial, since the corresponding phase density is the Maxwellian, which yields the Euler equations (4.21).

We shall consider the Grad method in detail for 13 and 26 moments, and then discuss its application to larger numbers of variables.

6.1.2 Grad's 13 moment equations

The 13 moment theory is based on the variables

$$u^{[13]} = \int \Psi_A^{[13]} f d\mathbf{c} = \left\{\rho, \ \rho v_i, \ \rho u = \frac{3}{2}\rho\theta, \ \sigma_{ij}, \ q_i\right\} ,$$

---

[1] A constitutive equation for the mass flux is not needed, since the mass flux equals the momentum density and thus is contained in $u_A^{[5]}$ already.

and the corresponding transport equations are the conservation laws (5.10, 5.12, 5.11),

$$\frac{D\rho}{Dt} + \rho \frac{\partial v_k}{\partial x_k} = 0 \,,$$

$$\rho \frac{Dv_i}{Dt} + \theta \frac{\partial \rho}{\partial x_i} + \rho \frac{\partial \theta}{\partial x_i} + \frac{\partial \sigma_{ik}}{\partial x_k} = \rho G_i \qquad (6.4)$$

$$\frac{3}{2} \rho \frac{D\theta}{Dt} + \rho \theta \frac{\partial v_k}{\partial x_k} + \frac{\partial q_k}{\partial x_k} + \sigma_{kl} \frac{\partial v_k}{\partial x_l} = 0 \,,$$

plus the moment equations for stress $\sigma_{ij} = u_{ij}^0$ and heat flux $q_i = \frac{1}{2} u_i^1$ which follow from (5.15, 5.16) as[2]

$$\frac{D\sigma_{ij}}{Dt} + \frac{\partial u_{ijk}^0}{\partial x_k} + \frac{4}{5} \frac{\partial q_{\langle i}}{\partial x_{j\rangle}} + 2\sigma_{k\langle i} \frac{\partial v_{j\rangle}}{\partial x_k} + \sigma_{ij} \frac{\partial v_k}{\partial x_k} + 2\rho\theta \frac{\partial v_{\langle i}}{\partial x_{j\rangle}} = \mathcal{P}_{ij}^0 \,, \qquad (6.5)$$

$$\frac{Dq_i}{Dt} + \frac{5}{2} \rho\theta \frac{\partial \theta}{\partial x_i} - \sigma_{ik} \frac{\partial \theta}{\partial x_k} - \sigma_{ik}\theta \frac{\partial \ln \rho}{\partial x_k} - \frac{\sigma_{ik}}{\rho} \frac{\partial \sigma_{kl}}{\partial x_l} - \frac{5}{2}\theta \frac{\partial \sigma_{ik}}{\partial x_k} + \frac{1}{2} \frac{\partial u_{ik}^1}{\partial x_k}$$
$$+ \frac{1}{6} \frac{\partial w^2}{\partial x_i} + u_{ikl}^0 \frac{\partial v_k}{\partial x_l} + \frac{7}{5} q_i \frac{\partial v_k}{\partial x_k} + \frac{7}{5} q_k \frac{\partial v_i}{\partial x_k} + \frac{2}{5} q_k \frac{\partial v_k}{\partial x_i} = \frac{1}{2} \mathcal{P}_i^1 \,. \qquad (6.6)$$

This set of 13 equations is not closed, since it contains not only the basic moments $u^{[13]}$, but in addition the higher moments

$$u_{ijk}^0 = m \int C_{\langle i} C_j C_{k\rangle} f d\mathbf{c} \,,$$

$$u_{ij}^1 = m \int C^2 C_{\langle i} C_{j\rangle} f d\mathbf{c} \,, \qquad (6.7)$$

$$w^2 = m \int C^4 (f - f_M) \, d\mathbf{c} \,,$$

as well as the production terms $\mathcal{P}_{ij}^0$ and $\frac{1}{2}\mathcal{P}_i^1$. We shall assume a gas of Maxwell molecules for which the production terms can be computed without knowledge of the distribution function as, see (5.23),

$$\mathcal{P}_{ij}^0 = -\frac{p}{\mu} \sigma_{ij} \quad \text{and} \quad \frac{1}{2}\mathcal{P}_i^1 = -\frac{2}{3} \frac{p}{\mu} q_i \,. \qquad (6.8)$$

In order to express the moments (6.7) through the variables $u^{[13]}$ the appropriate Grad distribution function (6.2) is required,

$$f_{|13} = f_M \sum \lambda_A \Psi_A^{[13]} = f_M \left(\lambda^0 + \lambda_i^0 C_i + \lambda^1 C^2 + \lambda_{\langle ij\rangle}^0 C_{\langle i} C_{j\rangle} + \lambda_i^1 C^2 C_i \right) \,. \qquad (6.9)$$

---

[2] Recall that $u_i^0 = w^0 = w^1 = 0$.

The coefficients $\lambda^a_{\langle i_1 \cdots i_n \rangle}$ must be chosen so that $f_{|13}$ reproduces the basic moments, that is

$$\rho = m \int f_{|13} d\mathbf{c} \; , \quad 0 = m \int C_i f_{|13} d\mathbf{c} \; , \quad \frac{3}{2}\rho\theta = \frac{m}{2} \int C^2 f_{|13} d\mathbf{c} \; ,$$

$$\sigma_{ij} = m \int C_{\langle i} C_{j \rangle} f_{|13} d\mathbf{c} \; , \quad q_i = \frac{m}{2} \int C^2 C_i f_{|13} d\mathbf{c} \; , \tag{6.10}$$

After straightforward computation, Grad's phase density for 13 moments follows as

$$f_{|13} = f_M \left[1 + \frac{\sigma_{ik}}{2p} \frac{C_{\langle i} C_{k \rangle}}{\theta} + \frac{2}{5} \frac{q_k}{p\theta} C_k \left(\frac{C^2}{2\theta} - \frac{5}{2} \right) \right] \; . \tag{6.11}$$

Evaluation of the additional moments (6.7) with $f_{|13}$ yields the constitutive equations for Grad's 13 moment theory as

$$u^0_{ijk|13} = 0 \; , \quad u^1_{ij|13} = 7\theta\sigma_{ij} \; , \quad w^2_{|13} = 0 \; . \tag{6.12}$$

Insertion of the results for production terms (6.8) and higher moments (6.12) into (6.5) and (6.6) finally gives the closed equations for 13 moments as the conservation laws (6.4) and

$$\frac{D\sigma_{ij}}{Dt} + \frac{4}{5} \frac{\partial q_{\langle i}}{\partial x_{j \rangle}} + 2\sigma_{k \langle i} \frac{\partial v_{j \rangle}}{\partial x_k} + \sigma_{ij} \frac{\partial v_k}{\partial x_k} + 2p\theta \frac{\partial v_{\langle i}}{\partial x_{j \rangle}} = -\frac{p}{\mu}\sigma_{ij} \; , \tag{6.13}$$

$$\frac{Dq_i}{Dt} + \frac{5}{2}\rho\theta \frac{\partial\theta}{\partial x_i} + \frac{5}{2}\sigma_{ik} \frac{\partial\theta}{\partial x_k} - \sigma_{ik}\theta \frac{\partial \ln\rho}{\partial x_k} - \frac{\sigma_{ik}}{\rho} \frac{\partial\sigma_{kl}}{\partial x_l} + \theta \frac{\partial\sigma_{ik}}{\partial x_k}$$
$$+ \frac{7}{5}q_i \frac{\partial v_k}{\partial x_k} + \frac{7}{5}q_k \frac{\partial v_i}{\partial x_k} + \frac{2}{5}q_k \frac{\partial v_k}{\partial x_i} = -\frac{2}{3}\frac{p}{\mu}q_i \; . \tag{6.14}$$

6.1.3 Grad's 26 moment equations

The 26 moment case adds the unknown moments from the 13 moment case to the list of variables, which considers the 13 equations for the 13 moments above (6.4–6.6) plus the moment equations for the other variables that appear in those, that is

$$u^{[26]} = \left\{ \rho, \; \rho v_i, \; \rho u = \frac{3}{2}\rho\theta, \; \sigma_{ij}, \; q_i, \; u^0_{ijk}, \; u^1_{ij}, \; w^2 \right\} \; . \tag{6.15}$$

The additional equations for u^0_{ijk}, u^1_{ij}, w^2 follow from Chapter 5 and are not printed here. They contain the additional moments

$$u^2_k \; , \; u^1_{ijk} \; , \; u^0_{ijkl} \; ,$$

which must, again, be related to the variables by means of Grad's distribution function. The appropriate Grad distribution for the 26 moment case is

$$f_{|26} = f_M \left(1 + \frac{w^2}{8p\theta} \left(1 - \frac{2}{3}\frac{C^2}{\theta} + \frac{1}{15}\frac{C^4}{\theta^2} \right) + \frac{2}{5}\frac{q_k}{p\theta}C_k \left(\frac{C^2}{2\theta} - \frac{5}{2} \right) \right.$$

$$\left. + \frac{2\theta\sigma_{ij}}{4p\theta^2}C_{\langle i}C_{j\rangle} + \frac{7\theta\sigma_{ij} - u_{ij}^1}{4p\theta^2}C_{\langle i}C_{j\rangle} \left(1 - \frac{1}{7}\frac{C^2}{\theta} \right) + \frac{u_{ijk}^0}{6p\theta^3}C_{\langle i}C_jC_{k\rangle} \right) .$$

$$(6.16)$$

This yields the constitutive equation for the 26 moment case as

$$u_{i|26}^2 = 28\theta q_i \ , \quad u_{ijk|26}^1 = 9\theta u_{ijk|26}^0 \ , \quad u_{ijkl|26}^0 = 0 \ . \tag{6.17}$$

For the production terms we chose again the results for Maxwell molecules (5.23), so that the equations finally read,

$$\frac{Dw^2}{Dt} + 8\theta\frac{\partial q_k}{\partial x_k} - 20\theta\sigma_{kl}\frac{\partial v_k}{\partial x_l} + 4u_{kl}^1\frac{\partial v_k}{\partial x_l} + 20q_k\frac{\partial\theta}{\partial x_k} - 8q_k\theta\frac{\partial\ln\rho}{\partial x_k}$$

$$- 8\frac{q_k}{\rho}\frac{\partial\sigma_{kl}}{\partial x_l} + \frac{7}{3}w^2\frac{\partial v_k}{\partial x_k} = -\frac{2}{3}\frac{p}{\mu}\left(w^2 + \frac{\sigma_{ij}\sigma_{ij}}{\rho} \right) , \quad (6.18)$$

$$\frac{Du_{ij}^1}{Dt} - 2u_{ijk}^0 \left(\frac{1}{\rho}\frac{\partial\sigma_{kl}}{\partial x_l} + \theta\frac{\partial\ln\rho}{\partial x_k} \right) + 7u_{ijk}^0\frac{\partial\theta}{\partial x_k} - \frac{28}{5}q_{\langle i}\left(\frac{1}{\rho}\frac{\partial\sigma_{j\rangle k}}{\partial x_k} + \theta\frac{\partial\ln\rho}{\partial x_{j\rangle}} \right)$$

$$+ \frac{28}{5}q_{\langle i}\frac{\partial\theta}{\partial x_{j\rangle}} + 9\theta\frac{\partial u_{ijk}^0}{\partial x_k} + \frac{56}{5}\theta\frac{\partial q_{\langle i}}{\partial x_{j\rangle}} + \frac{6}{7}u_{\langle ij}^1\frac{\partial v_{k\rangle}}{\partial x_k} + \frac{4}{5}u_{k\langle i}^1\frac{\partial v_k}{\partial x_{j\rangle}} + 2u_{k\langle i}^1\frac{\partial v_{j\rangle}}{\partial x_k}$$

$$+ u_{ij}^1\frac{\partial v_k}{\partial x_k} + \frac{14}{15}w^2\frac{\partial v_{\langle i}}{\partial x_{j\rangle}} + 14p\theta^2\frac{\partial v_{\langle i}}{\partial x_{j\rangle}} = -\frac{7}{6}\frac{p}{\mu}\left(u_{ij}^1 - \theta\sigma_{ij} + \frac{4}{7}\frac{1}{\rho}\sigma_{k\langle i}\sigma_{j\rangle k} \right) ,$$

$$(6.19)$$

$$\frac{Du_{ijk}^0}{Dt} - 3\sigma_{\langle ij}\left(\frac{1}{\rho}\frac{\partial\sigma_{k\rangle l}}{\partial x_l} + \theta\frac{\partial\ln\rho}{\partial x_{k\rangle}} + \frac{\partial\theta}{\partial x_{k\rangle}} \right) + \frac{3}{7}\frac{\partial u_{\langle ij}^1}{\partial x_{k\rangle}} + 3u_{\langle ij}^0\frac{\partial v_{k\rangle}}{\partial x_l}$$

$$+ u_{ijk}^0\frac{\partial v_l}{\partial x_l} + \frac{12}{5}q_{\langle i}\frac{\partial v_j}{\partial x_{k\rangle}} = -\frac{3}{2}\frac{p}{\mu}u_{ijk}^0 . \quad (6.20)$$

Of course, the other equations for the 26 moment case are (6.4–6.6).

6.1.4 Extended moment sets

Now we consider general sets of moment equations for the moments

$$w^a = m\int C^{2a}\left(f_{|G} - f_M \right) d\mathbf{c} \ , \quad u_{i_1\cdots i_n}^a = m\int C^{2a}C_{\langle i_1}C_{i_2}\cdots C_{i_n\rangle}f_{|G}d\mathbf{c} \ ,$$

$$(6.21)$$

where $a = 0, \ldots, A_n$, and $n = 1, \ldots, n_{\max}$. The numbers A_n count the moments of tensorial rank n that are taken into account; they correspond to the number of Sonine polynomials in the Chapman-Enskog expansion.

The appropriate Grad phase density reads

$$f_{|G} = f_M (1 + \Phi) = f_M \left[1 + \sum_{n=0}^{n_{\max}} \sum_{a=0}^{A_n} \lambda_{i_1 \cdots i_n}^a C^{2a} C_{\langle i_1} C_{i_2} \cdots C_{i_n \rangle} \right] \qquad (6.22)$$

where $\lambda_{i_1 \cdots i_n}^a$ are expansion coefficients that follow from plugging the Grad phase density (6.22) into the definition of the moments (6.21). This yields at first

$$w^a = \rho \sum_{b=0}^{A_0} \theta^{a+b} \mathcal{B}_{ab}^{(0)} \lambda^b \ , \quad u_{i_1 \cdots i_n}^a = n! \rho \sum_{b=0}^{A_r} \theta^{a+b+n} \mathcal{B}_{ab}^{(n)} \lambda_{i_1 \cdots i_n}^b \ ,$$

with the matrices of numbers

$$\mathcal{B}_{ab}^{(n)} = \prod_{k=n+1}^{a+b+n} (2k+1) \quad (a, b \in [0, A_r]) \ .$$

The relations between moments and the expansion coefficients are linear, and we find by inversion

$$\lambda_{i_1 \cdots i_n}^a = \sum_b \left[\mathcal{B}_{ab}^{(n)} \right]^{-1} \frac{\bar{u}_{i_1 \cdots i_n}^b}{n! \rho \theta^{a+b+n}} \quad \text{with} \quad \bar{u}^a = w^a, \ \bar{u}_{i_1 \cdots i_n}^a = u_{i_1 \cdots i_n}^a \ ,$$

$$\tag{6.23}$$

so that the Grad phase density is explicitly given. With this, the phase density is known, and *all* of its moments can be computed,

$$w^a = \sum_{b,c=0}^{A_0} \theta^{a-c} \left[\prod_{k=r+1}^{a+b} (2k+1) \right] \left[\mathcal{B}_{bc}^{(0)} \right]^{-1} w^c ,$$

$$u_{i_1 \cdots i_n}^a = \sum_{b,c=0}^{A_r} \theta^{a-c} \left[\prod_{k=n+1}^{a+b+n} (2k+1) \right] \left[\mathcal{B}_{bc}^{(n)} \right]^{-1} u_{i_1 \cdots i_n}^c ,$$

$$u_{i_1 \cdots i_{n_{\max}} j_i \cdots j_m}^a = 0 \ , \ m = 1, 2, \cdots$$

We note that all trace-free moments of rank higher than $n_{\max}$ vanish, and that the equations for $n \leq n_{\max}$ reduce to identities for $a \in [0, A_n]$. For more details on the above calculations the reader is referred to [24].

6.2 Remarks on Grad-type equations

6.2.1 Large moment numbers

The main question is, of course, *how many moments should be considered?* Certainly, it would be nice to know the answer to this questions before simulating a process, e.g. in dependence of the Knudsen number. However, until recently, the relation between moment equations and Knudsen number was not clear, and the assessment of the required moment number was done as follows: A process is simulated with a certain set of moment equations, and then again with a higher number of moments. If the increase of the moment number does change the results noticeably, the number of moments is increased again, and so on, until a further increase of the moment number does not change the results. The required number of moments is the number used before the last test.

Obviously, this strategy requires an easy access to Grad moment systems with arbitrary number of moments. By introducing formal notation one can develop algorithms that produce the closed set of moment equations for large moment numbers. This was pioneered by Weiss in his dissertation [75], and further developed by the author [76] and Au [77], see [24] for a good summary. Modern algebraic computer programs, e.g. Mathematica® [78], allow to implement these algorithms, and to generate output that can be further used for numerical solutions of the equations [76][77][24].

Large moment numbers can be considered in this manner, and this was done for a number of one-dimensional problems in a gas of Maxwell molecules, e.g., sound propagation [75][17], heat transfer at large Knudsen numbers [76], shock waves [77], and light scattering [75][17]. In extreme cases, up to 506 one-dimensional equations were considered [75][17], which would correspond to 15180 moments in three-dimensional settings. Two- or three-dimensional processes were never considered with large moment numbers, however.

These large numbers of moments are necessary since the convergence with increasing moment number is rather slow. Altogether the cited studies, and others, show that the number of moments must be large in case of steep gradients, fast oscillations, or strong shocks—all cases that are related to relatively large Knudsen numbers. At low and moderate Knudsen numbers a lower number of moments can be sufficient. When the Knudsen number is sufficiently small, the moment solutions stand in agreement with the Navier-Stokes-Fourier equations. At larger Knudsen numbers, when convergence is observed, the results stand in agreement with solutions of the Boltzmann equation, e.g. DSMC simulations.

Chapter 8 will discuss the relations between number of moments and Knudsen number.

All applications of large moment numbers rely on Maxwellian molecules or BGK models to compute the production terms. Eigenfunctions of the linearized Boltzmann collision term for Maxwell molecules are known [87][21][13], and the eigenvalues can be computed numerically. The trace-free moments used in this text can be considered as linear combinations of the eigenfunctions and thus the production terms for the moments can be computed [24]. Non-Maxwellian interaction potentials, or the non-linear collision term, were, to my best knowledge, never considered for large moment numbers.

Applications of moment equations with large moment numbers to quantum gases include radiative transfer [79]–[82], phonon transport in crystals [83], and electron transport in semi-conductors [84]–[86]. Here, the same observation was made, namely that steep gradients or fast changes, i.e. large Knudsen numbers, require a larger number of moments.

6.2.2 Moment systems and its competitors

Altogether it is safe to say that Grad-type moment equations can reproduce solutions of the Boltzmann equation, if the number of moments is large enough. However, the number of moments required might be huge, and then the use of moment equations has no advantage against direct simulations, or numerical solutions of the Boltzmann equation by discrete velocity schemes.

The idea of using moment equations should be to save computational time while producing reliable results in accordance with the Boltzmann equation. Large moment numbers require a fair amount of computational time, and the equations are difficult to generate, and implement, at least for two- or three-dimensional geometry. All advantages that the moment systems can have are lost if the number of moments must be large, and one will prefer DSMC simulations, or other solutions of the Boltzmann equation itself.

A lower number of moments, say 13 or 26, can be solved much faster, and, as long as the results are good, might be a useful tool for the computation of rarefied gas flows at moderate Knudsen numbers.

The Burnett and super-Burnett equations were derived by means of the Chapman-Enskog expansion to derive processes at moderate Knudsen numbers, and they are the natural competition for small moment systems. Indeed, Burnett and super-Burnett equations are unstable, and therefore useless for many problems, and a better alternative is desirable. We shall show below that Burnett and super-Burnett equations follow from a Chapman-Enskog expansion of the 13 respectively 26 moment case, and thus these must be considered as of similar accuracy.

We also note that moment methods and the higher order CE expansions share the problem of how to prescribe boundary conditions for higher moments, or higher derivatives, see Chapter 12.

6.2.3 Mathematical properties

The closed sets of Grad-type moment equations contain only first order derivatives in space and time, and form, at least for most values of the moments, a symmetric hyperbolic system of the general form

$$\frac{\partial u_A}{\partial t} + \mathcal{A}^k_{AB}\left(u_C\right)\frac{\partial u_B}{\partial x_k} = \mathcal{C}_A\left(u_C\right)\,,$$

where the matrix $\mathcal{A}^k_{AB}n_k$ has positive eigenvalues for each unit vector n_k. This follows from the strong relation between extended thermodynamics [17] and the Grad method, which we shall explore deeper in Sec. 6.6. Two particular features of the equations is that they are stable, and that they yield discontinuous shocks, that is solutions with a jump, see Chapter 11.

Shocks in gases can be observed, and a closer look, e.g. by experiment [88] or DSMC simulation [1], shows that gas properties like density, temperature, etc. change continuously over a distance of only few mean free paths. The computation of these shock structures is an important benchmark problem for theories on rarefied gas flows.

Due to their hyperbolic character, Grad-type moment methods fail to describe smooth shock structures when the inflow velocity lies above a critical Mach number [8][89].[3] For instance, in the 13 moment case the critical Mach number is Ma = 1.65. The critical Mach number grows with the number of moments [89], but only slowly. This is a severe handicap, and an important reason why Grad-type equations are seen critically.

Another argument put forward occasionally against Grad's method is that the phase density $f_{|G}$ will become negative for large velocities, while by definition the phase density is positive. A similar argument could be put forward against the phase density in the CE method, $f_{|CE}$. Both methods are approximations, and one cannot expect that they meet reality in all aspects. The phase densities $f_{|G}$ and $f_{|CE}$ are polynomials in C_i times the Maxwellian. The polynomials will assume negative values for large values of the velocity vector, but the Maxwellian, which is an exponential in $-C^2$, will ensure that the actual values of the phase densities for large C_i are very small. In other words, where the approximations to the phase densities are negative, they will be very small, so that the negative values have negligible influence on the results.

Chapters 7 and 8 will present regularizations, where the mathematical character of the equations is changed in a rational manner, and the resulting equations yield smooth shocks.

6.3 CE expansion of Grad's moment equations

6.3.1 Grad and CE phase density for Maxwell molecules and BGK model

Since the full (infinite) moment system is equivalent to the Boltzmann equation, one will expect that the Chapman-Enskog expansion of the moment equations will give the same results as the Chapman-Enskog expansion of the Boltzmann equation itself. However, the truncated moment systems of the Grad method already contain enough elements of the Boltzmann equation to

---

[3] The Mach number is defined as the ratio of inflow velocity and the local speed of sound. The critical velocities are the eigenvalues of $\mathcal{A}_{AB}^k n_k$, evaluated in equilibrium.

reproduce the Navier-Stokes-Fourier, Burnett, and super-Burnett equations for Maxwell molecules.

Comparison of the distribution function for the first order Chapman-Enskog expansion for Maxwell molecules or ES-BGK model, (4.25), and the 13 moment distribution (6.11) shows that both have a very similar form. Indeed, using the laws of Navier-Stokes and Fourier (4.22), the Chapman-Enskog phase density to first order can be written as

$$f_{|CE} = f^{(0)} + f^{(1)} = f_M \left[1 + \frac{\sigma_{ik}^{(1)}}{2p} \frac{C_{\langle i} C_{k \rangle}}{\theta} + \frac{2}{5} \frac{q_k^{(1)}}{p\theta} C_k \left(\frac{C^2}{2\theta} - \frac{5}{2} \right) \right] , \quad (6.24)$$

which has just the same form as Grad's 13 moment distribution $f_{|13}$ (6.11). The difference lies in heat flux and stress: The CE phase density contains only the first approximations to stress and heat flux, $\sigma_{ij}^{(1)}$ and $q_i^{(1)}$, while the Grad distribution contains both as independent variables, σ_{ij} and q_i.

This resemblance explains why the Grad 13 moment equations will reduce to the NSF equations in the Chapman-Enskog limit. The transport equations for stress and heat flux contain additional elements of the Boltzmann equations and that probably is the reason why the 13 moment system also contains the Burnett equations, although its phase density is considerably simpler than the Burnett phase density.

6.3.2 13 moments

For the CE expansion of the moment equations we first introduce the smallness parameter ε to scale the production terms in (6.13, 6.14), so that these read

$$\frac{D\sigma_{ij}}{Dt} + \cdots = -\frac{1}{\varepsilon} \frac{p}{\mu} \sigma_{ij} \; , \quad \frac{Dq_i}{Dt} + \cdots = -\frac{1}{\varepsilon} \frac{2}{3} \frac{p}{\mu} q_i \; ,$$

and then use the expansions (4.11)

$$\sigma_{ij} = \varepsilon \sigma_{ij}^{(1)} + \varepsilon^2 \sigma_{ij}^{(2)} \; , \quad q_i = \varepsilon q_i^{(1)} + \varepsilon^2 q_i^{(2)} \; .$$

Equating terms of the same power in the smallness parameter ε we find at first order the laws of Navier-Stokes and Fourier,

$$\sigma_{ij}^{(1)} = -2\mu \frac{\partial v_{\langle i}}{\partial x_{j \rangle}} = -2\mu S_{ij} \text{ and } q_i^{(1)} = -\frac{15}{4} \mu \frac{\partial \theta}{\partial x_i} \; , \quad (6.25)$$

and the second order equations read at first

$$\sigma_{ij}^{(2)} = -\frac{\mu}{p} \left[\frac{D_0 \sigma_{ij}^{(1)}}{Dt} + \frac{4}{5} \frac{\partial q_{\langle i}^{(1)}}{\partial x_{j \rangle}} + 2\sigma_{k \langle i}^{(1)} \frac{\partial v_{j \rangle}}{\partial x_k} + \sigma_{ij}^{(1)} \frac{\partial v_k}{\partial x_k} \right] ,$$

$$q_i^{(2)} = -\frac{3}{2}\frac{\mu}{p}\left[\frac{D_0 q_i^{(1)}}{Dt} + \frac{5}{2}\sigma_{ik}^{(1)}\frac{\partial\theta}{\partial x_k} - \sigma_{ik}^{(1)}\theta\frac{\partial\ln\rho}{\partial x_k} + \theta\frac{\partial\sigma_{ik}^{(1)}}{\partial x_k}\right.$$

$$\left. +\frac{7}{5}q_i^{(1)}\frac{\partial v_k}{\partial x_k} + \frac{7}{5}q_k^{(1)}\frac{\partial v_i}{\partial x_k} + \frac{2}{5}q_k^{(1)}\frac{\partial v_k}{\partial x_i}\right].$$

For further evaluation we chose a generalized viscosity of the form $\mu = \mu_0 (\theta/\theta_0)^\omega$ (4.24) and insert the first order approximations. The time derivatives in the equations have to be replaced by

$$\frac{D_0\theta}{Dt} = -\frac{2}{3}\theta\frac{\partial v_k}{\partial x_k},$$

$$\frac{D_0}{Dt}\frac{\partial\theta}{\partial x_i} = -\frac{2}{3}\theta\frac{\partial^2 v_k}{\partial x_i\partial x_k} - \frac{2}{3}\frac{\partial v_k}{\partial x_k}\frac{\partial\theta}{\partial x_i} - \frac{\partial\theta}{\partial x_k}\frac{\partial v_k}{\partial x_i}, \qquad (6.26)$$

$$\frac{D_0}{Dt}\frac{\partial v_{\langle i}}{\partial x_{j\rangle}} = \frac{\partial}{\partial x_{\langle i}}\left(-\frac{1}{\rho}\frac{\partial p}{\partial x_{j\rangle}}\right) - \frac{\partial v_{\langle i}}{\partial x_r}\frac{\partial v_r}{\partial x_{j\rangle}}.$$

The resulting equations are just the Burnett equations (4.26, 4.27)

$$\sigma_{ij}^{(2)} = \frac{\mu^2}{p}\left[\varpi_1\frac{\partial v_k}{\partial x_k}S_{ij} - \varpi_2\left(\frac{\partial}{\partial x_{\langle i}}\left(\frac{1}{\rho}\frac{\partial p}{\partial x_{j\rangle}}\right) + \frac{\partial v_k}{\partial x_{\langle i}}\frac{\partial v_{j\rangle}}{\partial x_k} + 2\frac{\partial v_k}{\partial x_{\langle i}}S_{j\rangle k}\right)\right.$$

$$\left. +\varpi_3\frac{\partial^2\theta}{\partial x_{\langle i}\partial x_{j\rangle}} + \varpi_4\frac{\partial\theta}{\partial x_{\langle i}}\frac{\partial\ln p}{\partial x_{j\rangle}} + \varpi_5\frac{1}{\theta}\frac{\partial\theta}{\partial x_{\langle i}}\frac{\partial\theta}{\partial x_{j\rangle}} + \varpi_6 S_{k\langle i}S_{j\rangle k}\right]$$

$$q_i^{(2)} = \frac{\mu^2}{\rho}\left[\theta_1\frac{\partial v_k}{\partial x_k}\frac{\partial\ln\theta}{\partial x_i} - \theta_2\left(\frac{2}{3}\frac{\partial^2 v_k}{\partial x_k\partial x_i} + \frac{2}{3}\frac{\partial v_k}{\partial x_k}\frac{\partial\ln\theta}{\partial x_i} + 2\frac{\partial v_k}{\partial x_i}\frac{\partial\ln\theta}{\partial x_k}\right)\right.$$

$$\left. +\theta_3 S_{ik}\frac{\partial\ln p}{\partial x_k} + \theta_4\frac{\partial S_{ik}}{\partial x_k} + 3\theta_5 S_{ik}\frac{\partial\ln\theta}{\partial x_k}\right]$$

with the Burnett coefficients (4.29),

$$\varpi_1 = \frac{4}{3}\left(\frac{7}{2} - \omega\right), \; \varpi_2 = 2, \; \varpi_3 = 3, \; \varpi_4 = 0, \; \varpi_5 = 3\omega, \; \varpi_6 = 8,$$

$$\theta_1 = \frac{15}{4}\left(\frac{7}{2} - \omega\right), \; \theta_2 = \frac{45}{8}, \; \theta_3 = -3, \; \theta_4 = 3, \; \theta_5 = \frac{35}{4} + \omega. \quad (6.27)$$

Therefore the Burnett equations are contained within Grad's 13 moment theory, and can be extracted by means of the Chapman-Enskog method.[4]

The super-Burnett equations, however, cannot be derived from the 13 moment system; their derivation requires the set of 26 moments. The close relation between Burnett and Grad's 13 moment equations suggests that both sets have the same range of applicability. This means in particular that those processes which are known to be best described by a large number of moments probably cannot be described accurately by the Burnett equations.

---

[4] Grad in [9] attributes this result to the 1956 edition of the classical textbook of Chapman and Cowling. However, nothing on this relation can be found in the last edition [2]

6.3.3 26 moments

The super-Burnett equations can be derived from the 26 moment equations. The computations are cumbersome, but straightforward, and were only performed for the linearized equations in three dimensions, and for the non-linear equations in one dimension, see the equations printed in Sec. 4.3.3. The derivation of the super-Burnett equations via the moment system is by far more convenient than deriving them from the Boltzmann equation. Programs for symbolic computation, e.g. Mathematica® or Maple®, can be used to facilitate the computations.

Here, we shall only consider the first order results to show that they stand in agreement with Grad's 13 moment closure (6.12). For this, we introduce the CE expansion of the moments,

$$\sigma_{ij} = \varepsilon \sigma_{ij}^{(1)} + \varepsilon^2 \sigma_{ij}^{(2)} \quad , \quad q_i = \varepsilon q_i^{(1)} + \varepsilon^2 q_i^{(2)}$$

$$u_{ijk}^0 = \varepsilon u_{ijk}^{0(1)} + \varepsilon^2 u_{ijk}^{0(2)} \quad , \quad w^2 = \varepsilon w^{2(1)} + \varepsilon^2 w^{2(2)} \quad , \quad u_{ij}^1 = \varepsilon u_{ij}^{1(1)} + \varepsilon^2 u_{ij}^{1(2)}$$

into the 26 moment equations, to obtain the NSF laws (6.25) and the first order expansion of (6.12), viz.

$$w^{2(1)} = u_{ijk}^{0(1)} = 0 \quad , \quad u_{ij}^{1(1)} = -14\mu \frac{\partial v_{\langle i}}{\partial x_{j\rangle}} = 7\theta \sigma_{ij}^{(1)} \; .$$

6.3.4 Maxwellian iteration

An alternative to extract the NSF and Burnett equations from moment systems is the Maxwellian iteration of Ikenberry and Truesdell [90][20]. As the CE method, the Maxwellian iteration can be applied to the infinite set of moment equations,[5] or to the closed sets of Grad-type equations.

Here we shall only consider the Maxwellian iteration applied to Grad's 13 moment system, which gives the Euler equations at zeroth iteration, the NSF laws at first iteration, and equations close to the Burnett equations for the second iteration [91]. The second iteration of the 13 moment system can be found in Müller's book [15], where it was used as a shortcut for an argument on the frame dependence of heat flux and stress tensor that he had given earlier based on the original Maxwell iteration [92]. Reinecke and Kremer recognized that his equations are basically equal to the Burnett equations [93].

The iteration procedure is as follows: The first iterates $\sigma_{ij}^{[1]}$, $q_i^{[1]}$ are obtained by inserting the equilibrium values, computed from the local Maxwellian, of stress and heat flux into the left hand sides of (6.13, 6.14) and solving for the first iterates which stand in the production terms on the right. Since the equilibrium values are $\sigma_{ij}^{[0]} = 0$ and $q_i^{[0]} = 0$ we obtain the first iterates as

---

[5] ... always assuming that the productions are known as explicit functions of the moments ...

$$\sigma_{ij}^{[1]} = -2\mu S_{ij} \quad , \quad q_i^{[1]} = -\frac{15}{4}\mu\frac{\partial\theta}{\partial x_i} \; , \tag{6.28}$$

that is, once again, the NSF laws.

For the second step of the iteration we insert the first iterates (6.28) into the left hand sides of (6.13, 6.14) and solve for the second iterates on the right hand side to obtain

$$\sigma_{ij}^{[2]} = -2\mu\frac{\partial v_{\langle i}}{\partial x_{j\rangle}} + 2\frac{\mu}{p}\left[\frac{D\mu S_{ij}}{Dt} + 2\mu S_{k\langle i}\frac{\partial v_{j\rangle}}{\partial x_k}\right]$$
$$+ 2\frac{\mu^2}{p}S_{ij}\frac{\partial v_k}{\partial x_k} + 3\frac{\mu}{p}\frac{\partial}{\partial x_{\langle i}}\left(\mu\frac{\partial\theta}{\partial x_{j\rangle}}\right) \; , \tag{6.29}$$

$$q_i^{[2]} = -\frac{15}{4}\mu\frac{\partial\theta}{\partial x_i} + \frac{45}{8}\frac{\mu}{p}\left[\frac{D}{Dt}\left(\mu\frac{\partial\theta}{\partial x_i}\right) + \mu\frac{\partial\theta}{\partial x_k}\frac{\partial v_i}{\partial x_k}\right] + \frac{75}{8}\frac{\mu^2}{p}\frac{\partial v_k}{\partial x_k}\frac{\partial\theta}{\partial x_i} +$$
$$+ 15\frac{\mu^2}{p}S_{ik}\frac{\partial\theta}{\partial x_k} + 3\frac{\mu}{\rho}\frac{\partial\mu S_{ik}}{\partial x_k} - 3\frac{\mu^2}{p\rho}S_{ik}\frac{\partial p}{\partial x_k} + 6\frac{\mu^2}{p\rho}S_{ik}\frac{\partial\mu S_{kn}}{\partial x_n} \; . \tag{6.30}$$

The second iteration adds quadratic and cubic terms in the viscosity, which plays here the role of a small parameter.[6] Higher order iterations will be obtained in the same manner, i.e. insertion of the last iterate on the left hand side, and solving for the new iterate on the right. Obviously, the higher iterates will contain terms with higher powers in μ. The third iterates, for instance, will recover the same second order terms that are present in the second iterates (6.29, 6.30), but will have additional third and higher order terms. Thus, the second iterates given above are only accurate within terms of second order, and we may drop third order terms in (6.29, 6.30).

Using the linear dependence of μ on temperature for Maxwell molecules, one can rewrite the second iterates without the obvious third order term in (6.30) as

$$\sigma_{ij}^{[2]} = -2\mu\frac{\partial v_{\langle i}}{\partial x_{j\rangle}} + \frac{\mu^2}{p}\frac{10}{3}S_{ij}\frac{\partial v_k}{\partial x_k} - 2\frac{\mu^2}{p}\left[-\frac{DS_{ij}}{Dt} + 2S_{k\langle i}\frac{\partial v_k}{\partial x_{j\rangle}}\right] +$$
$$+ 3\frac{\mu^2}{p}\frac{\partial^2\theta}{\partial x_{\langle i}\partial x_{j\rangle}} + 3\frac{\mu^2}{p\theta}\frac{\partial\theta}{\partial x_{\langle i}}\frac{\partial\theta}{\partial x_{j\rangle}} + 8\frac{\mu^2}{p}S_{k\langle i}S_{j\rangle k} + 2\frac{\mu^2}{p\theta}S_{ij}\left[\frac{D\theta}{Dt} + \frac{2}{3}\theta\frac{\partial v_k}{\partial x_k}\right] , \tag{6.31}$$

---

[6] Indeed, when the scaling parameter ε is introduced, μ would be replaced by $\varepsilon\mu$, or by the Knudsen number Kn in a dimensionless formulation. The Maxwell iteration is less formal than the CE method, and does not employ formal smallness parameters.

$$q_i^{[2]} = -\frac{15}{4}\mu\frac{\partial\theta}{\partial x_i} + \frac{75}{8}\frac{\mu^2}{p}\frac{\partial v_k}{\partial x_k}\frac{\partial\theta}{\partial x_i} + \frac{45}{8}\frac{\mu^2}{p}\left[\frac{D}{Dt}\frac{\partial\theta}{\partial x_i} - \frac{\partial v_k}{\partial x_i}\frac{\partial\theta}{\partial x_k}\right] - 3\frac{\mu^2}{p\rho}S_{ik}\frac{\partial p}{\partial x_k}$$

$$+ 3\frac{\mu^2}{\rho}\frac{\partial S_{ik}}{\partial x_k} + 3\frac{39}{4}\frac{\mu^2}{p}S_{ik}\frac{\partial\theta}{\partial x_k} + \frac{45}{8}\frac{\mu^2}{p\theta}\frac{\partial\theta}{\partial x_i}\left[\frac{D\theta}{Dt} + \frac{2}{3}\theta\frac{\partial v_k}{\partial x_k}\right], \quad (6.32)$$

Both equations contain hidden third order terms in the time derivatives. To make this visible, we use the conservation laws for mass and energy to write the time derivatives as

$$\frac{D\theta}{Dt} = \frac{D_0\theta}{Dt} - \frac{\partial q_k}{\partial x_k},$$

$$\frac{D}{Dt}\frac{\partial\theta}{\partial x_i} = \frac{D_0}{Dt}\frac{\partial\theta}{\partial x_i} - \frac{\partial^2 q_k}{\partial x_i\partial x_k},$$

$$\frac{DS_{ij}}{Dt} = \frac{D}{Dt}\frac{\partial v_{\langle i}}{\partial x_{j\rangle}} = \frac{D_0}{Dt}\frac{\partial v_{\langle i}}{\partial x_{j\rangle}} - \frac{\partial}{\partial x_{\langle i}}\left(\frac{1}{\rho}\frac{\partial\sigma_{j\rangle k}}{\partial x_k}\right),$$

with D_0/Dt as in (6.26). Obviously, within second order accuracy it is sufficient to replace the time derivatives by their zeroth order approximation, $\frac{D}{Dt} \to \frac{D_0}{Dt}$, and then (6.31, 6.32) reduce to the Burnett equations for Maxwell molecules (4.26, 4.27).

The Maxwellian iteration gives the same results as the CE method only when some higher order contributions are removed, as explained above. As was pointed out, the iteration at order $[\alpha]$ contains some terms from higher order iterations $[\alpha+1]$, but not all of these. The occurrence of terms of higher order is not disturbing, as they should play only a marginal role when the equations are applied in their proper range, and thus the higher order terms could be left in the equations. Indeed, the higher order terms could help in stabilizing the Burnett equations (which are unstable), but that seems not to happen.

Altogether the CE expansion gives a more direct relation between the order in the Knudsen number than the Maxwell iteration, and that is a reason to prefer it. Both methods yield the stable NSF laws to first order, but their higher order equations (Burnett, super-Burnett) are unstable, which puts both methods into question.

6.4 Reinecke-Kremer-Grad method

The previous sections clarified the relation between the methods of Grad and Chapman-Enskog for the special case of Maxwell molecules. In particular it became clear that the Chapman-Enskog expansion of moment systems yields the same results as the expansion of the Boltzmann equation, and is simpler by far.

The obvious next step is to consider other types of interaction, i.e. non-Maxwellian potentials. The general CE expansion of the Boltzmann equation was discussed in Sec. 4.3, and gave the phase density as an series in Sonine polynomials,

$$f_{|CE} = f_M \left(1 + \sum_{r=0}^{n_b} b_r S_{\frac{5}{2}}^{(r)} \left(\xi^2 \right) \xi_{\langle i} \xi_{j \rangle} \frac{\partial v_{\langle i}}{\partial x_{j \rangle}} + \sum_{r=1}^{n_a} a_r S_{\frac{3}{2}}^{(r)} \left(\xi^2 \right) \xi_i \sqrt{2\theta} \frac{\partial \ln \theta}{\partial x_i} \right) .$$

(6.33)

Reinecke and Kremer must have noticed that this phase density has the same structure as a Grad phase density, and can be written formally as [54][93][94]

$$f_{|G} = f_M \left(1 + \sum_{r=0}^{n_b} \lambda_{\langle ij \rangle}^r C^{2r} C_{\langle i} C_{j \rangle} + \sum_{r=0}^{n_a} \lambda_i^r C^{2r} C_i \right) .$$

The moments associated with this phase density are[7]

$$u_{\langle ij \rangle}^r = m \int C^{2r} C_{\langle i} C_{j \rangle} f_{|G} d\mathbf{c} \ , \ u_i^s = m \int C^{2s} C_i f_{|G} d\mathbf{c} \ (s = 0, \ldots, n_a) \ ,$$

(6.34)

for $r = 0, \ldots, n_b$; $s = 0, \ldots, n_a$. Reinecke and Kremer's idea is to *first* compute the moment equations for the variables (6.34), and *then* perform the CE expansion on the moment system. The classical CE method performs the expansion on the Boltzmann equation, and then computes the moments. Since the computation of moments, and moment equations, involves integration, we can say that in the classical method expansion comes first, followed by integration, while in the Reinecke-Kremer-Grad method integration comes first, followed by the expansion.

In this manner, one avoids the integral equations that appear in the CE expansion of the Boltzmann equation, which in any case must be solved by expansions in polynomials. In Reinecke and Kremer's method, the polynomial expansion is an integral part of the method.

Using Maxwellian iteration, Reinecke and Kremer used this method to compute the NSF laws and the Burnett equations for hard spheres and power potentials, and report, not surprisingly, results that stand in perfect agreement with the Chapman-Enskog method.

6.5 Moments of the collision term

Motivated by the success of the Reinecke-Kremer-Grad method in computing the NSF and Burnett equations, we use a similar procedure to compute the production terms $\mathcal{P}_{i_1 \cdots i_n}^a$ (5.19) in terms of moments.

Essentially, this method uses the Grad method to compute the phase density as a function of the moments and the microscopic velocity c_i. That phase

---

[7] Reinecke and Kremer also include the moments $u^r = m \int C^{2r} f d\mathbf{c}$, which, however, are not necessary to compute the NSF and Burnett equations.

density is then used to compute the production terms (5.19). One problem that arises here, is that this method requires the restriction to a *finite* number of moments.

With the Grad phase density (6.22), the collision moments are given by (5.19) as

$$\mathcal{P}_{i_1\cdots i_n}^a = m \int\int \left(C'^{2a} C'_{\langle i_1} C'_{i_2} \cdots C'_{i_n\rangle} - C^{2a} C_{\langle i_1} C_{i_2} \cdots C_{i_n\rangle} \right)$$
$$\times f_{|G} f_{|G,1} \sigma g \sin\Theta d\Theta d\varepsilon d\mathbf{c}_1 d\mathbf{c} \;.$$

$f_{|G} = f_M (1 + \Phi)$ is used to obtain

$$\mathcal{P}_{i_1\cdots i_n}^a = m \int\int \left(C'^{2a} C'_{\langle i_1} C'_{i_2} \cdots C'_{i_n\rangle} - C^{2a} C_{\langle i_1} C_{i_2} \cdots C_{i_n\rangle} \right)$$
$$\times (\Phi_1 + \Phi + \Phi\Phi_1) f_M f_{M,1} \sigma g \sin\Theta d\Theta d\varepsilon d\mathbf{c}_1 d\mathbf{c} \;.$$

Without evaluating the integral further, it is obvious that $\mathcal{P}_{i_1\cdots i_n}^a$ is at most quadratic in the moments, and thus must be of the general form

$$\mathcal{P}_{i_1\cdots i_n}^a = -\sum_b \mathcal{C}_{ab}^{(n)} \frac{\bar{u}_{i_1\cdots i_n}^b}{\tau\theta^{b-a}} - \sum_{r,m}\sum_{b,c} \mathcal{Y}_{a,bc}^{n,r,m} \frac{\bar{u}_{j_1\cdots j_r\langle i_1\cdots i_m}^b \bar{u}_{i_{m+1}i_{m+2}\cdots i_n\rangle j_1\cdots j_r}^c}{\tau\rho\theta^{b+c+r-a}} \;.$$

$$(6.35)$$

where the matrices $\mathcal{C}_{ab}^{(n)}$, $\mathcal{Y}_{a,bc}^{n,r,m}$ contain pure numbers and τ is a measure for the mean free time.

There is some freedom in the choice of τ, which could, e.g., be chosen as the inverse mean collision frequency (3.50),

$$\frac{1}{\tau} = \bar{\nu} = \frac{1}{n} \int\int f_M f_M^1 \, g\,\sigma \sin\Theta \, d\Theta \, d\varepsilon \, d\mathbf{c}^1 d\mathbf{c} \;. \qquad (6.36)$$

In the case of inverse power potentials, where $\sigma g = g^{\frac{\gamma-5}{\gamma-1}} F(\Theta)$, τ is best defined through the first approximation to the viscosity (4.45) as

$$\frac{1}{\tau} = \frac{p}{\mu^{[1]}} = \frac{p}{m\theta^\omega} \frac{3\chi^{(2,3)} \Gamma\left(\frac{4\nu-6}{\nu-1}\right)}{2^{\frac{5-\nu}{\nu-1}} \Gamma\left(\frac{7}{2}\right)} \qquad (6.37)$$

with

$$\chi^{(2,3)} = 2\pi \int_0^{\pi/2} \cos^2\Theta \sin^3\Theta F(\Theta) \, d\Theta \;.$$

For now, we shall not compute the matrices $\mathcal{C}_{ab}^{(n)}$, $\mathcal{Y}_{a,bc}^{n,r,m}$ which depend on the particularities of the molecular interaction, as they are reflected in the collisional cross section σ. Obviously, their computation is cumbersome, and one will try to compute only those matrices that are necessary indeed. One of the goals in the following will be to give arguments in order to determine which matrices must be considered for a theory of second order. It will be assumed that the matrices $\mathcal{C}_{ab}^{(n)}$ are invertible, which is the case as long as the numbers A_n remain finite.

6.6 Entropy maximization/Extended Thermodynamics

Grad's moment method is strongly related to the theory of Rational Extended Thermodynamics [17], and we shall discuss this without much detail.

Rational Extended Thermodynamics is a phenomenological field theory, and its counterpart in kinetic theory is the method of maximizing entropy [95][17], see also [96]. The idea of maximization of entropy was already discussed—and criticized—in [13], and gained new momentum due to its equivalence to extended thermodynamics, and due to the nice mathematical properties that the resulting moment systems should have.

It is now known, however, that the method suffers from severe difficulties, and cannot be used as designed. We shall first roughly sketch some of the main ideas, then summarize the main points of criticism, and finally discuss the relation to Grad's method.

6.6.1 Brief outline of entropy maximization

For the arguments in this section we consider a finite set of convective moments defined as

$$F_A = F^\alpha_{i_1 \cdots i_n} = m \int c^{2\alpha} c_{\langle i_1} c_{i_2} \cdots c_{i_n\rangle} f d\mathbf{c} = m \int c_A f d\mathbf{c} , \qquad (6.38)$$

where A is a suitable multi-index. The corresponding system of moment equations can be written as

$$\frac{\partial F_A}{\partial t} + \frac{\partial \mathcal{F}_{Ak}}{\partial x_k} = \Pi_A \qquad (6.39)$$

where

$$\mathcal{F}_{Ak} = m \int c_A c_k f d\mathbf{c} , \quad \Pi_A = m \int c_A \mathcal{S} d\mathbf{c} .$$

The second law reads (3.40)

$$\frac{\partial \eta}{\partial t} + \frac{\partial \Phi_k}{\partial x_k} = \Sigma \geq 0$$

where entropy density, entropy flux and entropy production are given as (3.38, 3.39),

$$\eta = -k \int f \ln \frac{f}{y} d\mathbf{c} , \quad \Phi_k = -k \int c_k f \ln \frac{f}{y} d\mathbf{c} , \quad \Sigma = -k \int \mathcal{S} \ln \frac{f}{y} d\mathbf{c} .$$

We recall Sec. 3.4.4 where it was shown that the Maxwellian maximizes the entropy in equilibrium. Now we ask which distribution function f_{max} maximizes the entropy under the constraints of given values of the moments F_A. As in Sec. 3.4.4 the constraints can be incorporated by means of Lagrange multipliers $\Lambda_A = \Lambda^\alpha_{\langle i_1 \cdots i_n \rangle}$ and the resulting phase density is[8]

---

[8] We use summation convention for the multi-indices $A, B, \ldots$

$$f_{\max} = y \exp[-1 - \frac{m}{k}\Lambda_A c_A] \, . \tag{6.40}$$

The Lagrange multipliers must be determined from the constraints (6.38). With the above definitions, it is straightforward to show that

$$\frac{\partial \eta}{\partial F_C} = \Lambda_A m \int c_A \frac{\partial f}{\partial \Lambda_B} d\mathbf{c} \frac{\partial \Lambda_B}{\partial F_C} = \Lambda_A \frac{\partial F_A}{\partial \Lambda_B} \frac{\partial \Lambda_B}{\partial F_C} = \Lambda_C \, ,$$

$$\frac{\partial \Phi_k}{\partial \mathcal{F}_{Cr}} = \Lambda_A m \int c_k c_A \frac{\partial f}{\partial \Lambda_B} d\mathbf{c} \frac{\partial \Lambda_B}{\partial \mathcal{F}_{Cr}} = \Lambda_A \frac{\partial \mathcal{F}_{Ak}}{\partial \Lambda_B} \frac{\partial \Lambda_B}{\partial \mathcal{F}_{Cr}} = \Lambda_C \delta_{kr}$$

This implies that entropy and entropy flux are potentials for the Lagrange multipliers in the sense that

$$d\eta = \Lambda_C dF_C \, , \quad d\Phi_k = \Lambda_C d\mathcal{F}_{Ck} \, .$$

Accordingly, the second law gives a relation between the entropy production and the production terms Π_A of the variables,

$$\frac{\partial \eta}{\partial F_C}\frac{\partial F_C}{\partial t} + \frac{\partial \Phi_k}{\partial \mathcal{F}_{Cr}}\frac{\partial \mathcal{F}_{Cr}}{\partial x_k} = \Lambda_C\left[\frac{\partial F_C}{\partial t} + \frac{\partial \mathcal{F}_{Ck}}{\partial x_k}\right] = \Lambda_C \Pi_C = \Sigma \geq 0 \, .$$

Introduction of $F_A = F_A(\Lambda_B)$, $\mathcal{F}_{Ak} = \mathcal{F}_{Ak}(\Lambda_B)$ in the moment equations (6.39) gives a coupled set of equations for the Lagrange multipliers,

$$\frac{\partial F_A}{\partial \Lambda_B}\frac{\partial \Lambda_B}{\partial t} + \frac{\partial \mathcal{F}_{Ak}}{\partial \Lambda_B}\frac{\partial \Lambda_B}{\partial x_k} = \Pi_A \, . \tag{6.41}$$

The matrices can be computed from to the definition of the moments, and the particular form of the maximized phase density, as

$$\frac{\partial F_A}{\partial \Lambda_B} = m \int c_A \frac{\partial f}{\partial \Lambda_B} d\mathbf{c} = -\frac{m^2}{k}\int c_A c_B f d\mathbf{c} \, ,$$

$$\frac{\partial \mathcal{F}_{Ak}}{\partial \Lambda_B} = m \int c_A c_k \frac{\partial f}{\partial \Lambda_B} d\mathbf{c} = -\frac{m^2}{k}\int c_A c_B c_k f d\mathbf{c} \, .$$

6.6.2 Properties and problems

Obviously, the matrices $\frac{\partial F_A}{\partial \Lambda_B}$ and $\frac{\partial \mathcal{F}_{Ak}}{\partial \Lambda_B} n_k$ are symmetric, and $\frac{\partial F_A}{\partial \Lambda_B}$ is negative definite. The definiteness ensures the invertibility of the relation between moments and Lagrange multipliers $F_A(\Lambda_B) \leftrightarrow \Lambda_A(F_B)$. The symmetry of the matrices shows that the system of moment equations forms a symmetric hyperbolic system [97], if the closure is performed by maximizing entropy. Hyperbolicity is a property of the system in any form (e.g. with the variables F_A), but becomes visible only when the Lagrange multipliers are considered as variables (which is why these are also known as the "main field" [98]). Systems of this type have welcome mathematical properties including stability and well-posedness of local Cauchy problems [99].

Another advantage of the method is that the maximized phase density (6.40) is strictly positive, while the Grad function (6.2) will become negative for large velocities (although the values will be small since the Maxwellian effectively suppresses polynomial contributions).

The method as outlined so far can work only, if the integrals over the infinite velocity space of the maximized phase density (6.40) exist. This requires the polynomial in the exponent to be negative in the limit $\mathbf{c} \to \infty$. Accordingly, the polynomial c_A with the highest power $A_{\max}$ must be even, and must have a negative coefficient, $\Lambda_{A_{\max}} > 0$.

Thus, the freedom of the choice of variables is reduced, since sets of polynomials where the highest power is odd are not allowed. For instance the 13 moment set $u^{[13]}$ is excluded, while the 26 moment set $u^{[26]}$ is allowed.

The Lagrange multipliers are the variables of the system, and should not be restricted. Their actual values will develop according to initial and boundary data and the differential equations, and it is unlikely that the condition $\Lambda_{A_{\max}} > 0$ will be fulfilled for all reasonable sets of initial and boundary data at all times.[9]

The space of variables Λ_A can be divided in an allowed and a forbidden region, and it turns out that the equilibria—the Maxwellians—just lie on the boundary between both domains [100]–[103]. Even the smallest fluctuations in the gas around its equilibrium are then restricted in order not to enter the forbidden region, which appears to be an artificial requirement. For a detailed discussion of these issues, the reader is referred to the papers cited above, which also show that the problems are related to the infinite particle velocities in kinetic theory.

Another feature of hyperbolic systems is the finite speed of disturbances. From the viewpoint of the theory of relativity one will be interested in excluding infinite speeds of propagation. In classical kinetic theory infinite particle speeds are allowed, and infinite propagation speeds therefore are not surprising, and cannot be seen as a valid point against a set of equations.

Finite propagation speeds lead to the occurrence of discontinuous shocks which are not observed in nature, and a different type of equations might be more desirable.

The elegant mathematical formulation of the theory of maximization of entropy is outweighed by far by the problems arising with the method, and its use cannot be recommended.

6.6.3 Linearization and Grad method

The Maxwellian can be taken out as an factor in (6.40), which then can be written as

$$f_{\max} = f_M \exp[-\lambda_A c_A]$$

---

[9] Obviously, initial data must be chosen so that $\Lambda_{A_{\max}} > 0$.

with new coefficients λ_A. The phase density reduces to the Maxwellian in equilibrium, which implies that the λ_A vanish in equilibrium. Then, for processes not to far from local equilibrium, the λ_A will be small, and the exponential can be expanded in a Taylor series about $\lambda_A = 0$, so that

$$f_{\max} \simeq f_M [1 - \lambda_A c_A] \ .$$

This is just the Grad distribution (6.2), which forms the base of Grad's method of moments. Obviously, the Grad function is closely related to the maximum entropy approach, but more important are the differences: (a) Grad's method can be considered with any set of moments, the restriction to even powers for the largest moment is not required. (b) The values of the coefficients λ_A are not restricted. (c) The Grad distribution can assume negative values.

Since $f_{\max}$ and $f_{|G}$ agree for small values of the coefficients λ_A, it follows that the Grad system is symmetric hyperbolic for not too large deviations from local equilibrium, that is when the coefficients λ_A are small enough. Müller and Ruggeri computed the region of hyperbolicity in [17] for the 13 moment case. The deviation from equilibrium can be quite substantial without violating the hyperbolicity of the equations, but a breakdown of solutions of Grad equations due to the loss of hyperbolicity is encountered, e.g. in strong shocks [104].

Problems

6.1. Euler equations
In the text it was stated that the Grad system with 5 moments is just the Euler equations. Proof this statement.

6.2. Grad phase density $f_{|13}$
(a) Use (6.10) to compute the coefficients $\lambda^a_{\langle i_1 \cdots i_n \rangle}$ in (6.9), and show that (6.11) is correct.
(b) Compute the moments $u^0_{ijk|13}$, $u^1_{ij|13}$, $w^2_{|13}$ (6.12).

6.3. Grad phase density $f_{|26}$
(a) Show that $f_{|26}$ (6.16) reproduces the moments $u^{[26]}$ (6.15).
(b) Show that $f_{|26}$ gives the constitutive laws (6.17).

6.4. Burnett equations
Proof (6.26), and then compute the Burnett equations (6.31, 6.32).

Regularization of Grad equations

The methods of Chapman-Enskog and Grad are the best known methods to produce macroscopic transport equations from the Boltzmann equation, but both come with problems: The Chapman-Enskog method has great success in deriving the Navier-Stokes and Fourier laws with explicit expressions for viscosity and heat conductivity, but the higher order expansions—the Burnett and super-Burnett equations—are unstable in transient processes, and their use cannot be recommended. The Grad equations on the other hand are always stable, but their hyperbolic character yields unphysical discontinuous shocks for large enough Mach numbers (Chapter 11).

In this chapter both methods are combined to derive a regularization of the Grad equations, which is non-hyperbolic, and yields continuous shock structures at all Mach numbers. The resulting set of equations, the R13 equations, combines the benefits of CE and Grad-type equations, while avoiding their shortcomings, as will be seen in later chapters. This chapter presents their original derivation [9][59] which requires the linearized production terms of Maxwell molecules. Chapter 8 presents an alternative, more general, method for their derivation. Furthermore, a regularization due to Jin and Slemrod is discussed and compared to the R13 equations.

This chapter is based on work with Manuel Torrilhon [59][60].

7.1 Grad distributions as pseudo-equilibrium manifolds

The local Maxwellians are the equilibrium solutions of the Boltzmann equation, and a gas left to itself in an initial non-equilibrium state will relax its phase density to a Maxwellian. The characteristic time for this process is of the order of the mean free time τ.

The Maxwellians define a manifold in phase space, and we can say that the gas will move towards this equilibrium manifold with the characteristic time τ. Processes on the manifold are solutions of the Euler equations, which were obtained by assuming that the phase density is always a Maxwellian, see, e.g., Sec. 4.2.2. This strong restriction holds when the Knudsen number approaches zero.

For non-zero Knudsen numbers we found the Navier-Stokes-Fourier laws as the first approximation, and we can say that these describe processes in the vicinity of the equilibrium manifold, that will ultimately end in an equilibrium state on the manifold.

The move from the Euler to the NSF equations changes the mathematical properties of the equations: The Euler equations are hyperbolic, and thus have discontinuous shock solutions, while the NSF equations are parabolic, and will always yield smooth shock structures (no discontinuities).

The higher order CE expansions, which give the Burnett and super-Burnett equations, aim at accessing a wider region in phase space around the equilibrium manifold. This fails, since the resulting equations are unstable.

The Grad distributions $f_{|G}$ (6.2), e.g. the 13 moment phase density $f_{|13}$ (6.11), define non-equilibrium manifolds in phase space [105][106]. The Maxwellians form a subset on these non-equilibrium manifolds. The Grad closure restricts the phase space so that the gas cannot access all states in phase space, but only those on the Grad non-equilibrium manifold. This strong restriction is inherent to Grad's closure, and has, as it seems, no physical foundation.

Indeed, the Grad non-equilibrium manifolds cannot be extracted from the Boltzmann equation as a special phase density that has a meaning. This is different, of course, for the Maxwellians, which are those phase densities that give a zero collision term. With no argument from physics to support the Grad distributions, it seems to be daring to restrict the gas on the Grad non-equilibrium manifolds. One way to relax the Grad assumption—at least somewhat—is to allow states in the *vicinity* of the Grad manifolds. This stands in analogy to the relation between Euler and NSF equations, which describe the equilibrium manifold, and its vicinity.

The NSF equations are derived by means of a first order Chapman-Enskog expansion around equilibrium, and, correspondingly, one will relax the Grad closure by performing a first order CE expansion around the Grad distribution. This will be done below for the 13 moment system. The Chapman-Enskog expansion requires a smallness parameter which is introduced through the following—artificial—assumption: It will be assumed that the gas, when in non-equilibrium, will approach the non-equilibrium manifold on a fast time scale τ_ϵ which is assumed to be smaller than the mean free time τ. The gas will then move through states in the vicinity of the manifold and move towards the equilibrium manifold with the characteristic time τ.

With this assumption the Grad distribution is considered as a pseudo-equilibrium manifold, in the sense that any non-equilibrium state will relax fast towards the pseudo-equilibrium, and than move towards the real equilibrium states on a slower time scale. It must be emphasized that there is no evidence in physics to support the existence of pseudo-equilibrium manifolds for the gas, and that the goal is merely to improve Grad moment systems, which restrict the gas on the pseudo-equilibrium manifold, by allowing states in the vicinity.

The resulting equations are not hyperbolic, and give smooth shock structures for all Mach numbers, and they are stable. Therefore, the relaxation of Grad's strong assumption yields a marked improvement over the original equations.

The regularization of Grad's 13 moment equations appears first as a side note in Grad's contribution to the Ecyclopedia of Physics [9], but—surprisingly—he did not consider the equations to be very useful. It seems that neither Grad nor other authors explored the equations further, so that [59][60] present the first in-depth study of the R13 equations.

7.2 Basic equations

The basic equations for Grad's 13 moment system are the conservation laws (6.4) and the moment equations for stress and heat flux (6.5, 6.6), which do not form a closed set of equations, since they contain the additional quantities w^2, u^1_{ij} and u^0_{ijk} (6.7) which must be related to the variables via the Grad phase density $f_{|13}$ (6.11). The result was given in (6.12) .

In order to regularize this system, we define the deviations of w^2, u^1_{ij} and u^0_{ijk} from their values obtained with the Grad closure (6.12) as

$$\Delta = w^2 - w^2_{|13} = w^2 ,$$
$$R_{ij} = u^1_{ij} - u^1_{ij|13} = u^1_{ij} - 7\theta\sigma_{ij} , \qquad (7.1)$$
$$m_{ijk} = u^0_{ijk} - u^0_{ijk|13} = u^0_{ijk} .$$

For the Grad closure we thus have $\Delta = R_{ik} = m_{ijk} = 0$. The goal is to compute non-zero approximations for these quantities from their respective moment equations. The equation for $\Delta = w^2$ follows directly from (5.14), the equation for $m_{ijk} = u^0_{ijk}$ follows directly from (5.17), and the equation for $R_{ij} = u^1_{ij} - 7\theta\sigma_{ij}$ follows from combining the equations for u^1_{ij}, $\sigma_{ij} = u^0_{ij}$ (5.16), and the energy balance. After introducing the linearized production terms for Maxwell molecules and the definitions of Δ, R_{ij}, and m_{ijk}, we obtain after some algebra[1]

$$\frac{D\Delta}{Dt} - 200\theta\frac{\partial q_k}{\partial x_k} + 8\theta\sigma_{kl}\frac{\partial v_k}{\partial x_l} + 4R_{kl}\frac{\partial v_k}{\partial x_l} - 8q_k\frac{\partial\theta}{\partial x_k} - 8q_k\theta\frac{\partial\ln\rho}{\partial x_k}$$
$$- 8\frac{q_k}{\rho}\frac{\partial\sigma_{kl}}{\partial x_l} + \frac{\partial u^2_k}{\partial x_k} + \frac{7}{3}\Delta\frac{\partial v_k}{\partial x_k} = -\frac{2}{3}\frac{p}{\mu}\Delta , \quad (7.2)$$

---

[1] ... including the use of $6\sigma_{\langle ij}\frac{\partial v_{k\rangle}}{\partial x_k} = 2\sigma_{ij}\frac{\partial v_k}{\partial x_k} + 4\sigma_{k\langle i}\frac{\partial v_{j\rangle}}{\partial x_k} - \frac{8}{5}\sigma_{r\langle i}\frac{\partial v_r}{\partial x_{j\rangle}}$.

$$\frac{DR_{ij}}{Dt} + \frac{2}{5}\frac{\partial u^2_{\langle i}}{\partial x_{j\rangle}} - \frac{28}{5}\theta\frac{\partial q_{\langle i}}{\partial x_{j\rangle}} - \frac{28}{5}q_{\langle i}\frac{\partial \theta}{\partial x_{j\rangle}} - \frac{28}{5}\theta q_{\langle i}\frac{\partial \ln \rho}{\partial x_{j\rangle}} - \frac{28}{5}\frac{q_{\langle i}}{\rho}\frac{\partial \sigma_{j\rangle k}}{\partial x_k}$$

$$+ 4\theta\left[\sigma_{k\langle i}\frac{\partial v_k}{\partial x_{j\rangle}} + \sigma_{k\langle i}\frac{\partial v_{j\rangle}}{\partial x_k} - \frac{2}{3}\sigma_{ij}\frac{\partial v_k}{\partial x_k}\right] - \frac{14}{3}\frac{1}{\rho}\sigma_{ij}\frac{\partial q_k}{\partial x_k} - \frac{14}{3}\frac{\sigma_{ij}\sigma_{kl}}{\rho}\frac{\partial v_k}{\partial x_l}$$

$$- 7\theta\frac{\partial m_{ijk}}{\partial x_k} - 2\frac{m_{ijk}}{\rho}\left(\frac{\partial \sigma_{kl}}{\partial x_l} + \frac{\partial \theta\rho}{\partial x_k}\right) + \frac{\partial u^1_{ijk}}{\partial x_k} + 2u^0_{ijkl}\frac{\partial v_k}{\partial x_l} + \frac{6}{7}R_{\langle ij}\frac{\partial v_{k\rangle}}{\partial x_k}$$

$$+ \frac{4}{5}R_{k\langle i}\frac{\partial v_k}{\partial x_{j\rangle}} + 2R_{k\langle i}\frac{\partial v_{j\rangle}}{\partial x_k} + R_{ij}\frac{\partial v_k}{\partial x_k} + \frac{14}{15}\Delta\frac{\partial v_{\langle i}}{\partial x_{j\rangle}} = -\frac{7}{6}\frac{p}{\mu}R_{ij}\,, \quad (7.3)$$

$$\frac{Dm_{ijk}}{Dt} - 3\frac{\sigma_{\langle ij}}{\rho}\frac{\partial \sigma_{k\rangle l}}{\partial x_l} - 3\sigma_{\langle ij}\theta\frac{\partial \ln \rho}{\partial x_{k\rangle}} + \frac{\partial u^0_{ijkl}}{\partial x_l} + \frac{3}{7}\frac{\partial R_{\langle ij}}{\partial x_{k\rangle}} + 3\theta\frac{\partial \sigma_{\langle ij}}{\partial x_{k\rangle}} + 3m_{l\langle ij}\frac{\partial v_{k\rangle}}{\partial x_l}$$

$$+ m_{ijk}\frac{\partial v_l}{\partial x_l} + \frac{12}{5}q_{\langle i}\frac{\partial v_j}{\partial x_{k\rangle}} = -\frac{3}{2}\frac{p}{\mu}m_{ijk}\,. \quad (7.4)$$

7.3 Expansion around $f_{|13}$

For the sake of the argument, we assume that the non-equilibrium moments of the 13 field case, σ_{ij} and q_i, change on the time scale defined by the mean free time $\tau = \mu/p$, while all other non-equilibrium moments change on a faster time scale $\tau_\epsilon = \epsilon\tau$ where ϵ is "small". Indeed, ϵ is a formal smallness parameter that will be set equal to unity at the end of the calculations, similar to the procedure in the classical CE expansion. The important difference to the Chapman-Enskog scaling is that only the production terms of higher moments are scaled by ϵ, while the production terms for stress and heat flux are not scaled.[2] Thus, the equations (7.2–7.4) are slightly changed on their right hand sides which now contain the parameter ϵ,

$$\frac{D\Delta}{Dt} + \{\cdots \text{space derivatives of moments} \cdots\} = -\frac{1}{\epsilon}\frac{2}{3}\frac{p}{\mu}\Delta\,,$$

$$\frac{DR_{ij}}{Dt} + \{\cdots \text{space derivatives of moments} \cdots\} = -\frac{1}{\epsilon}\frac{7}{6}\frac{p}{\mu}R_{ij}\,, \quad (7.5)$$

$$\frac{Dm_{ijk}}{Dt} + \{\cdots \text{space derivatives of moments} \cdots\} = -\frac{1}{\epsilon}\frac{3}{2}\frac{p}{\mu}m_{ijk}\,.$$

Similar equations can be written for all higher moments. Next, the equations are expanded in terms of the parameter ϵ as

$$\Delta = \Delta^{(0)} + \epsilon\Delta^{(1)} + \cdots\,,$$

$$R_{ij} = R_{ij}^{(0)} + \epsilon R_{ij}^{(1)} + \cdots\,, \quad (7.6)$$

$$m_{ijk} = m_{ijk}^{(0)} + \epsilon m_{ijk}^{(1)} + \cdots\,,$$

---

[2] Or rather, they are scaled by a different parameter $\varepsilon \gg \epsilon$.

where only terms up to order $\mathcal{O}\left(\epsilon^1\right)$ are accounted for. Obviously, this is a Chapman-Enskog expansion with ϵ as smallness parameter.

The constitutive laws for Δ, R_{ik} and m_{ijk} result from inserting the ansatz (7.6) into the equations (7.5), and balancing terms of the same order in ϵ. It is straightforward to see that the zeroth order approximation (balancing terms with the factor $1/\epsilon$) just gives Grad's 13 moment case, corresponding to

$$\Delta^{(0)} = R_{ij}^{(0)} = m_{ijk}^{(0)} = 0 . \tag{7.7}$$

The first order corrections result from balancing terms of order ϵ^0 and can be written as

$$\left[\frac{D\Delta}{\partial t} + \{\cdots \text{space derivatives of moments} \cdots\}\right]_{|f_{13}} = -\frac{2}{3}\frac{p}{\mu}\Delta^{(1)} ,$$

$$\left[\frac{DR_{ij}}{Dt} + \{\cdots \text{space derivatives of moments} \cdots\}\right]_{|f_{13}} = -\frac{7}{6}\frac{p}{\mu}R_{ij}^{(1)} ,$$

$$\left[\frac{Dm_{ijk}}{Dt} + \{\cdots \text{space derivatives of moments} \cdots\}\right]_{|f_{13}} = -\frac{3}{2}\frac{p}{\mu}m_{ijk}^{(1)} .$$

The notation $[\cdot]_{|f_{13}}$ indicates that *all* moments in *all* terms inside the square brackets must be evaluated with the phase density $f_{|13}$ as given by (6.11), that is

$$u_{ijkl|13}^0 = u_{ijk|13}^1 = 0 , \quad u_{k|13}^2 = 28\theta q_k \quad \text{and} \quad m_{ijk|13} = R_{ij|13} = \Delta_{|13} = 0 .$$

The result can be summarized as follows: The equations for pressure deviator and heat flux vector are

$$\frac{D\sigma_{ij}}{Dt} + \sigma_{ij}\frac{\partial v_k}{\partial x_k} + \frac{4}{5}\frac{\partial q_{\langle i}}{\partial x_{j\rangle}} + 2p\frac{\partial v_{\langle i}}{\partial x_{j\rangle}} + 2\sigma_{k\langle i}\frac{\partial v_{j\rangle}}{\partial x_k} + \frac{\partial m_{ijk}}{\partial x_k} = -\frac{p}{\mu}\sigma_{ij} , \tag{7.8}$$

$$\frac{Dq_i}{Dt} + \frac{5}{2}p\frac{\partial\theta}{\partial x_i} + \frac{5}{2}\sigma_{ik}\frac{\partial\theta}{\partial x_k} + \theta\frac{\partial\sigma_{ik}}{\partial x_k} - \theta\sigma_{ik}\frac{\partial\ln\rho}{\partial x_k} + \frac{7}{5}q_k\frac{\partial v_i}{\partial x_k} + \frac{2}{5}q_k\frac{\partial v_k}{\partial x_i}$$
$$+ \frac{7}{5}q_i\frac{\partial v_k}{\partial x_k} + \frac{1}{2}\frac{\partial R_{ik}}{\partial x_k} + \frac{1}{6}\frac{\partial\Delta}{\partial x_i} + m_{ijk}\frac{\partial v_j}{\partial x_k} - \frac{\sigma_{ij}}{\rho}\frac{\partial\sigma_{jk}}{\partial x_k} = -\frac{2}{3}\frac{p}{\mu}q_i . \tag{7.9}$$

The corrections to Grad's 13 moment equations read (after setting $\epsilon = 1$)

$$\Delta = -12\frac{\mu}{p}\left[\theta\frac{\partial q_k}{\partial x_k} + \frac{5}{2}q_k\frac{\partial\theta}{\partial x_k} - \theta q_k\frac{\partial\ln\rho}{\partial x_k} - \frac{q_k}{\rho}\frac{\partial\sigma_{kl}}{\partial x_l} + \theta\sigma_{kl}\frac{\partial v_k}{\partial x_l}\right] , \tag{7.10}$$

$$R_{ij} = -\frac{24}{5}\frac{\mu}{p}\left[\theta\frac{\partial q_{\langle i}}{\partial x_{j\rangle}} + q_{\langle i}\frac{\partial\theta}{\partial x_{j\rangle}} - \theta q_{\langle i}\frac{\partial\ln\rho}{\partial x_{j\rangle}} - \frac{q_{\langle i}}{\rho}\frac{\partial\sigma_{j\rangle k}}{\partial x_k}\right.$$
$$\left. + \frac{5}{7}\theta\left[\sigma_{k\langle i}\frac{\partial v_{j\rangle}}{\partial x_k} + \sigma_{k\langle i}\frac{\partial v_k}{\partial x_{j\rangle}} - \frac{2}{3}\sigma_{ij}\frac{\partial v_k}{\partial x_k}\right] - \frac{5}{6}\frac{\sigma_{ij}}{\rho}\frac{\partial q_k}{\partial x_k} - \frac{5}{6}\frac{\sigma_{ij}\sigma_{kl}}{\rho}\frac{\partial v_k}{\partial x_l}\right] , \tag{7.11}$$

$$m_{ijk} = -2\frac{\mu}{p}\left[\theta\frac{\partial\sigma_{\langle ij}}{\partial x_{k\rangle}} - \theta\sigma_{\langle ij}\frac{\partial\ln\rho}{\partial x_{k\rangle}} + \frac{4}{5}q_{\langle i}\frac{\partial v_j}{\partial x_{k\rangle}} - \frac{\sigma_{\langle ij}}{\rho}\frac{\partial\sigma_{k\rangle l}}{\partial x_l}\right].\qquad(7.12)$$

With these equations, we have derived the complete set of regularized equations for the 13 variables $\rho, v_i, \theta, \sigma_{ij}, q_i$ that consists of the equations (6.4, 7.8–7.12).

These equations are denoted as the R13 equations, where "R" stands for "regularized", and "13" denotes the number of variables [59][60]. They were first presented by Grad [9], who, however, omitted some non-linear terms, and did not replace convective time derivatives of v_i and T.

7.4 Euler and Navier-Stokes-Fourier equations

The procedure in the last section was less formal than our treatment of the classical CE expansion. In this section, we use the same method of regularization in order to derive the equations of Navier-Stokes and Fourier which can be considered as the regularization of the Euler equations. In this case the method reduces to the first order CE expansion of the 13 moment system, which means that in the present framework the Navier-Stokes-Fourier equations might be denoted as the R5 equations.

The first member of the meaningful sets of Grad type moment equations is the 5 moments case where $\Psi_A = m\left\{1, c_i, \frac{1}{2}C^2\right\}$, corresponding to the five variables $\rho_A = \left\{\rho, \rho v_i, \rho\varepsilon = \frac{3}{2}\rho\theta\right\}$ or $\{\rho, v_i, \theta\}$. Accordingly, the relevant moment equations are the conservation laws for mass, momentum and energy (6.4). Of course, these five equations do not form a closed set for the five variables $\{\rho, v_i, \theta\}$, since they contain the pressure deviator σ_{ij} and the heat flux vector q_i. Straightforward application of the Grad closure (6.2) reveals that the corresponding distribution function for the closure is the Maxwellian, $f_{|5} = f_M$, which yields

$$\sigma_{ij|5} = 0 \quad\text{and}\quad q_{i|5} = 0.\qquad(7.13)$$

The resulting equations are the well-known Euler equations for an ideal gas (4.21). The Euler equations are of hyperbolic nature and exhibit discontinuous shocks in supersonic flow problems.

For the regularization the smallness parameter ϵ is introduced in the right hand sides of the balances for σ_{ij} and q_i, (6.5, 6.6), which then read

$$\frac{D\sigma_{ij}}{Dt} + \{\cdots\text{space derivatives of moments}\cdots\} = -\frac{1}{\epsilon}\frac{p}{\mu}\sigma_{ij},$$

$$\frac{Dq_i}{Dt} + \{\cdots\text{space derivatives of moments}\cdots\} = -\frac{1}{\epsilon}\frac{2}{3}\frac{p}{\mu}q_i.$$

Next, σ_{ij} and q_i are expanded as

$$\sigma_{ij} = \sigma_{ij}^{(0)} + \epsilon\sigma_{ij}^{(1)}, \quad q_i = q_i^{(0)} + \epsilon q_i^{(1)},$$

and terms of equal powers in ε are equated. As can be expected, this gives

$$\sigma_{ij}^{(0)} = q_i^{(0)} = 0 \,,$$

so that the zeroth order expansion results in the Euler equations. The first order expansion yields

$$\left[\frac{D\sigma_{ij}}{Dt} + \{ \cdots \text{space derivatives of moments} \cdots \} \right]_{\mid f_5} = -\frac{1}{\epsilon} \frac{p}{\mu} \sigma_{ij}^{(1)} \,,$$

$$\left[\frac{Dq_i}{\partial t} + \{ \cdots \text{space derivatives of moments} \cdots \} \right]_{\mid f_5} = -\frac{1}{\epsilon} \frac{2}{3} \frac{p}{\mu} q_i^{(1)} \,,$$

where all moments in the square brackets must be replaced by their values computed with the Maxwellian $f_{\mid 5} = f_M$, that is from the Grad function for the case under consideration. These values are easily computed as

$$\sigma_{ij\mid 5} = q_{i\mid 5} = u_{ijk\mid 5}^0 = u_{ij\mid 5}^1 = w_{\mid 5}^2 = 0 \,,$$

and it follows that the two equations simply reduce to the Navier-Stokes law for the stress tensor and the Fourier law for the heat flux, (4.22).

7.5 Linearized equations

We consider the R13 equations for small deviations from an equilibrium state given by ρ_0, θ_0, $v_{i,0} = 0$. Dimensionless variables $\hat{\rho}$, $\hat{\theta}$, $\hat{v}_i$, $\hat{\sigma}_{ij}$, $\hat{q}$ are introduced as

$$\rho = \rho_0 \left(1 + \hat{\rho} \right) \,, \quad \theta = \theta_0 \left(1 + \hat{\theta} \right) \,, \quad p = \rho_0 \theta_0 \left(1 + \hat{\rho} + \hat{\theta} \right) \,,$$

$$v_i = \sqrt{\theta_0} \hat{v}_i \,, \quad \sigma_{ij} = \rho_0 \theta_0 \hat{\sigma}_{ij} \,, \quad q_i = \rho_0 \sqrt{\theta_0}^3 \hat{q}_i \,.$$

Moreover, a relevant length scale L of the process is used to non-dimensionalize the space and time variables according to

$$x_i = L \hat{x}_i \,, \quad t = \frac{L}{\sqrt{\theta_0}} \hat{t} \,.$$

The corresponding dimensionless collision time is then given by the Knudsen number, which is here defined as

$$\text{Kn} = \frac{\mu_0 \sqrt{\theta_0}}{p_0 L} \quad \text{where } \mu_0 = \mu \left(\theta_0 \right) \,.$$

Linearization in the deviations from equilibrium $\hat{\varrho}$, $\hat{T}$, $\hat{v}_i$, $\hat{\sigma}_{ij}$, $\hat{q}$ yields the dimensionless linearized system in three dimensions as

$$\frac{\partial \hat{\rho}}{\partial \hat{t}} + \frac{\partial \hat{v}_k}{\partial \hat{x}_k} = 0 \,,$$

$$\frac{\partial \hat{v}_i}{\partial \hat{t}} + \frac{\partial \hat{\rho}}{\partial \hat{x}_i} + \frac{\partial \hat{\theta}}{\partial \hat{x}_i} + \frac{\partial \hat{\sigma}_{ik}}{\partial \hat{x}_k} = 0 \,,$$

$$\frac{3}{2}\frac{\partial \hat{\theta}}{\partial \hat{t}} + \frac{\partial \hat{q}_k}{\partial \hat{x}_k} + \frac{\partial \hat{v}_k}{\partial \hat{x}_k} = 0 \,, \tag{7.14}$$

$$\frac{\partial \hat{\sigma}_{ij}}{\partial \hat{t}} + \frac{4}{5}\frac{\partial \hat{q}_{\langle i}}{\partial \hat{x}_{j\rangle}} + 2\frac{\partial \hat{v}_{\langle i}}{\partial \hat{x}_{j\rangle}} - 2\mathrm{Kn}\frac{\partial}{\partial \hat{x}_k}\frac{\partial \hat{\sigma}_{\langle ij}}{\partial \hat{x}_{k\rangle}} = -\frac{\hat{\sigma}_{ij}}{\mathrm{Kn}} \,,$$

$$\frac{\partial \hat{q}_i}{\partial \hat{t}} + \frac{5}{2}\frac{\partial \hat{\theta}}{\partial \hat{x}_i} + \frac{\partial \hat{\sigma}_{ik}}{\partial \hat{x}_k} - \frac{12}{5}\mathrm{Kn}\frac{\partial}{\partial \hat{x}_k}\frac{\partial \hat{q}_{\langle i}}{\partial \hat{x}_{k\rangle}} - 2\mathrm{Kn}\frac{\partial}{\partial \hat{x}_i}\frac{\partial \hat{q}_k}{\partial \hat{x}_k} = -\frac{2}{3}\frac{\hat{q}_i}{\mathrm{Kn}} \,.$$

This set of equations is equivalent to equations proposed by Karlin et al. [105], who, however, did not give explicit numerical expressions for the factors that multiply the second derivatives of $\hat{\sigma}_{ij}$ and $\hat{q}_i$, but presented them as integrals over the linearized collision operator which are not further evaluated.

Karlin et al. found their equations not through manipulations of the moment equations, but by considering a modification of the linearized Boltzmann equation which forces the phase density towards the pseudo-equilibrium manifold $f_{|13}$.

The approach through the moment equations is faster, easier, and more transparent, and allows us to derive the non-linear regularization for the 13 moment equations.

7.6 Discussion

It is well known that the relaxation times for moments grow only slowly [21], so that the relaxation times of the first 26 moments are of the same magnitude $\tau = \mu/p$. This becomes obvious by a glance at the right hand sides of (7.2–7.4) where the factors $\left\{ \frac{3}{2}\frac{p}{\mu}, \frac{7}{6}\frac{p}{\mu}, \frac{2}{3}\frac{p}{\mu} \right\}$ define the respective time scales for the moments w^2, u^1_{ij}, u^0_{ijk}, while the corresponding time scales for σ_{ij} and q_i are $\left\{ \frac{p}{\mu}, \frac{2}{3}\frac{p}{\mu} \right\}$. Accordingly, the parameter ϵ is not small, but of order unity. Thus, the basic assumption in the derivation—the smallness of ϵ—is not well justified. Nevertheless, the argument shows that the assumption behind the R13 equations, which require ϵ as "small", is less restrictive than the the assumption behind Grad's 13 moment system, which requires $\epsilon = 0$.

The R13 equations follow from the 26 moment equations, for which we have shown that they include the Burnett and super-Burnett equations, and contain the complete 13 moment equations, which include the Burnett equations. Thus, it is clear that the R13 equations contain the Burnett equations. Since the regularization changes or removes some terms in the equations, it must be checked again whether the R13 equations include the super-Burnett equations as well. As was shown in [59][60], this is indeed the case. The alternative derivation of the R13 equations in the next chapter will make clear why this is so.

We therefore see that the regularization improves the Chapman-Enskog order: The regularization of the Euler equations which are $\mathcal{O}\left(\varepsilon^0\right)$ adds terms of order $\mathcal{O}\left(\varepsilon\right)$ to yield the Navier-Stokes-Fourier equations. The regularization of the 13 moment equations which are accurate up to Burnett order $\mathcal{O}\left(\varepsilon^2\right)$ adds third order terms $\mathcal{O}\left(\varepsilon^3\right)$.

The CE expansion of the R13 equations can be continued ad infinitum, with $\sigma_{ij}^{(\alpha)} \neq 0$, $q_i^{(\alpha)} \neq 0$ for all α. Of course, the $\sigma_{ij}^{(\alpha)}$, $q_i^{(\alpha)}$ for $\alpha \geq 4$ do not yield the correct expressions of the Chapman-Enskog expansion to order α, since terms from other (higher) moment equations will enter the proper expressions. Nevertheless, there are contributions of all orders $\mathcal{O}\left(\varepsilon^\alpha\right)$, $\alpha = 1, \ldots, \infty$, present in the R13 equations. It is in these higher order contributions that the R13 equations differ from the Burnett and super-Burnett equations.

The presented regularization method is based on a Chapman-Enskog expansion of only *first order* but gives a set of equations which agrees with the Boltzmann equation to a *higher order*. The first order Chapman-Enskog expansion yields stable equations, while usually expansions to higher order yield unstable equations, see [56][57][107] and Sec. 10.3.

The regularization of the moment equations is a straightforward procedure that can be performed similarly on moment systems with more than 13 moments. A natural choice for the next system to consider is to add the moments w^2, u_{ij}^1 and u_{ijk}^0 to the list of variables, and then find the regularized closure conditions from even higher moment equations. This procedure would yield the R26 equations, and would need 45 moment equations to be derived.

The R13 equations have several advantages above the Burnett and super-Burnett equations: (a) They can be derived much easier and faster, so that errors can be excluded with higher certainty. (b) The R13 equations contain only space derivatives of first and second order while the super-Burnett equations contain derivatives of up to fourth order. Thus, the R13 equations fit more conveniently to existing analytical and numerical methods. Note that their mathematical structure is very similar to the NSF equations, so that methods for these can be used as well for solving the R13 equations. (c) Most important, however, is the fact that the R13 equations are linearly stable as was shown in [59] (see Sec. 10.1), while the Burnett and super-Burnett equations are linearly unstable [56], and that they give smooth shock structures.

We finish the discussion with a remark on the non-linear contributions to the production terms. If these are included, the equations for the deviations m, R_{ij}, Δ (7.5) change to

$$\frac{D\Delta}{Dt} + \{\cdots \text{space derivatives} \cdots\} = -\frac{1}{\epsilon}\frac{2}{3}\frac{p}{\mu}\left(\Delta + \frac{\sigma_{ij}\sigma_{ij}}{\rho}\right),$$

$$\frac{DR_{ij}}{Dt} + \{\cdots \text{space derivatives} \cdots\} = -\frac{1}{\epsilon}\frac{7}{6}\frac{p}{\mu}\left(R_{ij} + \frac{4}{7}\frac{1}{\rho}\sigma_{k\langle i}\sigma_{j\rangle k}\right),$$

$$\frac{Dm_{ijk}}{Dt} + \{\cdots \text{space derivatives} \cdots\} = -\frac{1}{\epsilon}\frac{3}{2}\frac{p}{\mu}m_{ijk} \ .$$

The expansion for small ϵ will then give non-zero values for the zeroth iteration,

$$\Delta^{(0)} = -\frac{\sigma_{ij}\sigma_{ij}}{\rho} \ , \quad R_{ij}^{(0)} = -\frac{4}{7}\frac{1}{\rho}\sigma_{k\langle i}\sigma_{j\rangle k} \ , \quad m_{ijk}^{(0)} = 0 \ .$$

These values are not related to the Grad distribution (which can only give linear expressions in σ_{ij}), and thus do not describe states on the Grad pseudo-equilibrium manifold. We have to conclude that the use of the Grad distribution, and the above regularization method, is restricted to processes where the non-linear terms in the production terms are not important. Note that their neglect leads to changes in the super-Burnett equations, as was discussed in Sec. 4.3.3.

The derivation of the R13 equations in the next chapter will incorporate the non-linear contributions in a straightforward manner.

7.7 Jin-Slemrod regularization

An alternative regularization was proposed by Jin and Slemrod [108][109], who set out to construct a set of equations that (a) gives the Burnett equations in a second order CE expansion, and (b) gives a positive entropy production for all values of the variables. Their equations share some similarity with the R13 equations.

7.7.1 Basic equations

Written in our notation, the equations of Jin and Slemrod are

$$\frac{D\sigma_{ij}}{Dt} - 2\frac{\partial v_k}{\partial x_{\langle i}}\sigma_{j\rangle k} = -\frac{2}{\varpi_2}\frac{p}{\mu}\left(\sigma_{ij} + 2\mu\frac{\partial v_{\langle i}}{\partial x_{j\rangle}} - P_{ij|2} - P_{ij|3}\right) \ ,$$

$$\frac{Dq_i}{Dt} - q_k\frac{\partial v_k}{\partial x_i} = -\frac{p\frac{3}{2}M}{\mu}\frac{}{\theta_2}\left(q_i + \frac{3}{2}M\mu\frac{\partial\theta}{\partial x_i} - q_{i|2} - q_{i|3}\right) \ , \quad (7.15)$$

with

$$P_{ij|2} = -\frac{\varpi_1}{2}\frac{\mu}{p}\frac{\partial v_k}{\partial x_k}\sigma_{ij} + \frac{\varpi_2}{2}\frac{\mu'}{p}\frac{D\theta}{Dt}\sigma_{ij} - \frac{\varpi_3}{\frac{3}{2}M}\frac{\mu^2}{p}\frac{\partial\frac{q_{\langle i}}{\mu}}{\partial x_{j\rangle}}$$
$$- \frac{\varpi_4}{\frac{3}{2}M}\frac{\mu}{p}q_{\langle i}\frac{\partial\ln p}{\partial x_{j\rangle}} - \frac{\mu}{p}\frac{\varpi_5}{\frac{3}{2}M}q_{\langle i}\frac{\partial\ln\theta}{\partial x_{j\rangle}} - \frac{\varpi_6}{2}\frac{\mu}{p}S_{k\langle i}\sigma_{j\rangle k} \ ,$$

$$P_{ij|3} = \frac{\mu^2}{p^2} \left[\hat{\omega}_2 S_{kl} S_{kl} + \hat{\omega}_3 \frac{1}{\theta} \frac{\partial \theta}{\partial x_k} \frac{\partial \theta}{\partial x_k} \right] \sigma_{ij} + \hat{\gamma}_1 \frac{\mu}{p} \frac{1}{\theta} \left(\frac{D\theta}{Dt} + \frac{2}{3} \theta \frac{\partial v_k}{\partial x_k} \right) \sigma_{ij}$$

$$+ \hat{\omega}_4 \frac{\partial}{\partial x_k} \left[\frac{\mu^3}{M\rho^2} \frac{\partial}{\partial x_k} \left(\frac{1}{2\mu\theta} \sigma_{ij} \right) \right] , \quad (7.16)$$

and

$$q_{i|2} = -\frac{\theta_1}{\frac{3}{2}M} \frac{\mu}{p} \frac{\partial v_k}{\partial x_k} q_i + \frac{\theta_2}{\frac{3}{2}M} \frac{\mu'}{p} \frac{D\theta}{Dt} q_i - \frac{\theta_3}{2} \frac{\mu}{\rho} \sigma_{ik} \frac{\partial \ln p}{\partial x_k}$$

$$- \frac{\theta_4}{2} \frac{\mu^2}{\rho} \frac{\partial \sigma_{ik} / (\mu)}{\partial x_k} - 3 \frac{\theta_5}{2} \frac{\mu}{\rho} \sigma_{ik} \frac{\partial \ln \theta}{\partial x_k} ,$$

$$q_{i|3} = \frac{\mu^2}{p^2} \left(\hat{\theta}_2 S_{kl} S_{kl} + \hat{\theta}_3 \frac{1}{\theta} \frac{\partial \theta}{\partial x_k} \frac{\partial \theta}{\partial x_k} \right) q_i + \frac{\hat{\lambda}_1}{\frac{3}{2}M} \frac{\mu}{p} \frac{1}{\theta} \left(\frac{D\theta}{Dt} + \frac{2}{3} \theta \frac{\partial v_k}{\partial x_k} \right) q_i$$

$$+ \frac{\hat{\theta}_4}{\frac{3}{2}M} \frac{\partial}{\partial x_k} \left(\frac{\mu^3 \theta}{\rho^2} \frac{\partial}{\partial x_k} \left(\frac{q_i}{\mu\theta^2} \right) \right) . \quad (7.17)$$

Here, ϖ_α and θ_α are the standard Burnett coefficients, $M = 5/(3\,\mathrm{Pr})$ is the Maxwell number, $\hat{\omega}_\alpha$, $\hat{\gamma}_1$, $\hat{\theta}_\alpha$, $\hat{\lambda}_1$ are additional coefficients that can be used for fitting, and $\mu' = \frac{d\mu}{d\theta}$.

These equations were apparently obtained from the super-Burnett expressions for stress and heat flux, $\sigma_{ij} = -2\mu S_{ij} + \sigma_{ij}^{(2)} + \sigma_{ij}^{(3)}$, $q_i = -q_i = -\frac{5}{2} \frac{\mu}{\mathrm{Pr}} \frac{\partial \theta}{\partial x_i} + q_i^{(2)} + q_i^{(3)}$ by the following modifications: (a) $\sigma_{ij}^{(2)}$ and $q_i^{(2)}$ are taken with the full time derivative, i.e. (4.47, 4.48). (b) In several, but not all terms the gradients of velocity and temperature are replaced by stress and heat flux according to the NSF laws, $S_{ij} \to -\frac{\sigma_{ij}}{2\mu}$ and $\frac{\partial \theta}{\partial x_k} \to -\frac{q_k}{\frac{3}{2}M\mu}$. (c) $\sigma_{ij}^{(3)}$ and $q_i^{(3)}$ are constructed to obtain a positive production of entropy.

7.7.2 Comparison with R13 equations

The structure of the Jin-Slemrod equations is very similar to that of the R13 equations, as can be seen best by rewriting the latter as

$$\frac{D\sigma_{ij}}{Dt} - 2\frac{\partial v_k}{\partial x_{\langle i}} \sigma_{j\rangle k} = -\frac{p}{\mu} \left(\sigma_{ij} + 2\mu S_{ij} - P_{ij|\mathrm{Grad}} - P_{ij|\mathrm{R13}} \right) , \quad (7.18)$$

$$P_{ij|\mathrm{Grad}} = -\frac{7}{3} \frac{\mu}{p} \frac{\partial v_k}{\partial x_k} \sigma_{ij} - 4\frac{\mu}{p} S_{k\langle i} \sigma_{j\rangle k} - \frac{4}{5} \frac{\mu}{p} \frac{\partial q_{\langle i}}{\partial x_{j\rangle}} ,$$

$$P_{ij|\mathrm{R13}} = -\frac{\mu}{p} \frac{\partial m_{ijk}}{\partial x_k} ,$$

and

$$\frac{Dq_i}{Dt} - q_k \frac{\partial v_k}{\partial x_i} = -\frac{2}{3}\frac{p}{\mu}\left(q_i + \frac{15}{4}\mu\frac{\partial \theta}{\partial x_i} - q_{i|Grad} - q_{i|R13}\right), \tag{7.19}$$

$$q_{i|Grad} = -\frac{7}{2}\frac{\mu}{p}q_i\frac{\partial v_r}{\partial x_r} + \frac{3}{2}\frac{\mu}{p}\theta\sigma_{ik}\frac{\partial \ln \rho}{\partial x_k} - \frac{15}{4}\frac{\mu}{\rho}\sigma_{ik}\frac{\partial \ln \theta}{\partial x_k} - \frac{3}{2}\frac{\mu}{p}\theta\frac{\partial \sigma_{ik}}{\partial x_k}$$
$$- \frac{21}{5}\frac{\mu}{p}q_k S_{ik} + \frac{3}{2}\frac{\mu}{p}\frac{\sigma_{ij}}{\rho}\frac{\partial \sigma_{jk}}{\partial x_k},$$

$$q_{i|R13} = -\frac{3}{4}\frac{\mu}{p}\frac{\partial R_{ik}}{\partial x_k} - \frac{1}{4}\frac{\mu}{p}\frac{\partial \Delta}{\partial x_i} - \frac{3}{2}\frac{\mu}{p}m_{ijk}\frac{\partial v_j}{\partial x_k}.$$

The R13 equations were derived for the special case of Maxwell molecules, and in order to compare with the Jin-Slemrod equations, we consider these for Maxwell molecules as well. Then, the Burnett coefficients are given by (4.29), and $M = \frac{5}{2}$. Moreover, we use the energy balance to replace the time derivative $\frac{D\theta}{Dt}$ to obtain

$$P_{ij|2} = -\frac{7}{3}\frac{\mu}{p}\frac{\partial v_k}{\partial x_k}\sigma_{ij} - 4\frac{\mu}{p}S_{k\langle i}\sigma_{j\rangle k} - \frac{4}{5}\frac{\mu}{p}\frac{\partial q_{\langle i}}{\partial x_{j\rangle}} - \frac{2}{3}\frac{\mu}{p}\frac{\sigma_{ij}\sigma_{kl}}{p}\frac{\partial v_k}{\partial x_l} - \frac{2}{3}\frac{\mu}{p}\frac{\sigma_{ij}}{p}\frac{\partial q_k}{\partial x_k} \tag{7.20}$$

$$q_{i|2} = -\frac{7}{2}\frac{\mu}{p}q_i\frac{\partial v_k}{\partial x_k} + \frac{3}{2}\frac{\mu}{\rho}\sigma_{ik}\frac{\partial \ln \rho}{\partial x_k} - \frac{15}{4}\frac{\mu}{\rho}\sigma_{ik}\frac{\partial \ln \theta}{\partial x_k} - \frac{3}{2}\frac{\mu}{\rho}\frac{\partial \sigma_{ik}}{\partial x_k}$$
$$- \frac{63}{8}\frac{\mu}{\rho}\sigma_{ik}\frac{\partial \ln \theta}{\partial x_k} - \underline{\frac{\mu}{p^2}\sigma_{kl}q_i\frac{\partial v_k}{\partial x_l}} - \underline{\frac{\mu}{p^2}q_i\frac{\partial q_k}{\partial x_k}} \tag{7.21}$$

The equations for $P_{ij|3}$ and $q_{i|3}$ remain unchanged, since the information about molecules types is hidden in the unknown parameters $\hat{\omega}_\alpha$, $\hat{\gamma}_1$, $\hat{\theta}_\alpha$, $\hat{\lambda}_1$.

Since stress and heat flux have no zeroth order contributions ($\sigma_{ij}^{(0)} = q_i^{(0)} = 0$), they are of first order in μ (or the Knudsen number, in a dimensionless formulation), and thus the underlined terms in the second order corrections (7.18–7.21) are of *third* order. Moreover, with the NSF laws we find that $\frac{21}{5}\frac{\mu}{p}q_k S_{ik} \simeq -\frac{21}{5}\frac{\mu}{p}\frac{15}{4}\mu\frac{\partial \theta}{\partial x_k}S_{ik} \simeq \frac{63}{8}\frac{\mu}{p}\frac{\partial \theta}{\partial x_k}\sigma_{ik}$ if higher order terms are ignored. Therefore the second order contributions agree within terms of higher order,

$$P_{ij|2} = P_{ij|Grad} + \mathcal{O}\left(\mu^3\right) \quad \text{and} \quad q_{i|2} = q_{i|Grad} + \mathcal{O}\left(\mu^3\right).$$

This leads to the following conclusion: The Jin-Slemrod equations agree with the Grad 13 equations up to terms of second order in the case of Maxwell molecules, and offer an interesting extension of the Grad 13 moment equations to other molecule models.

Jin and Slemrod found their equation in an indirect manner, by first expanding the Boltzmann equation in a CE series, and then reversing the expansion intuitively by strategically replacing $S_{ij} \to -\frac{\sigma_{ij}}{2\mu}$ and $\frac{\partial \theta}{\partial x_k} \to -\frac{q_k}{\frac{3}{2}M\mu}$. The next chapter will present a more rigorous approach to derive a generalization of the 13 moment equations for arbitrary molecules models.

Next, we compare the higher order contributions $P_{ij|3}$, $q_{i|3}$ and $P_{ij|R13}$, $q_{i|R13}$. For this it is important to note that the equations (7.18, 7.19) follow directly from the Boltzmann equation by integration over $C_{\langle i}C_{j\rangle}$ and $C^2 C_i$, without further assumptions. Closure assumptions are required to close the system, and these affect only the expressions for Δ, R_{ik}, and m_{ijk}. Grad's traditional closure gives $\Delta = R_{ik} = m_{ijk} = 0$, and the regularization procedure gives the corrections (7.10–7.12).

But independently of the closure assumption, the principal structure of the equations follows from the Boltzmann equation, and thus is mandatory. In particular this implies that the third order correction for stress is the divergence of a trace-free tensor, $\frac{p}{\mu}P_{ij|R13} = -\frac{\partial m_{ijk}}{\partial x_k}$. This is, of course, the case for the R13 equations, where m_{ijk} is given by (7.12), but not for the Jin-Slemrod correction (7.16)$_2$, which has a tensorial structure in disagreement to the Boltzmann equation. This becomes visible in the linear case, where

$$P_{ij|R13}^{lin} = \frac{\mu^2}{\rho p}\left(\frac{2}{3}\frac{\partial^2\sigma_{ij}}{\partial x_k\partial x_k} + \frac{4}{5}\frac{\partial^2\sigma_{k\langle i}}{\partial x_{j\rangle}\partial x_k}\right) \quad,\quad P_{ij|3}^{lin} = \frac{\hat\omega_4}{5}\frac{\mu^2}{\rho p}\frac{\partial^2\sigma_{ij}}{\partial x_k\partial x_k}$$

and

$$q_{i|R13}^{lin} = \frac{9}{5}\frac{\mu^2}{\rho p}\left(\frac{\partial^2 q_i}{\partial x_k\partial x_k} + 2\frac{\partial q_k}{\partial x_i\partial x_k}\right) \quad,\quad q_{i|3}^{lin} = \frac{\hat\theta_4}{\frac{3}{2}M}\frac{\mu^2}{\rho p}\frac{\partial^2 q_i}{\partial x_k\partial x_k}\;.$$

The corrections $P_{ij|3}$ and $q_{i|3}$ are of super-Burnett order, and one might ask whether the super-Burnett equations can be extracted from the Jin-Slemrod equations by means of a third order CE expansion. The R13 equations do contain the super-Burnett equations, and the difference between the third order corrections $P_{ij|3}$, $P_{ij|R13}q_{i|3}$ and $q_{i|3}$, $q_{i|R13}$ in the linear case already shows that the super-Burnett equations cannot be contained in the Jin-Slemrod equations.

A comparison of the one-dimensional equations between both theories allows us to identify the unknown coefficients of Jin and Slemrod as

$$\hat\omega_4 = 6 \quad,\quad \hat\theta_4 = \frac{81}{4}\;.$$

For the comparison of the R13 and the Jin-Slemrod equations we note: (a) The R13 equations are accurate to super-Burnett order while the regularized Burnett equations are accurate only to Burnett order. (b) No unknown coefficients appear in the R13 equations, while the Jin-Slemrod equations contain 8 unknown parameters. (c) The Jin-Slemrod equations were specifically constructed to give an entropy law with non-negative production terms. No entropy law is currently known for the R13 equations. (d) Both systems are stable. (e) Both systems give smooth shock structures, but the shock structures from the R13 equations fit experiments better (Chapter 11). (f) The R13 equations are restricted to Maxwellian molecules, while the Jin-Slemrod equations are available for arbitrary interaction potentials.

8
Order of magnitude approach

This chapter introduces a method for accounting for the order of magnitude of moments and terms in moment equations, in order to derive transport equations up to third order in the Knudsen number [74][110]. For Maxwell molecules this procedure yields the Euler and NSF equations in zeroth and first order, Grad's 13 moment equations (minus a non-linear term) in second order, and a variation of the R13 equations in third order. For non-Maxwellian molecules the method is developed to second order only, where it gives a generalization to Grad's 13 moment equations.

Since the method is based on accounting orders in the Knudsen number, it provides a direct link between Grad type equations, and the Knudsen number. Moreover, it reproduces the zeroth and first order results—Euler and NSF—of the Chapman-Enskog method, but not the unstable higher order Chapman-Enskog results, that is the Burnett and super-Burnett equations. Thus, the new method provides a common umbrella for sets of equations that up to now were thought to stem from very different arguments. Moreover, all sets of equations derived are stable for disturbances of all wavelengths and frequencies.

8.1 Introduction

The first attempt to derive Grad equations by means of arguments on the Knudsen number is due to Müller, Reitebuch and Weiss, termed as "Consistently Ordered Extended Thermodynamics" (COET) [111]. They considered the infinite system of coupled moment equations of the BGK equation [29]. The order of magnitude approach shares some similarity with the COET method, but is distinctly different in detail.

As does COET, the method is based on considering not the Boltzmann equation, but the infinite system of moment equations, that was discussed already in the last chapters. In particular we have seen that the classical

Chapman-Enskog expansion can be performed either on the Boltzmann equation, or on the moment equations. The method discussed now was developed so far only for the moment equations, and it is not clear whether it can be formulated for the Boltzmann equation directly.

The method of finding the proper equations with *order of accuracy* λ_0 in the Knudsen number consists of the following three steps:

1. Determination of the *order of magnitude* λ of the moments.
2. Construction of a moment set with minimum number of moments at order λ.
3. Deletion of all terms in all equations that would lead only to contributions of orders $\lambda > \lambda_0$ in the conservation laws for energy and momentum.

Step 1 is based on a Chapman-Enskog expansion where a moment ϕ is expanded according to

$$\phi = \phi_0 + \varepsilon\phi_1 + \varepsilon^2\phi_2 + \varepsilon^3\phi_3 + \cdots,$$

and the leading order of ϕ is determined by inserting this ansatz into the complete set of moment equations. A moment is said to be of leading order λ if $\phi_\beta = 0$ for all $\beta < \lambda$. It must be emphasized that only the leading order of the moments is of interest, and that the coefficients ϕ_β in the expansion will not be determined. Indeed, the latter is the approach of the Chapman-Enskog method, which aims at computing the coefficients ϕ_β in terms of gradients of mass density, velocity and temperature. This first step agrees with the ideas of COET [111], where, however, the authors do not perform a Chapman-Enskog expansion, but a Maxwellian iteration [20] (see Sec. 6.3.4).

In Step 2, new variables are introduced by linear combination of the moments originally chosen. The new variables are constructed such that the number of moments at a given order λ is minimal. This step does not only simplify the later discussion, but gives an unambiguous set of moments at order λ. This ensures that the final result will be independent of the initial choice of moments.

Step 3 follows from the definition of the order of accuracy λ_0: A set of equations is said to be accurate of order λ_0, when the pressure deviator σ_{ij} and the heat flux q_i are known within the order $\mathcal{O}\left(\varepsilon^{\lambda_0}\right)$.

The evaluation of this condition is based on the fact that all moment equations are strongly coupled. This implies that each term in any of the moment equations has some influence on all other equations, in particular on the conservation laws. The influence of each term can be weighted by powers in the Knudsen number, and is related, but not equal to, the order of magnitude of the moments appearing in the term. A theory of order λ_0 will consider only those terms in all equations whose leading order of influence in the conservation laws is $\lambda \leq \lambda_0$. Luckily, in order to evaluate this condition, it suffices to start with the conservation laws, and step by step, order by order, add the relevant terms that are required: We start with the $\mathcal{O}\left(\varepsilon^0\right)$ equations

(Euler), then add the relevant terms to obtain the $\mathcal{O}\left(\varepsilon^1\right)$ equations (NSF), and so on.

The accounting for the order of accuracy is the main difference between the order of magnitude approach and COET, which assumes that *all* terms in *all* moment equations that are of leading order $\lambda \leq \lambda_0$ or smaller must be retained. This is quite different, e.g. in order to compute the heat flux with third order accuracy, as is necessary in a third order theory, the present method requires some moments only with second order accuracy, while others can be ignored completely. COET, on the other hand, would require higher order accuracy for these moments, and a larger number of moments. The order of magnitude approach leads to smaller systems of equations for a given order, and can be performed for the full three dimensional and time dependent equations, while [111] presents the equations only for one-dimensional steady state processes.

The basic equations for the procedure are the moment equations (5.10–5.17) as introduced in Chapter 5, with the production terms computed by means of the Grad distribution in Sec. 6.5, where the production terms are scaled by the scaling parameter ε.

The equations are not printed again in full, but in an abbreviated fashion,[1]

$$\frac{Dw^a}{Dt} + \cdots = -\frac{1}{\varepsilon}\left(\sum_b C_{ab}^{(0)}\frac{w^b}{\tau\theta^{b-a}} + \sum_{r,b,c} y_{a,bc}^{0,r,0}\frac{\bar{u}_{j_1\cdots j_r}^b\bar{u}_{j_1\cdots j_r}^c}{\tau\rho\theta^{b+c+r-a}}\right)$$

$$\frac{Du_i^a}{Dt} + \cdots = -\frac{1}{\varepsilon}\left(\sum_b C_{ab}^{(1)}\frac{u_i^b}{\tau\theta^{b-a}} + \sum_{r,b,c} y_{a,bc}^{1,r,1}\frac{\bar{u}_{ij_1\cdots j_r}^b\bar{u}_{j_1\cdots j_r}^c}{\tau\rho\theta^{b+c+r-a}}\right)$$

$$\frac{Du_{ij}^a}{Dt} + \cdots = -\frac{1}{\varepsilon}\left(-\sum_b C_{ab}^{(2)}\frac{u_{ij}^b}{\tau\theta^{b-a}} - \sum_{r,b,c} y_{a,bc}^{2,r,2}\frac{\bar{u}_{k_1\cdots k_r ij}^b\bar{u}_{k_1\cdots k_r}^c}{\tau\rho\theta^{b+c+r-a}}\right.$$
$$\left. - \sum_{r,b,c} y_{a,bc}^{2,r,1}\frac{\bar{u}_{k_1\cdots k_r}^b\langle_i\bar{u}_{j\rangle k_1\cdots k_r}^c}{\tau\rho\theta^{b+c+r-a}}\right)$$

$$\frac{Du_{i_1\cdots i_n}^a}{Dt}\cdots = -\frac{1}{\varepsilon}\left(\sum_b C_{ab}^{(n)}\frac{\bar{u}_{i_1\cdots i_n}^b}{\tau\theta^{b-a}}\right.$$
$$\left. + \sum_{r,b,c} y_{a,bc}^{n,r,m}\frac{\bar{u}_{j_1\cdots j_r\langle i_1\cdots i_m}^b\bar{u}_{i_{m+1}i_{m+2}\cdots i_n\rangle j_1\cdots j_r}^c}{\tau\rho\theta^{b+c+r-a}}\right)$$

$$(8.1)$$

For filling in the dots, the reader is referred to (5.14-5.17).

---

[1] The conservation laws are not shown again.

8.2 The order of magnitude of moments

We shall now assign orders of magnitude to the moments, and then construct new sets of moments, such that at each order of magnitude we have the minimal number of variables.

The discussion is based on a Chapman-Enskog expansion of the moments, with ε as smallness parameter. All moments are expanded according to

$$u^a_{i_1 \cdots i_n} = \sum_{\beta=0} \varepsilon^\beta u^a_{i_1 \cdots i_n | \beta} = \varepsilon^0 u^a_{i_1 \cdots i_n | 0} + \varepsilon^1 u^a_{i_1 \cdots i_n | 1} + \varepsilon^2 u^a_{i_1 \cdots i_n | 2} + \varepsilon^3 u^a_{i_1 \cdots i_n | 3} + \cdots$$

(8.2)

and a similar series for the w^a.

We shall say that $u^a_{i_1 \cdots i_n}$ is of leading order λ if $u^a_{i_1 \cdots i_n | \beta} = 0$ for all $\beta < \lambda$. Note that we are not interested in computing the expansion coefficients $u^a_{i_1 \cdots i_n | \beta}$, but only in finding the leading order of the moments.

8.2.1 Zeroth and first order expansion

For the evaluation of the order of magnitude, it is important to note that the production terms are multiplied by the factor $\frac{1}{\varepsilon}$. If the above expansion is inserted into the moment equations for the non-conserved quantities, it becomes immediately clear that

$$w^a_{|0} = u^a_{i_1 \cdots i_n | 0} = 0$$

for all moments. This follows from balancing the factors of $\frac{1}{\varepsilon}$ on both sides of the equations—there are none of these on the left, and thus the above result. In other words, all quantities that are not conserved are at least of first order.

In the next step, the factors of ε^0 in the equations are balanced, to find

$$0 = -\sum_b C^{(0)}_{ab} \frac{w^b_{|1}}{\tau \theta^{b-a}} \,,$$

$$\frac{a(2a+3)!!}{3} \rho \theta^a \frac{\partial \theta}{\partial x_i} = -\sum_b C^{(1)}_{ab} \frac{u^b_{i|1}}{\tau \theta^{b-a}} \,,$$

$$\frac{2}{15}(2a+5)!! \rho \theta^{a+1} \frac{\partial v_{\langle i}}{\partial x_{j\rangle}} = -\sum_b C^{(2)}_{ab} \frac{u^b_{ij|1}}{\tau \theta^{b-a}} \,,$$

$$0 = -\sum_b C^{(n)}_{ab} \frac{u^b_{i_1 \cdots i_n | 1}}{\tau \theta^{b-a}} \,.$$

(8.3)

Note that all quadratic terms in the production terms are at least of second order, since all moments appearing in them are at least of first order, so that no quadratic terms appear in (8.3).

It follows that the leading order of vectors and rank-2 tensors, u^a_i and u^a_{ij}, is the first order, while the non-equilibrium parts of the scalar moments w^a, and the higher moments $u^a_{i_1 \cdots i_n}$ ($n \geq 3$) are at least of second order.

8.2.2 Second order

Next, we have a look at the second order quantities. Since the vectors and rank-2 tensors are already known to be of first order in ε, only the other moments must be considered. We make the equations for the scalars w^a and tensors of rank 3 and 4 explicit. Keeping only factors of order ε^1 in the equations (note that e.g. u_i^a is a $\mathcal{O}(\varepsilon)$ contribution) we find

$$
-\frac{2a}{3}(2a+1)!!\theta^{a-1}\frac{\partial q_k}{\partial x_k} - \frac{2a}{3}(2a+1)!!\theta^{a-1}\sigma_{kl}\frac{\partial v_k}{\partial x_l} + 2au_{kl}^{a-1}\frac{\partial v_k}{\partial x_l}
$$

$$
- 2au_k^{a-1}\frac{\partial \theta}{\partial x_k} - 2au_k^{a-1}\theta\frac{\partial \ln \rho}{\partial x_k} + \frac{\partial u_k^a}{\partial x_k}
$$

$$
= -\sum_b \mathcal{C}_{ab}^{(0)}\frac{w_{|2}^b}{\tau\theta^{b-a}} - \sum_{b,c} \mathcal{Y}_{a,bc}^{0,1,0}\frac{u_j^b u_j^c}{\tau\rho\theta^{b+c+1-a}} - \sum_{b,c} \mathcal{Y}_{a,bc}^{0,2,0}\frac{u_{jk}^b u_{jk}^c}{\tau\rho\theta^{b+c+2-a}} , \quad (8.4)
$$

so that the w^a are of second order.

For the moments of rank three and four it follows that

$$
-\frac{3}{7}(2a+7)u_{\langle ij}^a\left[\theta\frac{\partial \ln \rho}{\partial x_{k\rangle}} + \frac{\partial \theta}{\partial x_{k\rangle}}\right] + \frac{3}{7}\frac{\partial u_{\langle ij}^{a+1}}{\partial x_{k\rangle}} + \frac{6}{35}(2a+7)u_{\langle i}^{a+1}\frac{\partial v_j}{\partial x_{k\rangle}}
$$

$$
= -\sum_b \mathcal{C}_{ab}^{(3)}\frac{u_{ijk|2}^b}{\tau\theta^{b-a}} - \sum_{b,c} \mathcal{Y}_{a,bc}^{3,0,2}\frac{u_{\langle ij}^b u_{k\rangle}^c}{\tau\rho\theta^{b+c-a}} , \quad (8.5)
$$

$$
\frac{12}{63}(2a+9)u_{\langle ij}^{a+1}\frac{\partial v_k}{\partial x_{l\rangle}} = -\sum_b \mathcal{C}_{ab}^{(4)}\frac{u_{ijkl|2}^b}{\tau\theta^{b-a}} - \sum_{b,c} \mathcal{Y}_{a,bc}^{4,0,2}\frac{u_{\langle ij}^b u_{kl\rangle}^c}{\tau\rho\theta^{b+c-a}}
$$

so that the u_{ijk}^b and the u_{ijkl}^b are also of second order. For all higher moments one finds

$$
0 = -\sum_b \mathcal{C}_{ab}^{(n)}\frac{u_{i_1\cdots i_n|2}^b}{\tau\theta^{b-a}} , \quad n \geq 5 .
$$

It follows that the non-equilibrium parts of the scalar moments w^a, and the tensors of rank 3 and 4 are second order quantities. All higher moments are *at least* of third order. We will not go further, but it can be expected that tensors of rank 5 and 6 are of third order, tensors of rank 7 and 8 are of fourth order, etc.

8.2.3 Minimal number of moments of order $\mathcal{O}(\varepsilon)$

From the first order result for the scalar and 2-tensors, (8.3), the first order terms $u_{i|1}^b$ and $u_{ij|1}^b$ are related to the gradients of temperature and velocity, respectively, and thus they are linearly dependent,

$$u^b_{i|1} = -\kappa_b \tau \rho \theta^b \frac{\partial \theta}{\partial x_i} \quad , \quad u^b_{ij|1} = -\mu_b \tau \rho \theta^{1+b} \frac{\partial v_{\langle i}}{\partial x_{j\rangle}} \ . \tag{8.6}$$

κ_b and μ_b are pure numbers, given by

$$\kappa_b = \sum_{a=1} \left[C^{(1)}_{ba} \right]^{-1} \frac{a(2a+3)!!}{3} \quad , \quad \mu_b = \sum_{a=0} \left[C^{(2)}_{ba} \right]^{-1} \frac{2}{15}(2a+5)!! \ . \tag{8.7}$$

The first few values of these coefficients for Maxwell molecules (MM) and the BGK model can be computed from the matrices $C^{(n)}_{ab}$ in Sec. 5.3 as

$$\text{MM:} \quad \kappa_1 = 15/2 \ , \ \kappa_2 = 105 \quad \text{and} \quad \mu_0 = 2 \ , \ \mu_1 = 14 \ , \tag{8.8}$$
$$\text{BGK:} \quad \kappa_1 = 5 \quad \ \ , \ \kappa_2 = 70 \quad \text{and} \quad \mu_0 = 2 \ , \ \mu_1 = 14 \ . \tag{8.9}$$

In particular, pressure deviator and heat flux are given to first order as

$$q_{i|1} = \frac{1}{2}u^1_{i|1} = -\frac{1}{2}\kappa_1 \tau \rho \theta \frac{\partial \theta}{\partial x_i} = -\kappa \frac{\partial \theta}{\partial x_i} \ ,$$
$$\sigma_{ij|1} = u^0_{ij|1} = -\mu_0 \tau \rho \theta \frac{\partial v_{\langle i}}{\partial x_{j\rangle}} = -2\mu \frac{\partial v_{\langle i}}{\partial x_{j\rangle}} \ . \tag{8.10}$$

where heat conductivity κ and viscosity μ were introduced. Thus, in first order the methods gives the the laws of Fourier and Navier-Stokes with explicit expressions for viscosity μ and heat conductivity κ. Note that the computation of μ and κ involves the inverses of the matrices $C^{(1)}_{ab}$, $C^{(2)}_{ab}$. Indeed, this is the result for viscosity and heat conductivity according to the Reinecke-Kremer-Grad method, see [54][93] and Sec. 6.4.

It follows from (8.6, 8.10) that

$$u^b_{i|1} = \frac{2\kappa_b}{\kappa_1}\theta^{b-1}q_{i|1} \ , \quad u^b_{ij|1} = \frac{\mu_b}{\mu_0}\theta^b \sigma_{ij|1} \ .$$

While these equations relate the first order contributions of the vector and 2-tensor moments, it is straightforward to introduce new moments w^a_i, w^a_{ij} that are of second order only,

$$w^a_i = u^a_i - \frac{2\kappa_a}{\kappa_1}\theta^{a-1}q_i \ (a \geq 2) \quad \text{and} \quad w^a_{ij} = u^a_{ij} - \frac{\mu_a}{\mu_0}\theta^a \sigma_{ij} \ (a \geq 1) \ , \tag{8.11}$$

so that

$$w^a_i = m \int \left(C^{2a-2} - \frac{2\kappa_a}{\kappa_1}\theta^{a-1} \right) C^2 C_i f d\mathbf{c} \ (a \geq 2) \ ,$$
$$w^a_{ij} = m \int \left(C^{2a-2} - \frac{\mu_a}{\mu_0}\theta^a \right) C_{\langle i}C_{j\rangle} f d\mathbf{c} \ (a \geq 1) \ .$$

This means that one can formulate a set of moments where only σ_{ij} and q_i are of first order, while all other moments are at least of second order (excluding the conserved moments, of course).

It is in principle possible to go to higher order with this: the second order terms of w_i^a (say), when expanded, will be linearly dependent, and again one can use that to obtain a minimal set of moments of second order, while the remaining ones can be constructed to be of third order, etc. This is not necessary for the levels of accuracy considered here, and thus this idea will not be pursued further.

8.3 The transport equations with 2nd order accuracy

In the previous section the order of magnitude of the various moments was established up to $\mathcal{O}\left(\varepsilon^2\right)$. Now we ask what equations are required in order to describe a flow process in a rarefied ideal gas with an accuracy of $\mathcal{O}\left(\varepsilon^\lambda\right)$.

From now on, the smallness parameter ε is used in a slightly different manner, namely as an indicator for the leading order of a quantity. Thus, in any equation $u_{i_1\cdots i_n}^a$ is replaced by $\varepsilon^\beta u_{i_1\cdots i_n}^a$ when β denotes the leading order of $u_{i_1\cdots i_n}^a$. This allows for a proper bookkeeping of the order of magnitude of all terms in an equation.

8.3.1 The conservation laws and the definition of $\lambda - th$ order accuracy

We start the argument by repeating the conservation laws for mass, momentum and energy, (5.10–5.12), which read, when the factor ε is assigned to the first order quantities σ_{ij} and q_i,

$$\frac{D\rho}{Dt} + \rho\frac{\partial v_k}{\partial x_k} = 0 \,,$$

$$\frac{3}{2}\rho\frac{D\theta}{Dt} + \rho\theta\frac{\partial v_k}{\partial x_k} + \varepsilon\left[\frac{\partial q_k}{\partial x_k} + \sigma_{kl}\frac{\partial v_k}{\partial x_l}\right] = 0 \,, \qquad (8.12)$$

$$\rho\frac{Dv_i}{Dt} + \rho\frac{\partial\theta}{\partial x_i} + \theta\frac{\partial\rho}{\partial x_i} + \varepsilon\left[\frac{\partial\sigma_{ik}}{\partial x_k}\right] = 0 \,.$$

These equations are not a closed set of equations for ρ, v_i, θ, but contain pressure deviator σ_{ij} and heat flux q_i as additional quantities, and equations for these are required to obtain a closed set of equations.

We shall speak of an theory of $\lambda - th$ order accuracy, when both, σ_{ij} and q_i, are known within the order $\mathcal{O}\left(\varepsilon^\lambda\right)$.

8.3.2 Zeroth order accuracy: Euler equations

The equations of order $\mathcal{O}\left(\varepsilon^0\right)$ result from (8.12) by setting $\sigma_{ij} = q_i = 0$, that is by ignoring the terms with the factor ε in the balance laws. This yields the Euler equations for ideal gases,

$$\frac{D\rho}{Dt} + \rho\frac{\partial v_k}{\partial x_k} = 0 \,, \quad \frac{3}{2}\rho\frac{D\theta}{Dt} + \rho\theta\frac{\partial v_k}{\partial x_k} = 0 \,, \quad \rho\frac{Dv_i}{Dt} + \theta\frac{\partial\rho}{\partial x_i} + \rho\frac{\partial\theta}{\partial x_i} = 0 \,.$$

8.3.3 Collision moments for vectors and tensors

For higher order accuracy, i.e. first order and higher, the moment equations for pressure deviator and heat flux must be considered together with the production terms for vectors and 2-tensors up to first order in ε.

For vectors one obtains from $(8.1)_2$, by making the first two terms of the r−summation explicit,

$$\frac{1}{\varepsilon}\mathcal{P}_i^a = -\frac{1}{\varepsilon}\sum_b C_{ab}^{(1)}\frac{\varepsilon u_i^b}{\tau\theta^{b-a}} - \frac{1}{\varepsilon}\sum_{b,c}\mathcal{Y}_{a,bc}^{1,0,1}\frac{\varepsilon u_i^b\varepsilon^2 w^c}{\tau\rho\theta^{b+c-a}}$$

$$-\frac{1}{\varepsilon}\sum_{b,c}\mathcal{Y}_{a,bc}^{1,1,1}\frac{\varepsilon u_{ij}^b\varepsilon u_j^c}{\tau\rho\theta^{b+c+1-a}} - \frac{1}{\varepsilon}\sum_{r=2}\sum_{b,c}\mathcal{Y}_{a,bc}^{1,r,1}\frac{\varepsilon^{2r-1}\bar{u}_{ij_1\cdots j_r}^b\,\bar{u}_{j_1\cdots j_r}^c}{\tau\rho\theta^{b+c+r-a}}\,.$$

Up to order $\mathcal{O}\left(\varepsilon\right)$, with (8.11), the above equation reduces to

$$\frac{1}{\varepsilon}\mathcal{P}_i^a = -\sum_b C_{ab}^{(1)}\frac{2\kappa_b}{\kappa_1}\frac{q_i}{\tau\theta^{1-a}} - \varepsilon\left[\sum_b C_{ab}^{(1)}\frac{w_i^b}{\tau\theta^{b-a}} + \sum_{b,c}\mathcal{Y}_{a,bc}^{1,1,1}\frac{2\mu_b\kappa_c}{\mu_0\kappa_1}\frac{\sigma_{ij}q_j}{\tau\rho\theta^{2-a}}\right]\,.$$

From $(8.1)_3$, the production term for 2-tensors is obtained in the same manner as

$$\frac{1}{\varepsilon}\mathcal{P}_{ij}^a = -\sum_b C_{ab}^{(2)}\frac{\mu_b}{\mu_0}\frac{\sigma_{ij}}{\tau\theta^{-a}}$$

$$-\varepsilon\left[\sum_b C_{ab}^{(2)}\frac{w_{ij}^b}{\tau\theta^{b-a}} + \sum_{b,c}\mathcal{Y}_{a,bc}^{2,0,1}\frac{4\kappa_b\kappa_c}{\kappa_1\kappa_1}\frac{q_{\langle i}q_{j\rangle}}{\tau\rho\theta^{2-a}} + \sum_{b,c}\mathcal{Y}_{a,bc}^{2,1,1}\frac{\mu_b\mu_c}{\mu_0\mu_0}\frac{\sigma_{k\langle i}\sigma_{j\rangle k}}{\tau\rho\theta^{1-a}}\right]\,.$$

8.3.4 Equations for pressure deviator and heat flux

Equation $(5.16)/(8.1)_3$ with $a = 0$ and the moments (8.11) reads, after the proper order of magnitude is assigned to the various terms,

$$\varepsilon\left[\frac{D\sigma_{ij}}{Dt} + \frac{4}{5}\frac{\partial q_{\langle i}}{\partial x_{j\rangle}} + 2\sigma_{k\langle i}\frac{\partial v_{j\rangle}}{\partial x_k} + \sigma_{ij}\frac{\partial v_k}{\partial x_k} + \sum_b C_{0b}^{(2)}\frac{w_{ik}^b}{\tau\theta^b}\right.$$

$$\left. + \sum_{b,c}\mathcal{Y}_{0,bc}^{2,0,1}\frac{4\kappa_b\kappa_c}{\kappa_1\kappa_1}\frac{q_{\langle i}q_{k\rangle}}{\tau\rho\theta^2} + \sum_{b,c}\mathcal{Y}_{0,bc}^{2,1,1}\frac{\mu_b\mu_c}{\mu_0\mu_0}\frac{\sigma_{j\langle i}\sigma_{k\rangle j}}{\tau\rho\theta}\right]$$

$$+ \varepsilon^2\left[\frac{\partial u_{ijk}^0}{\partial x_k} + \cdots\right] = -\rho\theta\left[\frac{\sigma_{ij}}{\mu} + 2\frac{\partial v_{\langle i}}{\partial x_{j\rangle}}\right]\,. \quad (8.13)$$

Here it was used that

$$\sum_{b=0} C_{0b}^{(2)} \mu_b = \sum_{a,b=0} C_{0b}^{(2)} \left[C_{ba}^{(2)}\right]^{-1} \frac{2}{15}(2a+5)!! = \sum_{a=0} \delta_{0a} \frac{2}{15}(2a+5)!! = 2$$

and

$$\frac{1}{\tau\mu_0} = \frac{\rho\theta}{2\mu}$$

where μ is the viscosity; the last two equations follow directly from (8.7) and (8.10).

The dots in the quadratic term (ε^2) stand for higher order contributions from the production terms which are not made explicit. No terms arise there for Maxwell molecules and BGK models.

The corresponding equation for the heat flux results from setting $a = 1$ in (5.15)/(8.1)$_2$ with the second order moments (8.11), and assigning the orders of magnitude, as (remember that $w^1 = 0$),

$$\varepsilon \left[\frac{Dq_i}{Dt} + \left[\frac{1}{2}\frac{\mu_1}{\mu_0} - 1\right]\sigma_{ik}\frac{\partial\theta}{\partial x_k} - \sigma_{ik}\theta\frac{\partial\ln\rho}{\partial x_k} + \left[\frac{1}{2}\frac{\mu_1}{\mu_0} - \frac{5}{2}\right]\theta\frac{\partial\sigma_{ik}}{\partial x_k}\right.$$

$$\left. + \frac{7}{5}q_i\frac{\partial v_k}{\partial x_k} + \frac{7}{5}q_k\frac{\partial v_i}{\partial x_k} + \frac{2}{5}q_k\frac{\partial v_k}{\partial x_i} + \frac{1}{2}\sum_b C_{1b}^{(1)}\frac{w_i^b}{\tau\theta^{b-1}} + \sum_{b,c} y_{1,bc}^{1,1,1}\frac{\mu_b\kappa_c}{\mu_0\kappa_1}\frac{\sigma_{ij}q_j}{\tau\rho\theta}\right]$$

$$+ \varepsilon^2\left[\frac{1}{2}\frac{\partial w_{ij}^1}{\partial x_k} + \frac{1}{6}\frac{\partial w^2}{\partial x_i} + u_{ikl}^0\frac{\partial v_k}{\partial x_l} - \frac{\sigma_{ik}}{\rho}\frac{\partial\sigma_{kl}}{\partial x_l} + \cdots\right] = -\frac{5}{2}\rho\theta\left[\frac{q_i}{\kappa} + \frac{\partial\theta}{\partial x_i}\right].$$

$$(8.14)$$

Here it was used that

$$\sum_{b=1} C_{1b}^{(1)}\kappa_b = \sum_{a,b=1} C_{1b}^{(1)}\left[C_{ba}^{(1)}\right]^{-1}\frac{a(2a+3)!!}{3} = \sum_{a=1}\delta_{1a}\frac{a(2a+3)!!}{3} = \frac{5!!}{3} = 5$$

and

$$\frac{1}{\tau\kappa_1} = \frac{1}{2}\frac{1}{\kappa}\rho\theta$$

where κ is the heat conductivity. The last two equations follow directly from (8.7) and (8.10).

The dots again refer to higher order contributions from the production terms; no such terms arise for Maxwell molecules and BGK models.

Note that μ and κ are $\mathcal{O}(\varepsilon)$, as are σ_{ij} and q_i, so that their respective ratios $\frac{\sigma_{ij}}{\mu}$, $\frac{q_i}{\kappa}$ are $\mathcal{O}(\varepsilon^0)$. Also τ is $\mathcal{O}(\varepsilon)$ and w_i^b, w_{ij}^b are $\mathcal{O}(\varepsilon^2)$ so that their respective ratio is $\mathcal{O}(\varepsilon)$.

8.3.5 First order accuracy: Navier-Stokes-Fourier equations

We recall the goal to provide the equations for pressure deviator σ_{ij} and heat flux q_i within an accuracy of a given order. For first order accuracy, only the leading terms in (8.13, 8.14)—those of $\mathcal{O}\left(\varepsilon^0\right)$—must be considered, which yields the laws of Navier-Stokes and Fourier,

$$\sigma_{ij} = -2\mu \frac{\partial v_{\langle i}}{\partial x_{j\rangle}} \quad , \quad q_i = -\kappa \frac{\partial \theta}{\partial x_i} \ ,$$

where viscosity μ and heat conductivity κ are given by (8.7, 8.10). The corresponding Prandtl number is $\mathrm{Pr} = \frac{5}{2}\frac{\mu_0}{\kappa_1}$.

The first order equations coincide with the first order Chapman-Enskog expansion. In first order the approach agrees with the Reinecke-Kremer-Grad method (Sec. 6.4) [54] where viscosity and heat conductivity are computed in accordance with the above formulae.

It is worthwhile to have a look at the corresponding phase density. At first order, the only relevant moments are the vectors and 2-tensors, so that the phase density (6.22, 6.23) reduces to

$$f_{|G} = f_M \left[1 + \sum_{a,b=0}^{A_1} \left[\mathcal{B}_{ab}^{(1)} \right]^{-1} \frac{u_i^b}{\rho \theta^{a+b+1}} C^{2a} C_i \right.$$

$$\left. + \sum_{a,b=0}^{A_2} \left[\mathcal{B}_{ab}^{(2)} \right]^{-1} \frac{u_{ij}^b}{2\rho \theta^{a+b+2}} C^{2a} C_{\langle i} C_{j\rangle} \right] \ .$$

But to first order the moments are given by (8.6, 8.10),

$$u_{i|1}^b = -2\kappa \frac{\kappa_b}{\kappa_1} \theta^b \frac{\partial \ln \theta}{\partial x_i} \quad , u_{ij|1}^b = -2\mu \frac{\mu_b}{\mu_0} \theta^b \frac{\partial v_{\langle i}}{\partial x_{j\rangle}} \ ,$$

so that the phase density for the NSF case can be written as

$$f_{|NSF} = f_M \left\{ 1 - \frac{2\kappa}{\rho \theta^{1/2}} \left[\sum_{a,b=0}^{A_1} \left[\mathcal{B}_{ab}^{(1)} \right]^{-1} \frac{\kappa_b}{\kappa_1} \frac{C^{2a}}{\theta^a} \right] \frac{C_i}{\sqrt{\theta}} \frac{\partial \ln \theta}{\partial x_i} \right.$$

$$\left. - \frac{\mu}{\rho \theta} \left[\sum_{a,b=0}^{A_2} \left[\mathcal{B}_{ab}^{(2)} \right]^{-1} \frac{\mu_b}{\mu_0} \frac{C^{2a}}{\theta^a} \right] \frac{C_{\langle i} C_{j\rangle}}{\theta} \frac{\partial v_{\langle i}}{\partial x_{j\rangle}} \right\} \ .$$

The terms in square brackets are polynomials in $\xi^2 = C^2/\theta$ and can be rewritten in terms of Sonine polynomials (4.40). Thus, it is evident that the present method yields the same phase density as the classical first order Chapman-Enskog method, see (6.33).

Heat conductivity κ and viscosity μ are given by

$$\kappa = \tau \rho \theta \sum_{a=1}^{A_1} \left[\mathcal{C}_{1a}^{(1)} \right]^{-1} \frac{a\,(2a+3)!!}{6} \quad , \quad \mu = \tau \rho \theta \sum_{a=0}^{A_2} \left[\mathcal{C}_{0a}^{(2)} \right]^{-1} \frac{(2a+5)!!}{15} \ ,$$

and thus depend on the numbers A_1, A_2 of moments considered. This gives a good criterion for the determination of these numbers: increasing A_1, A_2 gives successive approximations for κ, μ and these will converge to the accurate value if A_1, A_2 go to infinity. One will chose these numbers such that an increase by one will lead only to a sufficiently small change in the values for κ, μ. This corresponds to the growing accuracy when higher order Sonine polynomials are taken into account in the Chapman-Enskog method [2].

Indeed, Reinecke and Kremer computed κ, μ essentially in this manner [54], and showed that a fourth order correction is already very accurate. Moreover, they showed that their values for κ, μ to order $A = A_1 = A_2$ agree with the corresponding values of the $A - th$ degree of approximation in Sonine polynomials, as given, e.g., in [2].

The difference to the approach of Reinecke and Kremer lies in the fact that Reinecke and Kremer chose the moments to base their approach on intuitively (they consider w^a, u_i^a, and u_{ij}^a), while here the full moment system (w^a, $u_{i_1 \cdots i_n}^a$) was reduced by discussion of the order of magnitude of moments and the order of accuracy of equations. This makes clear that at first order the scalar moments w^a need not to be considered. Moreover, Reinecke and Kremer perform a Maxwell iteration on the full moment equations. For a second order theory, they perform the second order Maxwell iteration in [93], which yields the Burnett equations, just as in the standard Chapman-Enskog expansion. These, as is well known, are unstable [56] (Chapter 10) and therefore should not be used. The order of magnitude approach does not yield the Burnett equations at second order, as will become clear in the next section.

8.3.6 Second order accuracy: 13 moment theory

In the next order, all terms in (8.13, 8.14) must be considered that have the factors ε^1 and ε^0,

$$
\frac{D\sigma_{ij}}{Dt} + \frac{4}{5}\frac{\partial q_{\langle i}}{\partial x_{j\rangle}} + 2\sigma_{k\langle i}\frac{\partial v_{j\rangle}}{\partial x_k} + \sigma_{ij}\frac{\partial v_k}{\partial x_k} + \sum_b C_{0b}^{(2)}\frac{w_{ij}^b}{\tau\theta^b}
$$
$$
+ \sum_{b,c} \mathcal{Y}_{0,bc}^{2,0,1} \frac{4\kappa_b\kappa_c}{\kappa_1\kappa_1} \frac{q_{\langle i}q_{j\rangle}}{\tau\rho\theta^2} + \sum_{b,c} \mathcal{Y}_{0,bc}^{2,1,1} \frac{\mu_b\mu_c}{\mu_0\mu_0} \frac{\sigma_{k\langle i}\sigma_{j\rangle k}}{\tau\rho\theta} = -\rho\theta\left[\frac{\sigma_{ij}}{\mu} + 2\frac{\partial v_{\langle i}}{\partial x_{j\rangle}}\right] .
$$
$$\tag{8.16}$$

$$
\frac{Dq_i}{Dt} + \left[\frac{1}{2}\frac{\mu_1}{\mu_0} - 1\right]\sigma_{ik}\frac{\partial\theta}{\partial x_k} - \theta\sigma_{ik}\frac{\partial\ln\rho}{\partial x_k} + \left[\frac{1}{2}\frac{\mu_1}{\mu_0} - \frac{5}{2}\right]\theta\frac{\partial\sigma_{ik}}{\partial x_k} + \frac{7}{5}q_i\frac{\partial v_k}{\partial x_k}
$$
$$
+ \frac{7}{5}q_k\frac{\partial v_i}{\partial x_k} + \frac{2}{5}q_k\frac{\partial v_k}{\partial x_i} + \sum_b C_{1b}^{(1)}\frac{w_i^b}{2\tau\theta^{b-1}} + \sum_{b,c}\mathcal{Y}_{1,bc}^{1,1,1}\frac{\mu_b\kappa_c}{\mu_0\kappa_1}\frac{\sigma_{ik}q_k}{\tau\rho\theta}
$$
$$
= -\frac{5}{2}\rho\theta\left[\frac{q_i}{\kappa} + \frac{\partial\theta}{\partial x_i}\right] . \tag{8.17}
$$

Before we proceed, we compare the above set to Grad's 13 moment theory (6.13, 6.14) [8][9]: In Grad's equations, the underlined terms in (8.16, 8.17) do not appear. Moreover, the equations contain the general expression $\frac{\mu_1}{\mu_0}$ which depends on the interaction potential (through $C_{ab}^{(2)}$), while in Grad's equations this value is set to $\frac{\mu_1}{\mu_0} = 7$. Finally, Grad's equations contain a term $-\frac{\sigma_{ik}}{\rho}\frac{\partial \sigma_{kl}}{\partial x_l}$ in the equation for heat flux, which does not appear here, since it is of order $\mathcal{O}\left(\varepsilon^2\right)$.

Indeed, for BGK model and Maxwell molecules, where (8.8, 8.9) hold, one finds $\frac{\mu_1}{\mu_0} = 7$, i.e. the value in the Grad equations. In addition, the underlined terms do not appear for Maxwell molecules, nor the BGK model. Thus, with the omission of the second order term, $-\frac{\sigma_{ik}}{\rho}\frac{\partial \sigma_{kl}}{\partial x_l}$, Grad's 13 moment equations are the proper equations of second order accuracy for the description of rarefied gas flows *only* for Maxwell molecules and the BGK model.[2]

For other interaction models, however, the above equations must be used in order to give second order agreement with the Boltzmann equation. Together with the conservation laws (8.12), the equations (8.16, 8.17) form a set of 13 equations for the 13 variables $\rho, v_i, \theta, \sigma_{ij}, q_i$. This system is not closed yet, since it contains the moments w_i^a, w_{ij}^a as well, and additional equations are required for these quantities. The w_i^a, w_{ij}^a are of order $\mathcal{O}\left(\varepsilon^2\right)$, and for the theory of second order accuracy considered here, it will be sufficient to know only the leading terms for these quantities. The calculations and results for this can be found in Appendix A.6, see (A.5, A.6).

The final equations for the pressure deviator and heat flux read

$$
\frac{D\sigma_{ij}}{Dt} + \sigma_{ij}\frac{\partial v_k}{\partial x_k} + 2\sigma_{k\langle i}\frac{\partial v_{j\rangle}}{\partial x_k} + \psi_1\frac{\partial q_{\langle i}}{\partial x_{j\rangle}} + 2\psi_2\sigma_{k\langle i}S_{j\rangle k} - \psi_3 q_{\langle i}\frac{\partial \ln \rho}{\partial x_{j\rangle}}
$$
$$
- \psi_4 q_{\langle i}\frac{\partial \ln \theta}{\partial x_{j\rangle}} + \psi_5\frac{q_{\langle i}q_{j\rangle}}{\mu\theta} + \psi_6\frac{\sigma_{k\langle i}\sigma_{j\rangle k}}{\mu} = -\psi_7\rho\theta\left[\frac{\sigma_{ij}}{\mu} + 2\frac{\partial v_{\langle i}}{\partial x_{j\rangle}}\right], \quad (8.18)
$$

$$
\frac{Dq_i}{Dt} + q_k\frac{\partial v_i}{\partial x_k} + \frac{5}{3}q_i\frac{\partial v_k}{\partial x_k} + \chi_1\sigma_{ik}\frac{\partial \theta}{\partial x_k} - \chi_2\theta\sigma_{ik}\frac{\partial \ln \rho}{\partial x_k}
$$
$$
+ \chi_3\theta\frac{\partial \sigma_{ik}}{\partial x_k} + 2\chi_4 q_k S_{ik} + \chi_5\frac{\sigma_{ik}q_k}{\mu} = -\chi_6\rho\theta\left[\frac{\mu_0}{\kappa_1}\frac{q_i}{\mu} + \frac{\partial \theta}{\partial x_i}\right]. \quad (8.19)
$$

In the above equations, the coefficients ψ_α and χ_α are pure numbers that are defined in (A.7, A.8). The coefficients must be computed from the matrices

$$
C_{ad}^{(2)} \;,\; \mathcal{Y}_{a,bc}^{2,0,1} \;,\; \mathcal{Y}_{a,bc}^{2,1,1} \;,\; C_{ad}^{(1)} \;,\; \mathcal{Y}_{a,bc}^{1,1,1} \;.
$$

These follow by using the reduced Grad function (6.22)

$$
f_{|G} = f_M\left(1 + \Phi\right) = f_M\left[1 + \sum_{a=0}^{A_1}\lambda_i^a C^{2a}C_i + \sum_{a=0}^{A_2}\lambda_{kl}^a C^{2a}C_{\langle k}C_{l\rangle}\right]
$$

---

[2] This statement includes the ES-BGK model for $\Pr = 2/3$.

to compute the collision moments $\mathcal{P}_i^a$, $\mathcal{P}_{ij}^a$ as outlined in Sec. 6.5. The numbers A_1, A_2 must be chosen such that their further increase does not change the values of the coefficients ψ_α, χ_α considerably.

For Maxwell molecules and the BGK model, the coefficients have the values

$$\psi_1 = \frac{4}{5} \ , \quad \psi_2 = \psi_3 = \psi_4 = \psi_5 = \psi_6 = 0 \ , \quad \psi_7 = 1$$

$$\chi_1 = \frac{5}{2} \ , \quad \chi_2 = 1 \ , \quad \chi_3 = 1 \ , \quad \chi_4 = \frac{2}{5} \ , \quad \chi_5 = 0 \ , \quad \chi_6 = \frac{5}{2} \ .$$

Furthermore, $\frac{\mu_0}{\kappa_1} = \frac{2}{5} \Pr = \frac{4}{15}$ for Maxwell molecules. This choice of coefficients results in Grad's original 13 moment equations [8][9].

8.3.7 Burnett equations

The Burnett equations follow from (8.18, 8.19) by means of the Chapman-Enskog expansion, just as in Sec. 6.3, where we considered Grad's 13 moment system to find the Burnett equations for Maxwell molecules. The Burnett coefficients are related to the coefficients ψ_α, χ_α as

$$\varpi_1 = \frac{1}{\psi_7} \frac{4}{3} \left(\frac{7}{2} - \omega \right) \ , \quad \varpi_2 = \frac{2}{\psi_7} \ , \quad \varpi_3 = \frac{5}{2} \frac{1}{\Pr} \frac{\psi_1}{\psi_7} \ , \quad \varpi_4 = -\frac{5}{2} \frac{1}{\Pr} \frac{\psi_3}{\psi_7} \ ,$$

$$\varpi_5 = \frac{1}{\psi_7} \frac{5}{2} \frac{1}{\Pr} \left(\psi_1 \omega + \psi_3 - \psi_4 - \frac{5}{2} \frac{1}{\Pr} \psi_5 \right) \ , \quad \varpi_6 = \frac{4}{\psi_7} (2 + \psi_2 - \psi_6) \ ;$$

$$\theta_1 = \frac{1}{\chi_6} \frac{25}{6} \frac{1}{\Pr^2} \left(\frac{7}{2} - \omega \right) \ , \quad \theta_2 = \frac{1}{\chi_6} \frac{25}{4} \frac{1}{\Pr^2} \ , \quad \theta_3 = -\frac{5}{\Pr} \frac{\chi_2}{\chi_6} \ ,$$

$$\theta_4 = 5 \frac{\chi_3}{\chi_6} \frac{1}{\Pr} \ , \quad \theta_5 = \frac{5}{3} \frac{1}{\chi_6} \frac{1}{\Pr} \left(\frac{5}{2} \frac{1}{\Pr} (1 + \chi_4 - \chi_5) + \chi_1 + \chi_2 + \chi_3 \omega \right) \ .$$

Here two well-known relations between the Burnett coefficients are recovered,

$$\theta_1 = \frac{2}{3} \left(\frac{7}{2} - \omega \right) \theta_2 \ , \quad \varpi_1 = \frac{2}{3} \left(\frac{7}{2} - \omega \right) \varpi_2 \ .$$

Unfortunately, the generalized 13 moment equations (8.18, 8.19) require more coefficients than the Burnett equations, and the ψ_α, χ_α cannot be completely determined from the known results for the Burnett coefficients. By inversion follows

$$\psi_1 = \frac{4}{5} \Pr \frac{\varpi_3}{\varpi_2} \ , \quad \psi_2 - \psi_6 = \frac{1}{2} \frac{\varpi_6}{\varpi_2} - 2 \ , \quad \psi_3 = -\frac{4}{5} \Pr \frac{\varpi_4}{\varpi_2} \ ,$$

$$\psi_4 = -\frac{5}{2} \frac{1}{\Pr} \psi_5 + \frac{4}{5} \Pr \left(\frac{\varpi_3}{\varpi_2} \omega - \frac{\varpi_5}{\varpi_2} - \frac{\varpi_4}{\varpi_2} \right) \ , \quad \psi_7 = \frac{2}{\varpi_2} \ ;$$

$$\chi_1 = -\frac{5}{2} \frac{1}{\Pr} (\chi_4 - \chi_5) + \frac{5}{2} \frac{1}{\Pr} \left(\frac{3}{2} \frac{\theta_5}{\theta_2} - \frac{1}{2} \frac{\theta_4}{\theta_2} \omega + \frac{1}{2} \frac{\theta_3}{\theta_2} - 1 \right) \ ,$$

$$\chi_2 = -\frac{5}{4} \frac{1}{\Pr} \frac{\theta_3}{\theta_2} \ , \quad \chi_3 = \frac{5}{4} \frac{1}{\Pr} \frac{\theta_4}{\theta_2} \ , \quad \chi_6 = \frac{1}{\theta_2} \frac{25}{4} \frac{1}{\Pr^2} \ .$$

The terms in (8.18) with ψ_4, ψ_5 and those with ψ_2, ψ_6 assume the same form in the Burnett limit, where σ_{ij} and q_i are replaced by the NSF laws. Similarly, the terms with χ_1, χ_4, χ_5 assume the same form when σ_{ij} and q_i are replaced by the NSF laws.

When the Burnett coefficients are introduced into (8.18, 8.19), these can be rewritten as

$$
\frac{D\sigma_{ij}}{Dt} + \sigma_{ij}\frac{\partial v_k}{\partial x_k} + 2\sigma_{k\langle i}\frac{\partial v_{j\rangle}}{\partial x_k} + \frac{4}{5}\Pr\frac{\varpi_3}{\varpi_2}\left(\frac{\partial q_{\langle i}}{\partial x_{j\rangle}} - \omega q_{\langle i}\frac{\partial \ln\theta}{\partial x_{j\rangle}}\right)
$$

$$
+ \frac{4}{5}\Pr\frac{\varpi_4}{\varpi_2}q_{\langle i}\frac{\partial \ln p}{\partial x_{j\rangle}} + \frac{4}{5}\Pr\frac{\varpi_5}{\varpi_2}q_{\langle i}\frac{\partial \ln\theta}{\partial x_{j\rangle}} + \frac{\varpi_6}{\varpi_2}\sigma_{k\langle i}S_{j\rangle k}
$$

$$
+ \underline{\frac{\psi_5}{\mu\theta}q_{\langle i}\left(q_{j\rangle} - q_{j\rangle}^{(1)}\right)} + \underline{\frac{\psi_6}{\mu}\sigma_{k\langle i}\left(\sigma_{j\rangle k} - \sigma_{j\rangle k}^{(1)}\right)} = -\frac{2}{\varpi_2}\frac{p}{\mu}\left[\sigma_{ij} + 2\mu\frac{\partial v_{\langle i}}{\partial x_{j\rangle}}\right],
$$

$$\tag{8.20}$$

$$
\frac{Dq_i}{Dt} + q_k\frac{\partial v_i}{\partial x_k} + \frac{5}{3}q_i\frac{\partial v_k}{\partial x_k} - \frac{5}{2}\frac{1}{\Pr}\sigma_{ik}\frac{\partial \theta}{\partial x_k}
$$

$$
+ \frac{5}{4}\frac{1}{\Pr}\frac{\theta_3}{\theta_2}\theta\sigma_{ik}\frac{\partial \ln p}{\partial x_k} + \frac{5}{4}\frac{1}{\Pr}\frac{\theta_4}{\theta_2}\theta\left(\frac{\partial \sigma_{ik}}{\partial x_k} - \omega\sigma_{ik}\frac{\partial \ln\theta}{\partial x_k}\right) + \frac{5}{2}\frac{1}{\Pr}\frac{3}{2}\frac{\theta_5}{\theta_2}\sigma_{ik}\frac{\partial \theta}{\partial x_k}
$$

$$
+ \underline{\frac{\chi_4}{\mu}\left(q_k^{(1)}\sigma_{ik} - q_k\sigma_{ik}^{(1)}\right)} + \underline{\frac{\chi_5}{\mu}\sigma_{ik}\left(q_k - q_k^{(1)}\right)} = -\frac{1}{\theta_2}\frac{5}{2}\frac{1}{\Pr}\frac{p}{\mu}\left[q_i + \frac{5}{2}\frac{\mu}{\Pr}\frac{\partial \theta}{\partial x_i}\right].
$$

$$\tag{8.21}$$

Here $\sigma_{ij}^{(1)}$ and $q_i^{(1)}$ are the NSF expressions, and therefore the underlined terms are of super-Burnett order (they do not contribute to the Burnett equations). These terms might be ignored for a theory of Burnett order.

It is not difficult to rewrite the last equations as

$$
\frac{D\sigma_{ij}}{Dt} - 2\frac{\partial v_k}{\partial x_{\langle i}}\sigma_{j\rangle k} = -\frac{2}{\varpi_2}\frac{p}{\mu}\left[\sigma_{ij} - \sigma_{ij}^{(1)} - P_{ij|2}\right],
$$

$$
\frac{Dq_i}{Dt} - q_k\frac{\partial v_k}{\partial x_i} = -\frac{1}{\theta_2}\frac{5}{2}\frac{1}{\Pr}\frac{p}{\mu}\left[q_i + \frac{5\mu}{2\Pr}\frac{\partial \theta}{\partial x_i} - q_{i|2}\right],
$$

where terms of super-Burnett order are ignored, and $P_{ij|2}$, $q_{i|2}$ are the Jin-Slemrod expressions (7.16, 7.17). The above equations are just the Jin-Slemrod equations (7.15) without the third order terms, $P_{ij|3} = q_{i|3} = 0$. Thus, the present derivation of second order transport equations gives a more rational background to the second order terms in the Jin-Slemrod equations, which were obtained from the Burnett equations in a rather intuitive manner.

8.4 Third order accuracy: R13 equations

The third order equations are discussed only for the case of Maxwell molecules where (8.13) and (8.14) simplify to

$$\varepsilon\left[\frac{D\sigma_{ij}}{Dt} + \frac{4}{5}\frac{\partial q_{\langle i}}{\partial x_{j\rangle}} + 2\sigma_{k\langle i}\frac{\partial v_{j\rangle}}{\partial x_k} + \sigma_{ij}\frac{\partial v_k}{\partial x_k}\right] + \varepsilon^2\frac{\partial u^0_{ijk}}{\partial x_k} = -\rho\theta\left[\frac{\sigma_{ij}}{\mu} + 2\frac{\partial v_{\langle i}}{\partial x_{j\rangle}}\right],$$
(8.22)

and

$$\varepsilon\left[\frac{Dq_i}{Dt} + \frac{5}{2}\sigma_{ik}\frac{\partial\theta}{\partial x_k} - \sigma_{ik}\theta\frac{\partial\ln\rho}{\partial x_k} + \theta\frac{\partial\sigma_{ik}}{\partial x_k} + \frac{7}{5}q_i\frac{\partial v_k}{\partial x_k} + \frac{7}{5}q_k\frac{\partial v_i}{\partial x_k} + \frac{2}{5}q_k\frac{\partial v_k}{\partial x_i}\right]$$
$$+ \varepsilon^2\left[\frac{1}{2}\frac{\partial w^1_{ij}}{\partial x_k} + \frac{1}{6}\frac{\partial w^2}{\partial x_i} + u^0_{ikl}\frac{\partial v_k}{\partial x_l} - \frac{\sigma_{ik}}{\rho}\frac{\partial\sigma_{kl}}{\partial x_l}\right] = -\frac{5}{2}\rho\theta\left[\frac{q_i}{\kappa} + \frac{\partial\theta}{\partial x_i}\right]. \quad (8.23)$$

To obtain the pressure deviator and the heat flux with third order accuracy, also the $\mathcal{O}\left(\varepsilon^2\right)$ terms in (8.13, 8.14) must be considered, so that the equations for σ_{ij} and q_i read (after setting $\varepsilon = 1$)

$$\frac{D\sigma_{ij}}{Dt} + \frac{4}{5}\frac{\partial q_{\langle i}}{\partial x_{j\rangle}} + 2\sigma_{k\langle i}\frac{\partial v_{j\rangle}}{\partial x_k} + \sigma_{ij}\frac{\partial v_k}{\partial x_k} + \frac{\partial u^0_{ijk}}{\partial x_k} = -\rho\theta\left[\frac{\sigma_{ij}}{\mu} + 2\frac{\partial v_{\langle i}}{\partial x_{j\rangle}}\right], \quad (8.24)$$

$$\frac{Dq_i}{Dt} + \frac{5}{2}\sigma_{ik}\frac{\partial\theta}{\partial x_k} - \sigma_{ik}\theta\frac{\partial\ln\rho}{\partial x_k} + \theta\frac{\partial\sigma_{ik}}{\partial x_k} + \frac{7}{5}q_i\frac{\partial v_k}{\partial x_k} + \frac{7}{5}q_k\frac{\partial v_i}{\partial x_k}$$
$$+ \frac{2}{5}q_k\frac{\partial v_k}{\partial x_i} + \frac{1}{2}\frac{\partial w^1_{ij}}{\partial x_k} + \frac{1}{6}\frac{\partial w^2}{\partial x_i} + u^0_{ikl}\frac{\partial v_k}{\partial x_l} - \frac{\sigma_{ik}}{\rho}\frac{\partial\sigma_{kl}}{\partial x_l} = -\frac{5}{2}\rho\theta\left[\frac{q_i}{\kappa} + \frac{\partial\theta}{\partial x_i}\right].$$
(8.25)

In addition to the 13 variables, these equations contain the quantities u^0_{ijk}, w^1_{ij}, w^2, and to obtain a closed set of equations, additional equations must be provided for these. As we have seen, the leading order of magnitude of u^0_{ijk}, w^1_{ij}, w^2 is $\mathcal{O}\left(\varepsilon^2\right)$, and if these are computed to second order accuracy, pressure deviator σ_{ij} and heat flux q_i are computed within *third order* accuracy. If u^0_{ijk}, w^1_{ij}, w^1 are computed up to third order accuracy, σ_{ij} and q_i will have fourth order accuracy, and so on.

The relevant equation for the proper third order theory for w^2 at $\mathcal{O}\left(\varepsilon^2\right)$ is (8.4) with $a = 2$ where we have to consider (8.11), so that

$$\left(\frac{2\kappa_2}{\kappa_1} - 20\right)\theta\frac{\partial q_k}{\partial x_k} + \left(4\frac{\mu_1}{\mu_0} - 20\right)\theta\sigma_{kl}\frac{\partial v_k}{\partial x_l} + \left(2\frac{\kappa_2}{\kappa_1} - 8\right)q_k\frac{\partial\theta}{\partial x_k}$$
$$- 8q_k\theta\frac{\partial\ln\rho}{\partial x_k} = -\frac{2}{3}\frac{p}{\mu}\left(w^2 + \frac{\sigma_{ij}\sigma_{ij}}{\rho}\right). \quad (8.26)$$

The second order equation for u^0_{ijk} is (8.5) for $a = 0$,

$$\frac{3}{7}\frac{\mu_1}{\mu_0}\theta\frac{\partial\sigma_{\langle ij}}{\partial x_{k\rangle}} - 3\sigma_{\langle ij}\theta\frac{\partial\ln\rho}{\partial x_{k\rangle}} + \frac{12}{5}q_{\langle i}\frac{\partial v_j}{\partial x_{k\rangle}} + 3\left(\frac{1}{7}\frac{\mu_1}{\mu_0} - 1\right)\sigma_{\langle ij}\frac{\partial\theta}{\partial x_{k\rangle}} = -\frac{3}{2}\frac{p}{\mu}u^0_{ijk}. \quad (8.27)$$

The equation for w^1_{ij} results from (5.16) with $a = 1$, after replacing u^1_{ij} by w^1_{ij} by means of (8.11), and subsequent elimination of the time derivatives of σ_{ij} and θ by means of their respective balance laws.[3] Keeping only the leading order terms we see, after some algebra,

$$\frac{4}{5}\left[\frac{\kappa_2}{\kappa_1} - \frac{\mu_1}{\mu_0}\right]\theta\frac{\partial q_{\langle i}}{\partial x_{j\rangle}} + \frac{4}{5}\left[\frac{\kappa_2}{\kappa_1} - 7\right]q_{\langle i}\frac{\partial\theta}{\partial x_{j\rangle}} - \frac{28}{5}\theta q_{\langle i}\frac{\partial\ln\rho}{\partial x_{j\rangle}}$$

$$+\frac{4}{7}\frac{\mu_1}{\mu_0}\theta\left(\sigma_{k\langle i}\frac{\partial v_{j\rangle}}{\partial x_k} + \sigma_{k\langle i}\frac{\partial v_k}{\partial x_{j\rangle}} - \frac{2}{3}\sigma_{ij}\frac{\partial v_k}{\partial x_k}\right)$$

$$+14\rho\theta^2\left(1 - \frac{1}{7}\frac{\mu_1}{\mu_0}\right)\left(\frac{\sigma_{ij}}{2\mu} + \frac{\partial v_{\langle i}}{\partial x_{j\rangle}}\right) = -\frac{7}{6}\frac{p}{\mu}\left(w^1_{ij} + \frac{4}{7}\frac{1}{\rho}\sigma_{k\langle i}\sigma_{j\rangle k}\right) . \quad (8.28)$$

It remains to use (8.8), which gives $\frac{\mu_1}{\mu_2} = 7$, $\frac{\kappa_2}{\kappa_1} = 14$, and the above equations for Maxwell molecules reduce to

$$w^2 = -\frac{\sigma_{ij}\sigma_{ij}}{\rho} - 12\frac{\mu}{p}\left[\theta\frac{\partial q_k}{\partial x_k} + \theta\sigma_{kl}\frac{\partial v_k}{\partial x_l} + \frac{5}{2}q_k\frac{\partial\theta}{\partial x_k} - q_k\theta\frac{\partial\ln\rho}{\partial x_k}\right] ,$$

$$u^0_{ijk} = -2\frac{\mu}{p}\left[\theta\frac{\partial\sigma_{\langle ij}}{\partial x_{k\rangle}} - \sigma_{\langle ij}\theta\frac{\partial\ln\rho}{\partial x_{k\rangle}} + \frac{4}{5}q_{\langle i}\frac{\partial v_j}{\partial x_{k\rangle}}\right] , \quad (8.29)$$

$$w^1_{ij} = -\frac{4}{7}\frac{1}{\rho}\sigma_{k\langle i}\sigma_{j\rangle k} - \frac{24}{5}\frac{\mu}{p}\left[\theta\frac{\partial q_{\langle i}}{\partial x_{j\rangle}} + q_{\langle i}\frac{\partial\theta}{\partial x_{j\rangle}} - \theta q_{\langle i}\frac{\partial\ln\rho}{\partial x_{j\rangle}}\right.$$

$$\left.+\frac{5}{7}\theta\left(\sigma_{k\langle i}\frac{\partial v_{j\rangle}}{\partial x_k} + \sigma_{k\langle i}\frac{\partial v_k}{\partial x_{j\rangle}} - \frac{2}{3}\sigma_{ij}\frac{\partial v_k}{\partial x_k}\right)\right] .$$

These are just the regularized 13 moment equations (7.8–7.12) [9][59], with the addition of the non-linear terms from the production terms (5.23), and the omission of some terms which are non-linear in σ_{ij} and q_k. These non-linear terms are not present here, since they add terms of third order to u^0_{ijk}, w^1_{ij}, w^1, which are ignored in the third order theory. However, these terms would be present in a 4th order theory, together with additional contributions from other moments.

For the BGK model the equations bear different factors,

$$w^2 = -8\tau\left[\cdots\right] , \quad u^0_{ijk} = -3\tau\left[\cdots\right] , \quad w^1_{ij} = -\tau\frac{28}{5}\left[\cdots\right]$$

where the the square brackets stand for their counterparts in (8.29).

8.5 Discussion

8.5.1 Higher order accuracy

We shall not go further with developing the equations at higher order accuracy. For Maxwell molecules it is clear that the next level of accuracy—fourth

---

[3] This is essentially (6.19) since $w^1_{ij} = R_{ij}$ for Maxwell molecules.

order—will contain full balance laws for u_{ijk}^0, w_{ij}^1, w^2, that is equations of the form $\frac{D\phi}{Dt} + [space\ derivatives] = -\frac{1}{\tau}\phi$. The resulting equations should then be equivalent to the Grad moment system with 26 moments of Sec. 6.1.3. The next order—the fifth—should be the regularization of the 26 moment system, similar as the third order is the regularization of the second order 13 moment case.

The (first order) NSF equations are the regularization of the (zeroth order) Euler equations. Thus, the following picture emerges: Equations at even order (zeroth, second, fourth, ...) are of the Grad moment type, and hyperbolic, while equations at odd orders (first, third, ...) form the regularization of these hyperbolic equations.

The complexity of the method increases substantially when one is not dealing with Maxwell molecules or the BGK model. Then, for the third order, full balance equations for the w_{ij}^b, w_i^b are required. For this, it will be necessary to construct new variables $\tilde{w}_{ij}^b$, $\tilde{w}_i^b$ such that a minimum number of these is of second order, similar to the procedure outlined in Sec. 8.2.3.

8.5.2 Comparison with Chapman-Enskog method

The most pressing question on the method introduced in this chapter is probably, where the differences are to the Chapman-Enskog method, and why the method gives stable and meaningful equations at each level of approximation, while the Chapman-Enskog method does not.

The Chapman-Enskog expansion aims at finding the coefficients in the expansion (8.2) expressed solely through gradients (of any order) of its basic variables, mass density ρ, temperature θ and velocity v_i. This is done by an iterative procedure, where the result of order n is used to compute the expressions at order $n + 1$.

The order of magnitude approach considers all moments as quantities in their own right, without aiming at expressions for the expansion coefficients. In particular, no iteration process occurs, so that the higher order contributions are independent of the lower order ones. Spiegel and Thiffeault use their own variant of an iterative expansion to compute transport equations up to second order of the Knudsen number [112]. While they argue that their method is different from, and better than, the Chapman-Enskog method, they nevertheless obtain unstable equations at second order. One might conclude that the independence of the higher order contributions of the lower ones is related to the stability (which is proven up to order three, and expected at any order).

The Chapman-Enskog method provides terms up to a definite order, while the equations found above contain higher order terms in the sense of the Chapman-Enskog expansion. This can be seen nicely in [107], where the authors perform a Chapman-Enskog expansion to infinite order of the linearized Grad equations. It seems that these higher order terms (in the Chapman-Enskog sense) are responsible for the stabilization, and the better agreement with experiments, when compared to the higher order Chapman-Enskog expansions, i.e. the Burnett and super-Burnett equations [59][60].

When this answer seems to be vague, the reason will be that no definite statement on why the higher order Chapman-Enskog expansions leads to unstable equations is available in the first place.

The Chapman-Enskog expansion can be performed on the third order equations, i.e. the R13 equations. The resulting equations are the Burnett equations at second order, and the super-Burnett equations at third order of the expansion. Evidently, the R13 equations are a more complete set of equations at third order.

To conclude, we summarize the advantages of the order of magnitude approach against the Chapman-Enskog method which are: (a) stable equations at any order, (b) better agreement with experiments, (c) a simpler derivation of the equations.

8.5.3 Comparison with the Grad method

In our discussion of the Grad method in Chapter 6, we have seen that it does not provide a statement on which sets of moments one should use, but experience shows that more moments lead to better results. The main feature of the Grad method is that it provides a certain phase density, $f_{|G}$, and allows only states that can be described by that function. As was argued in the last chapter, one can say that the Grad method assumes a non-equilibrium manifold, and forces the gas to stay on that manifold. However, there is no argument from physics on why the gas should be restricted to that non-equilibrium manifold.

The problem is centered in the question why a certain set of moments should be just enough to describe the gas properly—after all, the gas is not aware of our choice of variables.

Only for the simplest system of Grad type, the Euler equations, is an entropy based argument valid: the Boltzmann collision term forces the gas towards the local Maxwellian, which forms the equilibrium manifold of the Boltzmann equation, and maximizes entropy. There is no intermediate non-equilibrium manifold for the Boltzmann collision term to which the gas would relax before reaching the final—Maxwellian—equilibrium. Intermediate equilibria are possible, however, in more complex systems where processes with distinct mean free times occur, e.g. in the phonon gas [83].

Extended Thermodynamics [17] has a strong relationship to the Grad method, and similar arguments can be applied.

The method presented above is quite different from the Grad method, although the equations obtained bear a strong similarity to Grad's 13 moment set, and the R13 equations. More significant are the differences: The equations are obtained from a thorough analysis of the *orders of magnitude* of moments, and the *order of accuracy* of the moment equations. The moments that appear in the equations of chosen order follow from that analysis. Moreover, the Grad phase density $f_{|G}$ was not required to obtain closed sets of equations when Maxwell molecules are considered.

Compared to the second order equations (8.16, 8.17), the Grad equations for 13 moments contain the term $-\frac{\sigma_{ik}}{\rho}\frac{\partial\sigma_{kl}}{\partial x_l}$ in the heat flux equation. This term is of second order, but contributes to third order in the heat flux. Accordingly, this terms appears in the third order equations (8.24, 8.25), but there are additional third order terms that must be accounted for, viz. $\frac{\partial u_{ijk}^0}{\partial x_k}$ in the equation for σ_{ij}, and $\frac{1}{2}\frac{\partial w_{ij}^1}{\partial x_k} + \frac{1}{6}\frac{\partial w^2}{\partial x_i} + u_{ikl}^0\frac{\partial v_k}{\partial x_l}$ in the equation for q_i. Accordingly, the original Grad 13 equations stand in between the orders of magnitude, since they contain some (in fact only one), but not all terms of third order.

8.5.4 Comparison with the original derivation of the R13 equations

Some relaxation of the strong requirements of the Grad method can be found in approximations which allow states also in the vicinity of the (assumed) non-equilibrium manifold, as discussed in Chapter 7, see also [105]–[107] and the references therein.

The original derivation of the R13 equations by Grad [9], and by Struchtrup and Torrilhon [59][60] is based on allowing small deviations from Grad's 13 moment manifold, which are taken into account by performing a first order Chapman-Enskog expansion centered in the 13 moment phase density, and not in the Maxwellian as in the standard Chapman-Enskog expansion. Thus, the original derivation of the R13 equations takes the Grad method for granted, and just derives a regularization by ignoring certain terms that are of higher order in the Knudsen number.

In the method outlined above, however, the R13 equations resulted from accounting for orders of magnitude and accuracy.

Indeed, the order of magnitude argument gives the equations (8.29), while the original R13 equations can be written as [59]

$$w_{|R13}^2 = -12\tau\left[\cdots - \frac{1}{\rho}q_j\frac{\partial\sigma_{jk}}{\partial x_k}\right] \quad, \quad u_{ijk|R13}^0 = -2\tau\left[\cdots - \frac{\sigma_{\langle ij}}{\rho}\frac{\partial\sigma_{k\rangle l}}{\partial x_l}\right] \quad,$$

$$w_{ij|R13}^1 = -\frac{24}{5}\tau\left[\cdots - \frac{1}{\rho}q_{\langle i}\frac{\partial\sigma_{j\rangle k}}{\partial x_k} - \frac{5}{6}\frac{\sigma_{ij}}{\rho}\frac{\partial q_k}{\partial x_k} - \frac{5}{6}\frac{\sigma_{ij}}{\rho}\sigma_{kl}\frac{\partial v_k}{\partial x_l}\right]$$

where the dots indicate the same terms as in (8.29), so that only those terms are explicitly shown above that are added in the original R13 equations. A short glance suffices to see that these additions are of third order, which means they contribute to the fourth order in σ_{ij} and q_i. However, a careful analysis of the full fourth order would reveal that these are not the only fourth order terms, so that the original R13 equations stand in between third and fourth order, just as the original Grad equations stand in between second and third order.

Moreover, the original derivation could only be performed by considering linearized production terms, while the derivation in this chapter allows to consider these without any problem.

8.5.5 Comparison with consistently ordered extended thermodynamics

The order of magnitude approach owes its foundation to the ideas of Müller et al., presented in [111]. These authors recognized that the order of magnitude of moments (in a Chapman-Enskog sense) can be used as the building block for a consistent hierarchy of equations in the orders of the Knudsen number.

The ideas presented above differ from COET mainly in the definition of the *order of accuracy* of the set of equations. COET assumes that all terms in all moment equations up to the order $\mathcal{O}\left(\varepsilon^\lambda\right)$ must be taken into account for a theory that aims to be accurate at $\lambda - th$ order.

At first, these are the moment equations for all moments of order $\beta \leq \lambda$ under omission of higher order terms. However, these equations split into two independent subsystems, and only a smaller number of equations (and variables) remain as equations of importance.

The order of magnitude approach does not ask for the order of terms in all moment equations, but for the order of magnitude of their influence in the conservations laws, i.e. on heat flux and stress tensor. As became clear, a third order accuracy in the equations does not require balance laws for all moments that have third order contributions. Rather it is sufficient to have σ_{ij} and q_i up to third order, and the moments u^0_{ijk}, w^1_{ij}, w^1 to only second order.

Müller et al., however, due to their different philosophy, require the moments u^0_{ijk}, w^1_{ij}, w^1 to third order accuracy, and need additional moments up to third order, as becomes clear from Equation (5.7) in [111]. Accordingly, their number of variables is higher for the third order than here.

In the present interpretation of the order of accuracy of equations, the third order COET will have higher than third order accuracy (but might stand in between orders of magnitude, similar to original Grad 13 and original R13 equations).

Müller et al. state that their method is independent of the phase density. This is right only as long as they use Maxwell molecules or the BGK model to describe collisions. The same is true for the new approach, and it was shown how the production terms $\mathcal{P}^a_{i_1 \dots i_n}$ (5.8) for arbitrary interaction potentials can be computed from the Reinecke-Kremer-Grad method.

Finally we note that COET chooses orthogonal polynomials similar to the eigenfunctions as the basic moments, while here the moment set was constructed to have the minimal number of moments at a given order. It seems that for the case of Maxwell molecules both sets of moments agree [113].

8.5.6 Comparison with Jin-Slemrod equations

For non-Maxwellian molecules the method was developed only to second order. Going to third order for arbitrary interaction models will be far more cumbersome than for Maxwell molecules. In particular, the third order theory will include full balance laws for the minimum number of vectors w_i^a and tensors w_{ij}^a that are of second order, and thus the number of variables will be higher than 13.

Above it was shown that the equations for arbitrary interaction potentials agree with the Jin-Slemrod equations up to second order. Moreover, it was shown earlier, in Chapter 7, that even for Maxwell molecules the Jin-Slemrod equations do not agree with the R13 equations, and that they are not accurate to super-Burnett order.

Thus, the order of magnitude approach can justify the Jin-Slemrod equations to Burnett order, while showing at the same time that they cannot be correct to super-Burnett order.

9

Macroscopic transport equations for rarefied gas flows

The previous chapters presented detailed derivations of various sets of transport equations, and here we take a step back and collect the equations on a few pages. The chapter begins with a brief overview of the equations, and their relations to the Boltzmann equation, and to each other. Then the full equations are printed, along with their forms for purely one-dimensional geometry, and their linearized form for small deviations from equilibrium.

9.1 Relations between the equations

The derivation of macroscopic equations from the Boltzmann equation, or the equivalent set of infinitely many moment equations, is simplest for the special case of Maxwell molecules. Accordingly, theories of higher orders in the Knudsen number, like the super-Burnett and the R13 equations, are—at present—only available for Maxwell molecules.

The Chapman-Enskog expansion of increasing order gives the Euler, Navier-Stokes-Fourier, Burnett, and super-Burnett equations.

The augmented Burnett equations contain terms of super-Burnett order, which are added ad hoc, and cannot be derived from the Boltzmann equation, or the infinite moment system.

Grad-type moment equations can be constructed for arbitrary moment sets, but the 13 moment system and the 26 moment system are particularly interesting, since they are equations of orders $\mathcal{O}\left(\varepsilon^2\right)$ and $\mathcal{O}\left(\varepsilon^4\right)$, respectively. They can be obtained also from the order of magnitude approach in the last chapter. Moreover, this method gave the R13 equations as the proper equations at order $\mathcal{O}\left(\varepsilon^3\right)$, and the NSF and Euler equations at orders $\mathcal{O}\left(\varepsilon^1\right)$ and $\mathcal{O}\left(\varepsilon^0\right)$, respectively.

Jin and Slemrod's equations are accurate to order $\mathcal{O}\left(\varepsilon^2\right)$, but contain terms of super-Burnett order, $\mathcal{O}\left(\varepsilon^3\right)$, which cannot be derived from the Boltzmann equation.

Table 9.1. The hierarchy of macroscopic equations for Maxwell molecules.

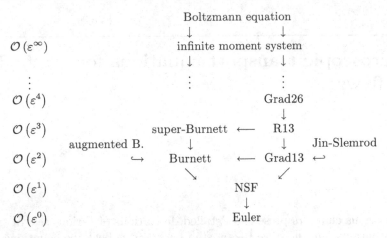

The relations between these sets of equations are depicted in Table 9.1, in which an arrow between two sets of equations indicates that one set can be derived from the other (e.g. the Burnett equations from the Grad13 equations). Note that at a given order the equations derived from the Chapman-Enskog method and from the order of magnitude approach are quite different, due to the marked differences in methodology. Indeed, the Chapman-Enskog based equations (e.g. the super-Burnett equations, at third order), contain less information than their counterparts (e.g. the R13 equations, also at third order), since the former can be derived from the latter, but not vice versa.

For other types of particles, e.g. those that interact through power potentials, accurate sets of equations are only available to order $\mathcal{O}\left(\varepsilon^2\right)$, namely the generalized 13 moment equations which were obtained from the order of magnitude method, and the Burnett equations from the CE method. The Euler and NSF equations form the proper equations at zeroth and first order, and the Jin-Slemrod equations are available as well. Table 9.2 shows the known equations as well as their order of accuracy, and their relations.

9.2 3-D non-linear equations

9.2.1 Conservation laws

The basic equations for all sets of equations are the conservation laws for mass, momentum and energy. The basic variables are mass density ρ, velocity v_i, and temperature in energy units $\theta = \frac{k}{m}T$. This text considers only monatomic ideal gases were the internal energy is given by $u = \frac{3}{2}\theta$, and the pressure p obeys the ideal gas law $p = \rho\theta$. The conservation laws read, see (3.35, 3.36),

Table 9.2. The hierarchy of macroscopic equations for molecules with arbitrary interaction potentials.

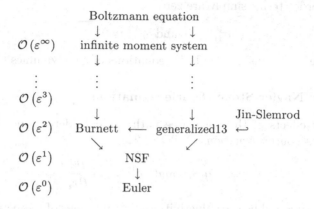

$$\frac{D\rho}{Dt} + \rho\frac{\partial v_k}{\partial x_k} = 0\,,$$

$$\rho\frac{Dv_i}{Dt} + \theta\frac{\partial\rho}{\partial x_i} + \rho\frac{\partial\theta}{\partial x_i} + \frac{\partial\sigma_{ik}}{\partial x_k} = \rho G_i\,, \qquad (9.1)$$

$$\frac{3}{2}\rho\frac{D\theta}{Dt} + \frac{\partial q_k}{\partial x_k} = -\rho\theta\frac{\partial v_k}{\partial x_k} - \sigma_{ij}S_{ij}\,.$$

where $\frac{D}{Dt} = \frac{\partial}{\partial t} + v_k\frac{\partial}{\partial x_k}$ is the convective time derivative, G_i is the external body force, and

$$S_{ij} = \frac{\partial v_{\langle i}}{\partial x_{j\rangle}} = \frac{1}{2}\frac{\partial v_i}{\partial x_j} + \frac{1}{2}\frac{\partial v_j}{\partial x_i} - \frac{1}{3}\frac{\partial v_k}{\partial x_k}\delta_{ij}\,.$$

Stress σ_{ij} and heat flux q_i depend on the method used, and will be given below.

9.2.2 Chapman-Enskog expansion

The Chapman-Enskog expansion gives stress and heat flux as a series in the formal smallness parameter ε that stands for the Knudsen number (4.11). Since ε must be set to unity in the final result, we have

$$\sigma_{ij} = \sum_{\alpha=0}\sigma_{ij}^{(\alpha)} = \sigma_{ij}^{(0)} + \sigma_{ij}^{(1)} + \sigma_{ij}^{(2)} + \sigma_{ij}^{(3)} + \cdots\,,$$

$$q_i = \sum_{\alpha=0}q_i^{(\alpha)} = q_i^{(0)} + q_i^{(1)} + q_i^{(2)} + q_i^{(3)} + \cdots\,.$$

where the $\sigma_{ij}^{(\alpha)}, q_i^{(\alpha)}$ are the contributions to order α, and the equations of order β will include all terms up to $\alpha = \beta$.

Zeroth order: Euler equations

The zeroth order terms simply are zero,

$$\sigma_{ij}^{(0)} = 0 \text{ and } q_i^{(0)} = 0 .$$

The resulting equations are the Euler equations of gas dynamics (4.21).

First order: Navier-Stokes-Fourier equations

If first order effects of the mean free path are considered, one obtains the Navier-Stokes-Fourier equations (NSF) (4.22),

$$\sigma_{ij}^{(1)} = -2\mu S_{ij} \text{ and } q_i^{(1)} = -\kappa \frac{\partial \theta}{\partial x_i} . \tag{9.2}$$

where viscosity μ and heat conductivity[1] κ are functions of temperature alone (4.24, 4.45). For power potentials for the interaction between the particles one finds

$$\mu = \mu_0 \left(\frac{\theta}{\theta_0}\right)^\omega \text{ and } \kappa = \frac{5}{2}\frac{\mu}{\text{Pr}} . \tag{9.3}$$

Here $\omega \in [0.5, 1]$ is a number that depends on the interaction potential, and Pr is the Prandtl number with $\text{Pr} = \frac{2}{3}$ for Maxwell molecules ($\omega = 1$), and a number very close to 2/3 for all other molecules, e.g. $\text{Pr} = 0.661$ for hard sphere molecules ($\omega = 0.5$). $\omega = \frac{\gamma+3}{2\gamma-2}$ for power potentials (3.9).

Second order: Burnett equations

The second order contributions are given by the Burnett equations, which read (4.26, 4.27)

$$\sigma_{ij}^{(2)} = \frac{\mu^2}{p}\left[\varpi_1 \frac{\partial v_k}{\partial x_k}S_{ij} - \varpi_2 \left(\frac{\partial}{\partial x_{\langle i}}\left(\frac{1}{\rho}\frac{\partial p}{\partial x_{j\rangle}}\right) + \frac{\partial v_k}{\partial x_{\langle i}}\frac{\partial v_{j\rangle}}{\partial x_k} + 2\frac{\partial v_k}{\partial x_{\langle i}}S_{j\rangle k}\right)\right.$$
$$\left. +\varpi_3\frac{\partial^2\theta}{\partial x_{\langle i}\partial x_{j\rangle}} + \varpi_4\frac{\partial\theta}{\partial x_{\langle i}}\frac{\partial\ln p}{\partial x_{j\rangle}} + \varpi_5\frac{1}{\theta}\frac{\partial\theta}{\partial x_{\langle i}}\frac{\partial\theta}{\partial x_{j\rangle}} + \varpi_6 S_{k\langle i}S_{j\rangle k}\right] , \tag{9.4}$$

$$q_i^{(2)} = \frac{\mu^2}{\rho}\left[\theta_1\frac{\partial v_k}{\partial x_k}\frac{\partial\ln\theta}{\partial x_i} - \theta_2\left(\frac{2}{3}\frac{\partial^2 v_k}{\partial x_k\partial x_i} + \frac{2}{3}\frac{\partial v_k}{\partial x_k}\frac{\partial\ln\theta}{\partial x_i} + 2\frac{\partial v_k}{\partial x_i}\frac{\partial\ln\theta}{\partial x_k}\right)\right.$$
$$\left. +\theta_3 S_{ik}\frac{\partial\ln p}{\partial x_k} + \theta_4\frac{\partial S_{ik}}{\partial x_k} + 3\theta_5 S_{ik}\frac{\partial\ln\theta}{\partial x_k}\right] . \tag{9.5}$$

The Burnett coefficients depend on the molecule type, and for power potentials (3.9) they are given in Table 9.3 [54].

---

[1] Recall that we employ energy units for temperature and heat conductivity; that is the usual heat conductivity $\hat{\kappa}$ is related to our heat conductivity as $\hat{\kappa} = \frac{k}{m}\kappa$.

Table 9.3. Burnett coefficients for power potentials ($\gamma = 5$ for Maxwell molecules, $\gamma = \infty$ for for hard sphere molecules) [54].

γ	ω	ϖ_1	ϖ_2	ϖ_3	ϖ_4	ϖ_5	ϖ_6	θ_1	θ_2	θ_3	θ_4	θ_5
5	1	10/3	2	3	0	3	8	75/8	45/8	-3	3	39/4
7	0.833	3.561	2.003	2.793	0.217	1.942	7.781	10.038	5.647	-3.010	2.793	9.113
7.66	0.8	3.600	2.004	2.761	0.254	1.784	7.748	10.160	5.656	-3.014	2.761	9.019
9	0.75	3.679	2.007	2.695	0.328	1.466	7.681	10.402	5.674	-3.023	2.695	8.829
17	0.625	3.863	2.016	2.553	0.500	0.814	7.543	10.995	5.736	-3.053	2.553	8.442
∞	0.5	4.056	2.028	2.418	0.681	0.219	7.424	11.644	5.822	-3.09	2.418	8.286

Table 9.4. First approximation to the Burnett coefficients for power potentials ($\gamma = 5$ for Maxwell molecules, $\gamma = \infty$ for for hard sphere molecules). The ES-BGK model with Pr=2/3 gives the same coefficients.

γ	ω	ϖ_1	ϖ_2	ϖ_3	ϖ_4	ϖ_5	ϖ_6	θ_1	θ_2	θ_3	θ_4	θ_5
5	1	10/3	2	3	0	3	8	75/8	45/8	-3	3	39/4
7	0.833	3.555	2	3	0	3	8	10.000	45/8	-3	3	9.583
7.66	0.8	3.600	2	3	0	3	8	10.125	45/8	-3	3	9.550
9	0.75	3.667	2	3	0	3	8	10.313	45/8	-3	3	9.500
17	0.625	3.833	2	3	0	3	8	10.781	45/8	-3	3	9.375
∞	0.5	4	2	3	0	3	8	11.25	45/8	-3	3	9.25

For the ES-BGK model the Burnett coefficients are related to the Prandtl number by (4.28),

$$\varpi_1 = \frac{4}{3}\left(\frac{7}{2} - \omega\right), \quad \varpi_2 = 2, \quad \varpi_3 = \frac{2}{\mathrm{Pr}}, \quad \varpi_4 = 0, \quad \varpi_5 = \frac{2\omega}{\mathrm{Pr}}, \quad \varpi_6 = 8,$$

$$\theta_1 = \frac{5}{3}\frac{1}{\mathrm{Pr}^2}\left(\frac{7}{2} - \omega\right), \quad \theta_2 = \frac{5}{2}\frac{1}{\mathrm{Pr}^2}, \quad \theta_3 = -\frac{2}{\mathrm{Pr}}, \quad \theta_4 = \frac{2}{\mathrm{Pr}},$$

$$\theta_5 = \frac{2}{3}\frac{1}{\mathrm{Pr}}\left(\frac{7}{2}\left(1 + \frac{1}{\mathrm{Pr}}\right) + \omega\right). \tag{9.6}$$

The first approximation for the Burnett coefficients (only the leading terms are considered in the Sonine expansion (4.42)), agrees with the coefficients for the ES-BGK model with Pr = 2/3, some of their numerical values are printed in Table 9.4.

The Burnett coefficients for the standard BGK model [29] are obtained from (9.6) by setting Pr = 1.

Third order: Super-Burnett equations

The third order contributions to stress and heat flux are given by the super-Burnett equations, which are only available for Maxwell molecules ($\omega = 1$), and only for one-dimensional problems, or in the linear case, see below in (9.25, 9.26) and (9.45).

Augmented Burnett equations

The instability of the original Burnett equations lead Zhong et al. to augment the Burnett equations by adding stabilizing terms of super-Burnett order in an ad-hoc manner [62][63]. These terms replace the super-Burnett expressions, so that

$$\sigma_{ij}^{(3)} \rightarrow \sigma_{ij}^{(A)} = \frac{\mu^3}{p^2} \left(\frac{1}{3} \theta \frac{\partial^2}{\partial x_k \partial x_k} \frac{\partial v_{\langle i}}{\partial x_{j\rangle}} \right),$$

$$q_i^{(3)} \rightarrow q_i^{(A)} = \frac{\mu^3}{p^2} \left(\frac{11}{16} \theta \frac{\partial^3 \theta}{\partial x_i \partial x_k \partial x_k} - \frac{5}{8} \frac{\theta^2}{\rho} \frac{\partial^3 \rho}{\partial x_i \partial x_k \partial x_k} \right) . \tag{9.7}$$

9.2.3 Moment equations for Maxwell molecules

Moment equations at higher order are most easily derived for Maxwell molecules, and we present these first. For higher moments, the notation of Chapter 7 is used, i.e.,

$$\Delta = w^2 , \quad R_{ij} = w_{ij}^1 = u_{ij}^1 - 7\theta\sigma_{ij} , \quad m_{ijk} = u_{ijk}^0 .$$

Basic equations for stress and heat flux

Moment methods consider the hydrodynamic variables ρ, v_i, θ plus additional higher moments as variables, and provide full balance equations for the additional variables as well. Stress σ_{ij} and heat flux q_i are the first moments added to the list of variables, and their balance equations are (7.8, 7.9),

$$\frac{D\sigma_{ij}}{Dt} + \sigma_{ij}\frac{\partial v_k}{\partial x_k} + \frac{4}{5}\frac{\partial q_{\langle i}}{\partial x_{j\rangle}} + 2p\frac{\partial v_{\langle i}}{\partial x_{j\rangle}} + 2\sigma_{k\langle i}\frac{\partial v_{j\rangle}}{\partial x_k} + \frac{\partial m_{ijk}}{\partial x_k} = -\frac{p}{\mu}\sigma_{ij} , \tag{9.9}$$

$$\frac{Dq_i}{Dt} + \frac{5}{2}p\frac{\partial\theta}{\partial x_i} + \frac{5}{2}\sigma_{ik}\frac{\partial\theta}{\partial x_k} + \theta\frac{\partial\sigma_{ik}}{\partial x_k} - \theta\sigma_{ik}\frac{\partial\ln\rho}{\partial x_k} + \frac{7}{5}q_k\frac{\partial v_i}{\partial x_k} + \frac{2}{5}q_k\frac{\partial v_k}{\partial x_i}$$
$$+ \frac{7}{5}q_i\frac{\partial v_k}{\partial x_k} + \frac{1}{2}\frac{\partial R_{ik}}{\partial x_k} + \frac{1}{6}\frac{\partial\Delta}{\partial x_i} + m_{ijk}\frac{\partial v_j}{\partial x_k} - \frac{\sigma_{ij}}{\rho}\frac{\partial\sigma_{jk}}{\partial x_k} = -\frac{2}{3}\frac{p}{\mu}q_i . \tag{9.10}$$

These equations contain the additional quantities Δ, R_{ij} and m_{ijk}, and equations for these depend on the method used, and will be given below.

Grad 13 moments

The quantities Δ, R_{ij} and m_{ijk} are chosen such that they vanish for Grad's classical 13 moment theory [8][9] which therefore gives

$$\Delta = 0 , \quad R_{ij} = 0 , \quad m_{ijk} = 0 . \tag{9.11}$$

R13 equations (original)

Corrections to the Grad13 equations were obtained with the Struchtrup-Torrilhon regularization method [59][60] as (7.10–7.12),

$$\Delta = -12\frac{\mu}{p}\left[\theta\frac{\partial q_k}{\partial x_k} + \frac{5}{2}q_k\frac{\partial\theta}{\partial x_k} - \theta q_k\frac{\partial\ln\rho}{\partial x_k} - \frac{q_k}{\rho}\frac{\partial\sigma_{kl}}{\partial x_l} + \theta\sigma_{kl}\frac{\partial v_k}{\partial x_l}\right]$$

$$R_{ij} = -\frac{24}{5}\frac{\mu}{p}\left[\theta\frac{\partial q_{\langle i}}{\partial x_{j\rangle}} + q_{\langle i}\frac{\partial\theta}{\partial x_{j\rangle}} - \theta q_{\langle i}\frac{\partial\ln\rho}{\partial x_{j\rangle}} - \frac{q_{\langle i}}{\rho}\frac{\partial\sigma_{j\rangle k}}{\partial x_k}\right.$$
$$\left. + \frac{10}{7}\theta\sigma_{k\langle i}S_{j\rangle k} - \frac{5}{6}\frac{\sigma_{ij}}{\rho}\frac{\partial q_k}{\partial x_k} - \frac{5}{6}\frac{\sigma_{ij}\sigma_{kl}}{\rho}\frac{\partial v_k}{\partial x_l}\right]$$

$$m_{ijk} = -2\frac{\mu}{p}\left[\theta\frac{\partial\sigma_{\langle ij}}{\partial x_{k\rangle}} - \theta\sigma_{\langle ij}\frac{\partial\ln\rho}{\partial x_{k\rangle}} + \frac{4}{5}q_{\langle i}\frac{\partial v_j}{\partial x_{k\rangle}} - \frac{\sigma_{\langle ij}}{\rho}\frac{\partial\sigma_{k\rangle l}}{\partial x_l}\right] \qquad (9.12)$$

These equations are only valid when the linearized production terms for Maxwell molecules (5.24) are considered.

R13 equations (order of magnitude)

The order of magnitude approach allows to account for non-linear terms in the production terms, and removes some terms from (9.12) which are of higher order, it gives (8.29),

$$\Delta = -\frac{\sigma_{ij}\sigma_{ij}}{\rho} - 12\frac{\mu}{p}\left[\theta\frac{\partial q_k}{\partial x_k} + \theta\sigma_{kl}\frac{\partial v_k}{\partial x_l} + \frac{5}{2}q_k\frac{\partial\theta}{\partial x_k} - q_k\theta\frac{\partial\ln\rho}{\partial x_k}\right]$$

$$R_{ij} = -\frac{4}{7}\frac{1}{\rho}\sigma_{k\langle i}\sigma_{j\rangle k} - \frac{24}{5}\frac{\mu}{p}\left[\theta\frac{\partial q_{\langle i}}{\partial x_{j\rangle}} + q_{\langle i}\frac{\partial\theta}{\partial x_{j\rangle}} - \theta q_{\langle i}\frac{\partial\ln\rho}{\partial x_{j\rangle}} + \frac{10}{7}\theta\sigma_{k\langle i}S_{j\rangle k}\right]$$

$$m_{ijk} = -2\frac{\mu}{p}\left[\theta\frac{\partial\sigma_{\langle ij}}{\partial x_{k\rangle}} - \sigma_{\langle ij}\theta\frac{\partial\ln\rho}{\partial x_{k\rangle}} + \frac{4}{5}q_{\langle i}\frac{\partial v_j}{\partial x_{k\rangle}}\right] \qquad (9.13)$$

Grad 26 moments

The 26 moment case, of order $\mathcal{O}\left(\varepsilon^4\right)$ according to the order of magnitude method, gives full balance equations for Δ, R_{ij} and m_{ijk} (6.18–6.20), which can be rewritten as

$$\frac{D\Delta}{Dt} + 8\theta\frac{\partial q_k}{\partial x_k} + 8\theta\sigma_{kl}\frac{\partial v_k}{\partial x_l} + 20q_k\frac{\partial\theta}{\partial x_k} - 8q_k\theta\frac{\partial\ln\rho}{\partial x_k} - 8\frac{q_k}{\rho}\frac{\partial\sigma_{kl}}{\partial x_l}$$
$$+ 4R_{kl}\frac{\partial v_k}{\partial x_l} + \frac{7}{3}\Delta\frac{\partial v_k}{\partial x_k} = -\frac{2}{3}\frac{p}{\mu}\left(\Delta + \frac{\sigma_{ij}\sigma_{ij}}{\rho}\right), \qquad (9.14)$$

$$\frac{DR_{ij}}{Dt} + \frac{28}{5}\theta\frac{\partial q_{\langle i}}{\partial x_{j\rangle}} + \frac{28}{5}q_{\langle i}\frac{\partial\theta}{\partial x_{j\rangle}} - \frac{28}{5}\theta q_{\langle i}\frac{\partial\ln\rho}{\partial x_{j\rangle}} - \frac{28}{5}\frac{q_{\langle i}}{\rho}\frac{\partial\sigma_{j\rangle k}}{\partial x_k} + 8\theta\sigma_{k\langle i}S_{j\rangle k}$$

$$- \frac{14}{3}\frac{\sigma_{ij}}{\rho}\frac{\partial q_k}{\partial x_k} - \frac{14}{3}\frac{\sigma_{ij}\sigma_{kl}}{\rho}\frac{\partial v_k}{\partial x_l} + \frac{14}{15}\Delta\frac{\partial v_{\langle i}}{\partial x_{j\rangle}} - 2\frac{m_{ijk}}{\rho}\frac{\partial\sigma_{kl}}{\partial x_l} - 2m_{ijk}\theta\frac{\partial\ln\rho}{\partial x_k}$$

$$+ 7m_{ijk}\frac{\partial\theta}{\partial x_k} + 2\theta\frac{\partial m_{ijk}}{\partial x_k} + \frac{8}{7}R_{k\langle i}S_{j\rangle k} + 2R_{k\langle i}\frac{\partial v_{j\rangle}}{\partial x_k} + \frac{5}{3}R_{ij}\frac{\partial v_k}{\partial x_k}$$

$$= -\frac{7}{6}\frac{p}{\mu}\left(R_{ij} + \frac{4}{7}\frac{1}{\rho}\sigma_{k\langle i}\sigma_{j\rangle k}\right) , \quad (9.15)$$

$$\frac{Dm_{ijk}}{Dt} + 3\theta\frac{\partial\sigma_{\langle ij}}{\partial x_{k\rangle}} - 3\sigma_{\langle ij}\theta\frac{\partial\ln\rho}{\partial x_{k\rangle}} + \frac{12}{5}q_{\langle i}\frac{\partial v_j}{\partial x_{k\rangle}} - 3\frac{\sigma_{\langle ij}}{\rho}\frac{\partial\sigma_{k\rangle l}}{\partial x_l}$$

$$+ \frac{3}{7}\frac{\partial R_{\langle ij}}{\partial x_{k\rangle}} + 3m_{l\langle ij}\frac{\partial v_{k\rangle}}{\partial x_l} + m_{ijk}\frac{\partial v_l}{\partial x_l} = -\frac{3}{2}\frac{p}{\mu}m_{ijk} . \quad (9.16)$$

The above equations are obtained from the classical Grad method. When the order of the terms in the equations is considered, one notices that the underlined term in (9.15) is of third order, and will lead to a fifth order contribution to the heat flux. When this term is removed, the equations are of fourth order accuracy, in the sense of the order of magnitude approach.

9.2.4 Moment equations for general molecule types

Generalized 13 moments

For non-Maxwellian molecules, and when some higher order terms are neglected, the order of magnitude approach gives as the equations of second order accuracy a variation of Grad's classical equations, (8.20, 8.21),

$$\frac{D\sigma_{ij}}{Dt} + \sigma_{ij}\frac{\partial v_k}{\partial x_k} + 2\sigma_{k\langle i}\frac{\partial v_{j\rangle}}{\partial x_k} + \frac{4}{5}\Pr\frac{\varpi_3}{\varpi_2}\left(\frac{\partial q_{\langle i}}{\partial x_{j\rangle}} - \omega q_{\langle i}\frac{\partial\ln\theta}{\partial x_{j\rangle}}\right)$$

$$+ \frac{4}{5}\Pr\frac{\varpi_4}{\varpi_2}q_{\langle i}\frac{\partial\ln p}{\partial x_{j\rangle}} + \frac{4}{5}\Pr\frac{\varpi_5}{\varpi_2}q_{\langle i}\frac{\partial\ln\theta}{\partial x_{j\rangle}} + \frac{\varpi_6}{\varpi_2}\sigma_{k\langle i}S_{j\rangle k}$$

$$= -\frac{2}{\varpi_2}\frac{p}{\mu}\left[\sigma_{ij} + 2\mu\frac{\partial v_{\langle i}}{\partial x_{j\rangle}}\right] , \quad (9.17)$$

$$\frac{Dq_i}{Dt} + q_k\frac{\partial v_i}{\partial x_k} + \frac{5}{3}q_i\frac{\partial v_k}{\partial x_k} - \frac{5}{2}\frac{1}{\Pr}\sigma_{ik}\frac{\partial\theta}{\partial x_k} + \frac{5}{4}\frac{1}{\Pr}\frac{\theta_3}{\theta_2}\theta\sigma_{ik}\frac{\partial\ln p}{\partial x_k}$$

$$+ \frac{5}{4}\frac{1}{\Pr}\frac{\theta_4}{\theta_2}\theta\left(\frac{\partial\sigma_{ik}}{\partial x_k} - \omega\sigma_{ik}\frac{\partial\ln\theta}{\partial x_k}\right) + \frac{5}{2}\frac{1}{\Pr}\frac{3}{2}\frac{\theta_5}{\theta_2}\sigma_{ik}\frac{\partial\theta}{\partial x_k}$$

$$= -\frac{1}{\theta_2}\frac{5}{2}\frac{1}{\Pr}\frac{p}{\mu}\left[q_i + \frac{5}{2}\frac{\mu}{\Pr}\frac{\partial\theta}{\partial x_i}\right] . \quad (9.18)$$

ϖ_α and θ_α are the Burnett coefficients of Table 9.3. For Maxwell molecules, the above equations reduce to Grad's 13 moment equations.

Jin-Slemrod equations

Jin and Slemrod's regularization gives the equations (7.15–7.17). Due to their similarity with the generalized 13 moment equations from above, they are obtained by adding terms to the right hand sides of the latter[2], i.e.

$$\frac{D\sigma_{ij}}{Dt} + \cdots = -\frac{2}{\varpi_2}\frac{p}{\mu}\left[\sigma_{ij} + 2\mu\frac{\partial v_{\langle i}}{\partial x_{j\rangle}} - P_{ij|3}\right],$$

$$\frac{Dq_i}{Dt} + \cdots = -\frac{1}{\theta_2}\frac{5}{2}\frac{1}{\Pr}\frac{p}{\mu}\left[q_i + \frac{5}{2}\frac{\mu}{\Pr}\frac{\partial\theta}{\partial x_i} - q_{i|3}\right]$$

with

$$P_{ij|3} = \frac{\mu^2}{p^2}\left[\hat{\omega}_2 S_{kl}S_{kl} + \hat{\omega}_3\frac{1}{\theta}\frac{\partial\theta}{\partial x_k}\frac{\partial\theta}{\partial x_k}\right]\sigma_{ij} + \hat{\gamma}_1\frac{\mu}{p}\frac{1}{\theta}\left(\frac{D\theta}{Dt} + \frac{2}{3}\theta\frac{\partial v_k}{\partial x_k}\right)\sigma_{ij}$$
$$+ \frac{3}{5}\Pr\hat{\omega}_4\frac{\partial}{\partial x_k}\left[\frac{\mu^3}{\rho^2}\frac{\partial}{\partial x_k}\left(\frac{\sigma_{ij}}{2\mu\theta}\right)\right], \quad (9.19)$$

$$q_{i|3} = \frac{\mu^2}{p^2}\left(\hat{\theta}_2 S_{kl}S_{kl} + \hat{\theta}_3\frac{1}{\theta}\frac{\partial\theta}{\partial x_k}\frac{\partial\theta}{\partial x_k}\right)q_i + \frac{2}{5}\Pr\hat{\lambda}_1\frac{\mu}{p}\frac{1}{\theta}\left(\frac{D\theta}{Dt} + \frac{2}{3}\theta\frac{\partial v_k}{\partial x_k}\right)q_i$$
$$+ \frac{2}{5}\Pr\hat{\theta}_4\frac{\partial}{\partial x_k}\left(\frac{\mu^3\theta}{\rho^2}\frac{\partial}{\partial x_k}\left(\frac{q_i}{\mu\theta^2}\right)\right). \quad (9.20)$$

$\hat{\omega}_2$, $\hat{\omega}_3$, $\hat{\omega}_4$, $\hat{\gamma}_1$, $\hat{\lambda}_1$, $\hat{\theta}_2$, $\hat{\theta}_3$, $\hat{\theta}_4$ are unknown dimesnionless coefficients.

9.3 One-dimensional equations

Many standard tests of macroscopic equations consider one-dimensional problems, where all variables depend only on the coordinate $x = x_1$, and all vectorial moments point only into the $x-$direction, so that, e.g.,

$$v_i = \{v(x,t),0,0\} \quad, \quad q_i = \{q(x,t),0,0\}.$$

We will use the notation

$$v_1 = v, \; q_1 = q, \; \sigma_{11} = \sigma, \; R_{11} = R, \; m_{111} = m.$$

The equations contain many trace-free tensors, and some care must be exercised when the 1-D equations are derived. For instance, the velocity gradient and its trace-free part are

$$\frac{\partial v_i}{\partial x_j} = \begin{bmatrix} \frac{\partial v}{\partial x} & 0 & 0 \\ 0 & 0 & 0 \\ 0 & 0 & 0 \end{bmatrix}_{ij}, \quad S_{ij} = \frac{\partial v_{\langle i}}{\partial x_{j\rangle}} = \begin{bmatrix} \frac{2}{3}\frac{\partial v}{\partial x} & 0 & 0 \\ 0 & -\frac{1}{3}\frac{\partial v}{\partial x} & 0 \\ 0 & 0 & -\frac{1}{3}\frac{\partial v}{\partial x} \end{bmatrix}_{ij} = \frac{2}{3}\frac{\partial v}{\partial x}\varsigma_{ij}$$

---

[2] There will be a small discrepancy, that plays no role.

where $\varsigma_{ij} = \text{diag}\left[1, -\frac{1}{2}, -\frac{1}{2}\right]_{ij}$. The stress will have only one significant component, $\sigma = \sigma_{11}$, but to be trace-free, the tensor must have the form $\sigma_{ij} = \sigma\varsigma_{ij}$. Additional equations for the derivation of the one-dimensional equations can be found in the problems at the end of the chapter.

9.3.1 Conservation laws

$$\frac{D\rho}{Dt} + \rho\frac{\partial v}{\partial x} = 0 \, ,$$

$$\rho\frac{Dv}{Dt} + \theta\frac{\partial\rho}{\partial x} + \rho\frac{\partial\theta}{\partial x} + \frac{\partial\sigma}{\partial x} = 0 \, , \qquad (9.21)$$

$$\frac{3}{2}\rho\frac{D\theta}{Dt} + \frac{\partial q}{\partial x} = -\rho\theta\frac{\partial v}{\partial x} - \sigma\frac{\partial v}{\partial x} \, .$$

where $\frac{D}{Dt} = \frac{\partial}{\partial t} + v\frac{\partial}{\partial x}$ is the convective time derivative, and the external body force is ignored.

9.3.2 Chapman-Enskog expansion

$$\sigma = \sigma^{(0)} + \sigma^{(1)} + \sigma^{(2)} + \sigma^{(3)} + \cdots \, , \quad q = q^{(0)} + q^{(1)} + q^{(2)} + q^{(3)} + \cdots \, .$$

Zeroth order: Euler equations

$$\sigma_{ij}^{(0)} = 0 \quad \text{and} \quad q^{(0)} = 0 \, .$$

First order: Navier-Stokes-Fourier equations

$$\sigma^{(1)} = -\frac{4}{3}\mu\frac{\partial v}{\partial x} \quad \text{and} \quad q_i^{(1)} = -\kappa\frac{\partial\theta}{\partial x} \, . \qquad (9.22)$$

Second order: Burnett equations

$$\sigma^{(2)} = \frac{\mu^2}{p}\left[\frac{2}{3}\left(\varpi_1 - \frac{7}{3}\varpi_2 + \frac{1}{3}\varpi_6\right)\left(\frac{\partial v}{\partial x}\right)^2 - \varpi_2\frac{2}{3}\frac{\partial}{\partial x}\left(\frac{1}{\rho}\frac{\partial p}{\partial x}\right)\right.$$

$$\left. + \frac{2}{3}\varpi_3\frac{\partial^2\theta}{\partial x^2} + \frac{2}{3}\varpi_4\frac{\partial\theta}{\partial x}\frac{\partial\ln p}{\partial x} + \frac{2}{3}\varpi_5\frac{1}{\theta}\frac{\partial\theta}{\partial x}\frac{\partial\theta}{\partial x}\right] \, , \quad (9.23)$$

$$q^{(2)} = \frac{\mu^2}{\rho}\left[\left(\theta_1 - \frac{8}{3}\theta_2 + 2\theta_5\right)\frac{\partial v}{\partial x}\frac{\partial\ln\theta}{\partial x} + \frac{2}{3}\left(\theta_4 - \theta_2\right)\frac{\partial^2 v}{\partial x^2} + \frac{2}{3}\theta_3\frac{\partial v}{\partial x}\frac{\partial\ln p}{\partial x}\right] \, .$$

$$(9.24)$$

Third order: Super-Burnett equations

$$\sigma^{(3)} = \frac{\mu^3}{p^2}\left[\frac{47}{3}\frac{1}{\rho}\frac{\partial\rho}{\partial x}\frac{\partial\theta}{\partial x}\frac{\partial v}{\partial x} - \frac{64}{9}\frac{\theta}{\rho^2}\left(\frac{\partial\rho}{\partial x}\right)^2\frac{\partial v}{\partial x} + \frac{40}{9}\frac{\theta}{\rho}\frac{\partial^2\rho}{\partial x^2}\frac{\partial v}{\partial x} - \frac{2}{3}\frac{\theta}{\rho}\frac{\partial\rho}{\partial x}\frac{\partial^2 v}{\partial x^2}\right.$$
$$\left. -7\frac{1}{\theta}\left(\frac{\partial\theta}{\partial x}\right)^2\frac{\partial v}{\partial x} - \frac{47}{9}\frac{\partial\theta}{\partial x}\frac{\partial^2 v}{\partial x^2} - \frac{31}{9}\frac{\partial^2\theta}{\partial x^2}\frac{\partial v}{\partial x} + \frac{2}{9}\theta\frac{\partial^3 v}{\partial x^3} + \frac{16}{27}\left(\frac{\partial v}{\partial x}\right)^3\right],$$

$$(9.25)$$

$$q^{(3)} = \frac{\mu^3}{p^2}\left[-\frac{8035}{336}\frac{\partial\theta}{\partial x}\left(\frac{\partial v}{\partial x}\right)^2 + \frac{166}{21}\frac{\theta}{\rho}\frac{\partial\rho}{\partial x}\left(\frac{\partial v}{\partial x}\right)^2 + \frac{949}{168}\theta\frac{\partial^2 v}{\partial x^2}\frac{\partial v}{\partial x}\right.$$
$$+\frac{917}{8}\frac{1}{\rho}\frac{\partial\rho}{\partial x}\left(\frac{\partial\theta}{\partial x}\right)^2 - \frac{1137}{16}\frac{\theta}{\rho^2}\left(\frac{\partial\rho}{\partial x}\right)^2\frac{\partial\theta}{\partial x} + \frac{397}{16}\frac{\theta}{\rho}\frac{\partial\rho}{\partial x}\frac{\partial^2\theta}{\partial x^2}$$
$$+\frac{701}{16}\frac{\theta}{\rho}\frac{\partial^2\rho}{\partial x^2}\frac{\partial\theta}{\partial x} - \frac{813}{16}\frac{1}{\theta}\left(\frac{\partial\theta}{\partial x}\right)^3 - \frac{1451}{16}\frac{\partial\theta}{\partial x}\frac{\partial^2\theta}{\partial x}$$
$$\left.-\frac{157}{16}\theta\frac{\partial^3\theta}{\partial x^3} - \frac{41}{8}\frac{\theta^2}{\rho^2}\frac{\partial^2\rho}{\partial x^2}\frac{\partial\rho}{\partial x} - \frac{5}{8}\frac{\theta^2}{\rho}\frac{\partial^3\rho}{\partial x^3} + \frac{23}{4}\frac{\theta^2}{\rho^3}\left(\frac{\partial\rho}{\partial x}\right)^3\right]. \quad (9.26)$$

Augmented Burnett equations

$$\sigma^{(A)} = \frac{\mu^3}{p^2}\left(\frac{2}{9}\theta\frac{\partial^3 v}{\partial x^3}\right) \;,\quad q^{(A)} = \frac{\mu^3}{p^2}\left(\frac{11}{16}\theta\frac{\partial^3\theta}{\partial x^3} - \frac{5}{8}\frac{\theta^2}{\rho}\frac{\partial^3\rho}{\partial x^3}\right). \quad (9.27)$$

9.3.3 Moment equations for Maxwell molecules

Basic equations for stress and heat flux

$$\frac{D\sigma}{Dt} + \sigma\frac{\partial v}{\partial x} + \frac{8}{15}\frac{\partial q}{\partial x} + \frac{4}{3}p\frac{\partial v}{\partial x} + \frac{4}{3}\sigma\frac{\partial v}{\partial x} + \frac{\partial m}{\partial x} = -\frac{p}{\mu}\sigma\,, \quad (9.28)$$

$$\frac{Dq}{Dt} + \frac{5}{2}p\frac{\partial\theta}{\partial x} + \frac{5}{2}\sigma\frac{\partial\theta}{\partial x} + \theta\frac{\partial\sigma}{\partial x} - \theta\sigma\frac{\partial\ln\rho}{\partial x} + \frac{16}{5}q\frac{\partial v}{\partial x}$$
$$+\frac{1}{2}\frac{\partial R}{\partial x} + \frac{1}{6}\frac{\partial\Delta}{\partial x} + m\frac{\partial v}{\partial x} - \frac{\sigma}{\rho}\frac{\partial\sigma}{\partial x_k} = -\frac{2}{3}\frac{p}{\mu}q\,. \quad (9.29)$$

Grad 13 moments

$$\Delta = 0 \;,\quad R = 0 \;,\quad m = 0\,. \quad (9.30)$$

R13 equations (original)

$$\frac{1}{2}R + \frac{1}{6}\Delta = -\frac{18}{5}\frac{\mu}{p}\left[\left(\theta - \frac{5}{9}\frac{\sigma}{\rho}\right)\frac{\partial q}{\partial x} - \frac{q}{\rho}\frac{\partial \sigma}{\partial x} + \frac{11}{6}q\frac{\partial \theta}{\partial x}\right.$$

$$\left. -\theta q\frac{\partial \ln \rho}{\partial x} + \frac{5}{9}\left(\frac{11}{7}\theta\sigma - \frac{\sigma^2}{\rho}\right)\frac{\partial v}{\partial x}\right] ,$$

$$m = -2\frac{\mu}{p}\left[\frac{3}{5}\theta\frac{\partial \sigma}{\partial x} - \frac{3}{5}\theta\sigma\frac{\partial \ln \rho}{\partial x} + \frac{4}{5}q_{\langle i}\frac{\partial v_j}{\partial x_{k\rangle}} - \frac{3}{5}\frac{\sigma}{\rho}\frac{\partial \sigma}{\partial x}\right] . \qquad (9.31)$$

R13 equations (order of magnitude)

$$\frac{1}{2}R + \frac{1}{6}\Delta = -\frac{13}{42}\frac{1}{\rho}\sigma^2 - \frac{18}{5}\frac{\mu}{p}\left[\theta\frac{\partial q}{\partial x} - \frac{q}{\rho}\frac{\partial \sigma}{\partial x} + \frac{11}{6}q\frac{\partial \theta}{\partial x} - \theta q\frac{\partial \ln \rho}{\partial x} + \frac{55}{63}\theta\sigma\frac{\partial v}{\partial x}\right]$$

$$m = -\frac{\mu}{p}\frac{6}{5}\left[\theta\frac{\partial \sigma}{\partial x} - \theta\sigma\frac{\partial \ln \rho}{\partial x} + \frac{8}{15}q\frac{\partial v}{\partial x}\right] . \qquad (9.32)$$

Grad 26 moments

$$\frac{D\Delta}{Dt} + 8\theta\frac{\partial q}{\partial x} + 8\theta\sigma\frac{\partial v}{\partial x} + 20q\frac{\partial \theta}{\partial x} - 8q\theta\frac{\partial \ln \rho}{\partial x_k} - 8\frac{q}{\rho}\frac{\partial \sigma}{\partial x}$$

$$+ 4R\frac{\partial v}{\partial x} + \frac{7}{3}\Delta\frac{\partial v}{\partial x} = -\frac{2}{3}\frac{p}{\mu}\left(\Delta + \frac{\sigma^2}{\rho}\right) , \qquad (9.33)$$

$$\frac{DR}{Dt} + \frac{56}{15}\theta\frac{\partial q}{\partial x} + \frac{56}{15}q\frac{\partial \theta}{\partial x} - \frac{56}{15}\theta q\frac{\partial \ln \rho}{\partial x} - \frac{56}{15}\frac{q}{\rho}\frac{\partial \sigma}{\partial x} + \frac{8}{3}\theta\sigma\frac{\partial v}{\partial x} - \frac{14}{3}\frac{\sigma}{\rho}\frac{\partial q}{\partial x}$$

$$- \frac{14}{3}\frac{\sigma^2}{\rho}\frac{\partial v}{\partial x} + \frac{28}{45}\Delta\frac{\partial v}{\partial x} - 2\frac{m}{\rho}\frac{\partial \sigma}{\partial x} - 2m\theta\frac{\partial \ln \rho}{\partial x} + 7m\frac{\partial \theta}{\partial x} + 2\theta\frac{\partial m}{\partial x}$$

$$+ \frac{71}{21}R\frac{\partial v}{\partial x} = -\frac{7}{6}\frac{p}{\mu}\left(R + \frac{2}{7}\frac{\sigma^2}{\rho}\right) , \qquad (9.34)$$

$$\frac{Dm}{Dt} + \frac{9}{5}\theta\frac{\partial \sigma}{\partial x} - \frac{9}{5}\theta\sigma\frac{\partial \ln \rho}{\partial x} + \frac{24}{25}q\frac{\partial v}{\partial x} - \frac{9}{5}\frac{\sigma}{\rho}\frac{\partial \sigma}{\partial x} + \frac{9}{35}\frac{\partial R}{\partial x}$$

$$+ \frac{14}{5}m\frac{\partial v}{\partial x} = -\frac{3}{2}\frac{p}{\mu}m . \qquad (9.35)$$

9.3.4 Moment equations for general molecule types

$$\frac{D\sigma}{Dt} + \frac{7}{3}\sigma\frac{\partial v}{\partial x} + \frac{4}{5}\Pr\frac{\varpi_3}{\varpi_2}\frac{2}{3}\left(\frac{\partial q}{\partial x} - \omega q\frac{\partial \ln\theta}{\partial x}\right) - \frac{8}{15}\Pr\frac{\varpi_4}{\varpi_2}q\frac{\partial \ln p}{\partial x}$$

$$+ \frac{8}{15}\Pr\frac{\varpi_5}{\varpi_2}q\frac{\partial \ln\theta}{\partial x} + \frac{\varpi_6}{\varpi_2}\frac{1}{3}\sigma\frac{\partial v}{\partial x} = -\frac{2}{\varpi_2}\frac{p}{\mu}\left[\sigma + \frac{4}{3}\mu\frac{\partial v}{\partial x} - P_{|3}\right], \quad (9.36)$$

$$\frac{Dq}{Dt} + \frac{8}{3}q\frac{\partial v}{\partial x} - \frac{5}{2}\frac{1}{\Pr}\sigma\frac{\partial \theta}{\partial x} + \frac{5}{4}\frac{1}{\Pr}\frac{\theta_3}{\theta_2}\theta\sigma\frac{\partial \ln p}{\partial x_k} + \frac{5}{4}\frac{1}{\Pr}\frac{\theta_4}{\theta_2}\theta\left(\frac{\partial\sigma}{\partial x} - \omega\sigma\frac{\partial \ln\theta}{\partial x}\right)$$

$$+ \frac{5}{2}\frac{1}{\Pr}\frac{3}{2}\frac{\theta_5}{\theta_2}\sigma\frac{\partial\theta}{\partial x} = -\frac{1}{\theta_2}\frac{5}{2}\frac{1}{\Pr}\frac{p}{\mu}\left[q + \frac{5}{2}\frac{\mu}{\Pr}\frac{\partial\theta}{\partial x} - q_{|3}\right]. \quad (9.37)$$

Generalized 13 moments

$$P_{|3} = q_{|3} = 0$$

Jin-Slemrod

$$P_{|3} = \frac{\mu^2}{p^2}\left[\frac{2}{3}\hat{\omega}_2\left(\frac{\partial v}{\partial x}\right)^2 + \hat{\omega}_3\frac{1}{\theta}\left(\frac{\partial\theta}{\partial x}\right)^2\right]\sigma + \hat{\gamma}_1\frac{\mu}{p}\frac{1}{\theta}\left(\frac{D\theta}{Dt} + \frac{2}{3}\theta\frac{\partial v}{\partial x}\right)\sigma$$

$$+ \frac{3}{5}\Pr\hat{\omega}_4\frac{\partial}{\partial x}\left[\frac{\mu^3}{\rho^2}\frac{\partial}{\partial x}\left(\frac{\sigma}{2\mu\theta}\right)\right], \quad (9.38)$$

$$q_{|3} = \frac{\mu^2}{p^2}\left(\frac{2}{3}\hat{\theta}_2\left(\frac{\partial v}{\partial x}\right)^2 + \hat{\theta}_3\frac{1}{\theta}\left(\frac{\partial\theta}{\partial x}\right)^2\right)q + \frac{2}{5}\Pr\hat{\lambda}_1\frac{\mu}{p}\frac{1}{\theta}\left(\frac{D\theta}{Dt} + \frac{2}{3}\theta\frac{\partial v}{\partial x}\right)q$$

$$+ \frac{2}{5}\Pr\hat{\theta}_4\frac{\partial}{\partial x}\left(\frac{\mu^3\theta}{\rho^2}\frac{\partial}{\partial x}\left(\frac{q_i}{\mu\theta^2}\right)\right). \quad (9.39)$$

9.4 Linear dimensionless equations

Next, all equations are presented for small deviations from an equilibrium state given by ρ_0, θ_0, $v_{i,0} = 0$. Dimensionless variables $\hat{\rho}$, $\hat{\theta}$, $\hat{v}_i$, $\hat{\sigma}_{ij}$, $\hat{q}$, $\hat{\Delta}$, $\hat{R}_{ij}$, $\hat{m}_{ijk}$, and dimensionless space and time variables $\hat{x}_i$, $\hat{t}$ are introduced as

$$\rho = \rho_0\left(1 + \hat{\rho}\right), \quad \theta = \theta_0\left(1 + \hat{\theta}\right), \quad p = \rho_0\theta_0\left(1 + \hat{\rho} + \hat{\theta}\right), \quad v_i = \sqrt{\theta_0}\hat{v}_i,$$

$$\sigma_{ij} = \rho_0\theta_0\hat{\sigma}_{ij}, \quad q_i = \rho_0\sqrt{\theta_0}^3\hat{q}_i, \quad \Delta = \rho_0\theta_0^2\hat{\Delta}, \quad R_{ij} = \rho_0\theta_0^2\hat{R}_{ij}, \quad (9.40)$$

$$m_{ijk} = \rho_0\sqrt{\theta_0}^3\hat{m}_{ijk}, \quad x_i = L\hat{x}_i, \quad t = \frac{L}{\sqrt{\theta_0}}\hat{t}.$$

L is a relevant length scale of the process, which also defines the Knudsen number,

$$\text{Kn} = \frac{\mu\sqrt{\theta}}{pL} = \frac{\mu_0\sqrt{\theta_0}}{p_0L} \ . \tag{9.41}$$

The definition of the Knudsen number varies in the literature; definitions frequently used are $\text{Kn}' = \sqrt{\frac{\pi}{2}}\text{Kn} = 1.25331\,\text{Kn}$ [3] and $\text{Kn}'' = \frac{8}{5}\sqrt{\frac{2}{\pi}}\text{Kn} = 1.27662\,\text{Kn}$ [1]. The definition (9.41) was chosen since it is most convenient to use in macroscopic equations.

To linearize, (9.40) is inserted into all equations, and only terms that are linear in the small quantities $\hat{\psi}$ are considered; the hats are omitted for better readability.

9.4.1 Conservation laws

$$\frac{\partial\rho}{\partial t} + \frac{\partial v_k}{\partial x_k} = 0 \ ,$$

$$\frac{\partial v_i}{\partial t} + \frac{\partial\theta}{\partial x_k} + \frac{\partial\rho}{\partial x_k} + \frac{\partial\sigma_{ik}}{\partial x_k} = G_i \ , \tag{9.42}$$

$$\frac{3}{2}\frac{\partial\theta}{\partial t} + \frac{\partial q_k}{\partial x_k} = -\frac{\partial v_k}{\partial x_k} \ .$$

9.4.2 Chapman-Enskog expansion

Zeroth order: Euler equations

$$\sigma_{ij}^{(0)} = 0 \ \text{and} \ q_i^{(0)} = 0 \ .$$

First order: Navier-Stokes-Fourier equations

$$\sigma_{ij}^{(1)} = -2\text{Kn}\frac{\partial v_{\langle i}}{\partial x_{k\rangle}} \ \text{and} \ q_i^{(1)} = -\frac{5}{2}\frac{\text{Kn}}{\text{Pr}}\frac{\partial\theta}{\partial x_i} \ . \tag{9.43}$$

Second order: Burnett equations

$$\sigma_{ij}^{(2)} = \text{Kn}^2\left[-\varpi_2\frac{\partial^2\rho}{\partial x_{\langle i}\partial x_{j\rangle}} + (\varpi_3 - \varpi_2)\frac{\partial^2\theta}{\partial x_{\langle i}\partial x_{j\rangle}}\right] \ ,$$

$$\tag{9.44}$$

$$q_i^{(2)} = \text{Kn}^2\left[\frac{\theta_4}{2}\frac{\partial^2 v_i}{\partial x_k\partial x_k} + \frac{2}{3}\left(\frac{\theta_4}{4} - \theta_2\right)\frac{\partial^2 v_k}{\partial x_i\partial x_k}\right] \ .$$

Third order: Super-Burnett equations

$$\sigma_{ij}^{(3)} = \text{Kn}^3 \left(\frac{5}{3} \frac{\partial^2}{\partial x_{\langle i} \partial x_{j\rangle}} \frac{\partial v_k}{\partial x_k} - \frac{4}{3} \frac{\partial^2}{\partial x_k \partial x_k} \frac{\partial v_{\langle i}}{\partial x_{j\rangle}} \right) ,$$

$$q_i^{(3)} = \text{Kn}^3 \left(-\frac{157}{16} \frac{\partial^3 \theta}{\partial x_i \partial x_k \partial x_k} - \frac{5}{8} \frac{\partial^3 \rho}{\partial x_i \partial x_k \partial x_k} \right) .$$

$$(9.45)$$

Augmented Burnett equations

$$\sigma_{ij}^{(3)} \rightarrow \sigma_{ij}^{(A)} = \text{Kn}^3 \left(\frac{1}{3} \frac{\partial^2}{\partial x_k \partial x_k} \frac{\partial v_{\langle i}}{\partial x_{j\rangle}} \right) ,$$

$$q_i^{(3)} \rightarrow q_i^{(A)} = \text{Kn}^3 \left(\frac{11}{16} \frac{\partial^3 \theta}{\partial x_i \partial x_k \partial x_k} - \frac{5}{8} \frac{\partial^3 \rho}{\partial x_i \partial x_k \partial x_k} \right) .$$

$$(9.46)$$

9.4.3 Moment equations for Maxwell molecules

Basis equations for stress and heat flux

$$\frac{\partial \sigma_{ij}}{\partial t} + \frac{4}{5} \frac{\partial q_{\langle i}}{\partial x_{j\rangle}} + 2 \frac{\partial v_{\langle i}}{\partial x_{j\rangle}} + \frac{\partial m_{ijk}}{\partial x_k} = -\frac{\sigma_{ij}}{\text{Kn}} ,$$

$$\frac{\partial q_i}{\partial t} + \frac{5}{2} \frac{\partial \theta}{\partial x_i} + \frac{\partial \sigma_{ik}}{\partial x_k} + \frac{1}{2} \frac{\partial R_{ik}}{\partial x_k} + \frac{1}{6} \frac{\partial \Delta}{\partial x_i} = -\frac{2}{3} \frac{q_i}{\text{Kn}} .$$

$$(9.47)$$

Grad 13 moments

$$m_{ijk} = 0 , \quad R_{ij} = 0 , \quad \Delta = 0 .$$

$$(9.48)$$

R13 equations (original/order of magnitude)

$$\Delta = -12\text{Kn} \frac{\partial q_k}{\partial x_k} , \quad R_{ij} = -\frac{24}{5} \text{Kn} \frac{\partial q_{\langle i}}{\partial x_{j\rangle}} , \quad m_{ijk} = -2\text{Kn} \frac{\partial \sigma_{\langle ij}}{\partial x_{k\rangle}} . \quad (9.49)$$

Grad 26 moments

$$\frac{\partial \Delta}{\partial t} + 8 \frac{\partial q_k}{\partial x_k} = -\frac{2}{3} \frac{\Delta}{\text{Kn}} ,$$

$$\frac{\partial R_{ij}}{\partial t} + \frac{28}{5} \frac{\partial q_{\langle i}}{\partial x_{j\rangle}} + 2 \frac{\partial m_{ijk}}{\partial x_k} = -\frac{7}{6} \frac{R_{ij}}{\text{Kn}} ,$$

$$\frac{\partial m_{ijk}}{\partial t} + 3 \frac{\partial \sigma_{\langle ij}}{\partial x_{k\rangle}} + \frac{3}{7} \frac{\partial R_{\langle ij}}{\partial x_{k\rangle}} = -\frac{3}{2} \frac{p}{\mu} m_{ijk} .$$

$$(9.50)$$

9.4.4 Moment equations for general molecule types

Generalized 13 moments/Jin-Slemrod

$$\frac{\partial \sigma_{ij}}{\partial t} + \frac{4}{5} \Pr \frac{\varpi_3}{\varpi_2} \frac{\partial q_{\langle i}}{\partial x_{j\rangle}} = -\frac{2}{\text{Kn}} \frac{1}{\varpi_2} \left[\sigma_{ij} + 2\text{Kn} \frac{\partial v_{\langle i}}{\partial x_{j\rangle}} - P_{ij|3} \right] ,$$

$$\tag{9.51}$$

$$\frac{\partial q_i}{\partial t} + \frac{5}{4} \frac{1}{\Pr} \frac{\theta_4}{\theta_2} \frac{\partial \sigma_{ik}}{\partial x_k} = -\frac{5}{2} \frac{1}{\text{Kn}} \frac{1}{\theta_2} \frac{1}{\Pr} \left[q_i + \frac{5}{2} \frac{\text{Kn}}{\Pr} \frac{\partial \theta}{\partial x_i} - q_{i|3} \right] .$$

For the generalized 13 moment case:

$$P_{ij|3} = q_{i|3} = 0 ,$$

for the Jin-Slemrod equations:

$$P_{ij|3} = \frac{3}{10} \text{Kn}^2 \Pr \hat{\omega}_4 \frac{\partial^2 \sigma_{ij}}{\partial x_k \partial x_k} , \quad q_{i|3} = \frac{2}{5} \text{Kn}^2 \Pr \hat{\theta}_4 \frac{\partial^2 q_i}{\partial x_k \partial x_k}$$

Problems

9.1. Trace-free 2-tensors in 1-D problems

Show that for one-dimensional problems the following relations hold

$$\frac{\partial v_k}{\partial x_{\langle i}} \frac{\partial v_{j\rangle}}{\partial x_k} = \frac{2}{3} \left(\frac{\partial v}{\partial x} \right)^2 \varsigma_{ij} , \quad \frac{\partial}{\partial x_{\langle i}} \left(\frac{1}{\rho} \frac{\partial p}{\partial x_{j\rangle}} \right) = \frac{2}{3} \frac{\partial}{\partial x} \left(\frac{1}{\rho} \frac{\partial p}{\partial x} \right) \varsigma_{ij} ,$$

$$\frac{\partial^2 \theta}{\partial x_{\langle i} \partial x_{j\rangle}} = \frac{2}{3} \frac{\partial^2 \theta}{\partial x^2} \varsigma_{ij} , \quad \frac{\partial v_k}{\partial x_{\langle i}} S_{j\rangle k} = \frac{4}{9} \left(\frac{\partial v}{\partial x} \right)^2 \varsigma_{ij} , \quad S_{k\langle i} S_{j\rangle k} = \frac{2}{9} \left(\frac{\partial v}{\partial x} \right)^2 \varsigma_{ij},$$

$$\sigma_{k\langle i} \frac{\partial v_{j\rangle}}{\partial x_k} = \frac{2}{3} \sigma \frac{\partial v}{\partial x} \varsigma_{ij} , \quad \sigma_{k\langle i} S_{j\rangle k} = \frac{1}{3} \sigma \frac{\partial v}{\partial x} \varsigma_{ij} , \quad \sigma_{k\langle i} \sigma_{j\rangle k} = \frac{1}{2} \sigma^2 \varsigma_{ij} .$$

9.2. Trace-free 3-tensors in 1-D problems

Show that for one-dimensional problems the following relations hold

$$\frac{\partial \sigma_{\langle 11}}{\partial x_{1\rangle}} = \frac{3}{5} \frac{\partial \sigma}{\partial x} , \quad \sigma_{\langle 11} \frac{\partial \ln \rho}{\partial x_{1\rangle}} = \frac{3}{5} \sigma \frac{\partial \ln \rho}{\partial x} , \quad \frac{\sigma_{\langle 11}}{\rho} \frac{\partial \sigma_{1\rangle l}}{\partial x_l} = \frac{3}{5} \frac{\sigma}{\rho} \frac{\partial \sigma}{\partial x} ,$$

$$\frac{\partial R_{\langle ij}}{\partial x_{k\rangle}} = \frac{3}{5} \frac{\partial R}{\partial x} , \quad m_{l\langle 11} \frac{\partial v_{1\rangle}}{\partial x_l} = \frac{3}{5} m \frac{\partial v}{\partial x} , \quad q_{\langle i} \frac{\partial v_j}{\partial x_{k\rangle}} = q_{\langle i} S_{jk\rangle} = \frac{3}{5} q S_{11} = \frac{2}{5} q \frac{\partial v}{\partial x} .$$

Hint: Use that, e.g.,

$$\frac{\partial \sigma_{\langle ij}}{\partial x_{k\rangle}} = \frac{1}{3} \left(\frac{\partial \sigma_{ij}}{\partial x_k} + \frac{\partial \sigma_{ik}}{\partial x_j} + \frac{\partial \sigma_{jk}}{\partial x_i} \right) - \frac{2}{15} \left(\frac{\partial \sigma_{ir}}{\partial x_r} \delta_{jk} + \frac{\partial \sigma_{jr}}{\partial x_r} \delta_{ik} + \frac{\partial \sigma_{kr}}{\partial x_r} \delta_{ij} \right)$$

10

Stability and dispersion

The last chapter summarized the equations which were derived so far in this text, and now it is time to study the properties of the equations, and discuss their relative merits. Stability problems of the Burnett and super-Burnett equations were mentioned several times and are now discussed in detail. Moreover, we briefly discuss speed of sound, and the dispersion and attenuation of sound waves.

10.1 Linear stability

10.1.1 Plane harmonic waves

For one-dimensional processes the linear transport equations of Sec. 9.4 can be written as

$$\frac{\partial u_A}{\partial t} + \mathcal{A}_{AB}\frac{\partial u_B}{\partial x} + \mathcal{B}_{AB}\frac{\partial^2 u_B}{\partial x^2} + \cdots = \mathcal{C}_{AB}u_B , \tag{10.1}$$

with constant matrices $\mathcal{A}_{AB}, \mathcal{B}_{AB}, \mathcal{C}_{AB}, \ldots$ For the solution, we assume plane waves of the form

$$u_A = \tilde{u}_A \exp\left[i\left(\Omega t - kx\right)\right]$$

where $\tilde{u}_A$ is the complex amplitude of the wave, Ω is its frequency, and k is its wave number. The use of complex quantities simplifies the treatment, and it is implicitly understood that only the real parts of the expressions for the u_A are relevant.

The equations (10.1) can then be written as

$$\mathcal{G}_{AB}\left(\Omega, k\right)\tilde{u}_B = 0 \text{ where } \mathcal{G}_{AB}\left(\Omega, k\right) = i\Omega\delta_{AB} - ik\mathcal{A}_{AB} - k^2\mathcal{B}_{AB} + \cdots - \mathcal{C}_{AB}$$

and nontrivial solutions require

$$\det\left[\mathcal{G}_{AB}\left(\Omega, k\right)\right] = 0 . \tag{10.2}$$

The resulting relation between Ω and k is the dispersion relation.

If a disturbance in space is considered, the wave number k is real, and the frequency is complex, $\Omega = \Omega_r(k) + i\Omega_i(k)$. Then, the wave solutions assume the form

$$u_A = \tilde{u}_A \exp[-\alpha t] \exp[ik(v_{ph}t - x)]$$

where phase velocity v_{ph} and damping α are given by

$$v_{ph} = \frac{\Omega_r(k)}{k} \quad \text{and} \quad \alpha = \Omega_i(k) .$$

Note that $\tilde{u}_A \exp[-\alpha t]$ describes the local amplitude as a function of time, while the exponential with imaginary argument, $\exp[ik(v_{ph}t - x)]$, describes harmonic waves moving with the velocity v_{ph}.[1]

Stability requires damping, and thus $\Omega_i(k) \geq 0$. If, however, $\Omega_i(k) < 0$ then a small disturbance in space will blow up in time.

If a disturbance in time at a given location is considered, the frequency Ω is real, while the wave number is complex, $k = k_r(\Omega) + ik_i(\Omega)$. Then, the wave solutions assume the form

$$u_A = \tilde{u}_A \exp[-\alpha x] \exp\left[i\Omega\left(t - \frac{x}{v_{ph}}\right)\right]$$

where phase velocity v_{ph} and damping α are now given by

$$v_{ph} = \frac{\Omega}{k_r(\Omega)} \quad \text{and} \quad \alpha = -k_i(\Omega) .$$

Note that $\tilde{u}_A \exp[-\alpha x]$ describes the amplitude at the point x, while the exponential with imaginary argument, $\exp\left[i\Omega\left(t - \frac{x}{v_{ph}}\right)\right]$, describes harmonic waves moving with the speed v_{ph}.[2]

For a wave traveling in positive x-direction ($k_r > 0$), the damping must be negative ($k_i < 0$), while for a wave traveling in negative x-direction ($k_r < 0$), the damping must be positive ($k_i > 0$). If this condition is not fulfilled, a disturbance in time at some point will lead to a much larger disturbance at other locations.

Thus, in order to test the stability of a given set of equations, one has to perform two tests, for stability in time, and stability in space. However, for the Burnett and super-Burnett equations, most authors only consider stability in time, and ignore stability in space [56][62]. Problem 10.1 gives a simple example for an equation that is stable in space but unstable in time.

---

[1] The complex amplitude can be written as $\tilde{u}_A = \bar{u}_A \exp[i\varphi]$, where φ is the phase shift, and $\bar{u}_A$ is the real amplitude. Then, the real part of u_A is $\mathrm{Re}[u_A] = \bar{u}_A \exp[-\alpha t] \cos[k(v_{ph}t - x) + \varphi]$.

[2] The real part is $\mathrm{Re}[u_A] = \bar{u}_A \exp[-\alpha x] \cos\left[\Omega\left(t - \frac{x}{v_{ph}}\right) + \varphi\right]$.

10.1.2 Linear one-dimensional equations

We ask for linear stability in purely one-dimensional processes, and treat the problem by means of the one-dimensional counterparts of the dimensionless linearized equations of Sec. 9.4.

The reference length L is chosen as the mean free path, and the reference time is the mean free time. Accordingly, we have to set $\mathrm{Kn} = 1$, which implies $L = \frac{\mu_0}{p_0}\sqrt{\theta_0}$. It follows that the wave number k is measured in units of the inverse mean free path, and the wave frequency in terms of the collision frequency $1/\tau$.

Thus, the proper definition for the Knudsen number for an oscillation with frequency ω is $\mathrm{Kn}_\Omega = \omega\tau$, that is $\mathrm{Kn}_\Omega = \Omega$ with the dimensionless measure used. Similarly, for a given wave number k, the proper Knudsen number is $\mathrm{Kn}_k = k$.

In the one-dimensional case, where all variables depend only on $x_1 = x$, and where $v_i = \{v(x,t), 0, 0\}$, the conservation laws (9.42) reduce to

$$\frac{\partial \rho}{\partial t} + \frac{\partial v}{\partial x} = 0\,,$$

$$\frac{\partial v}{\partial t} + \frac{\partial \rho}{\partial x} + \frac{\partial \theta}{\partial x} + \frac{\partial \sigma}{\partial x} = 0\,,$$

$$\frac{3}{2}\frac{\partial \theta}{\partial t} + \frac{\partial q}{\partial x} + \frac{\partial v}{\partial x} = 0\,,$$

where $\sigma = \sigma_{11}$ and $q = q_1$.

10.1.3 Euler equations, speed of sound

For the Euler equations stress and heat flux vanish, $\sigma = q = 0$, and this gives

$$\mathcal{G}_{AB} = \begin{bmatrix} i\Omega & -ik & 0 \\ -ik & i\Omega & -ik \\ 0 & -ik & \frac{3}{2}i\Omega \end{bmatrix}_{AB}$$

with the dispersion relation

$$\det \mathcal{G}_{AB} = \frac{3}{2}i^3\Omega^3 - \frac{5}{2}i^3 k^2\Omega = 0\,.$$

This implies

$$\Omega_{1,2} = \pm\sqrt{\frac{5}{3}}k \quad \text{and} \quad \Omega_3 = 0\,,$$

or, when solved for k,

$$k_{1,2} = \pm\sqrt{\frac{3}{5}}\Omega\,.$$

Both cases lead to the absolute phase velocity $c_{ph} = \sqrt{\frac{5}{3}}$, or, when dimensions are re-inserted,

$$c_{ph} = a = \sqrt{\frac{5}{3}\theta_0} = \sqrt{\frac{5}{3}\frac{k}{m}T_0} \ . \tag{10.3}$$

a denotes the speed of sound of an ideal monatomic gas.

The Euler equations are frictionless, that is the damping vanishes, $\alpha = 0$.

10.1.4 Linear stability in time

First, we test the stability against a disturbance of given wave length, or wave number k.

The equations from the CE-expansion are $(9.43$–$9.45)$[3]

$$\sigma_{CE} = -\varepsilon\frac{4}{3}\frac{\partial v}{\partial x} - \varepsilon^2\left[\frac{4}{3}\frac{\partial^2\rho}{\partial x^2} - \frac{2}{3}\frac{\partial^2\theta}{\partial x^2}\right] + \varepsilon^3\frac{2}{9}\frac{\partial^3 v}{\partial x^3} + \cdots \ ,$$

$$q_{CE} = -\varepsilon\frac{15}{4}\frac{\partial\theta}{\partial x} - \varepsilon^2\frac{7}{4}\frac{\partial^2 v}{\partial x^2} - \varepsilon^3\left[\frac{157}{16}\frac{\partial^3\theta}{\partial x^3} + \frac{5}{8}\frac{\partial^3\rho}{\partial x^3}\right] + \cdots \ ,$$

and these are the NSF equations if only terms of order $\mathcal{O}(\varepsilon)$ are considered, the Burnett equations when the quadratic terms in ε are added, and the super-Burnett equations when also the third order terms are considered. For the augmented Burnett equations (9.46), the factor $(-157/16)$ in the super-Burnett contribution to the heat flux must be replaced by $(11/16)$.

The linear 13 moment equations $(9.47,\ 9.48)$, and their regularization (9.49) can be written together as

$$\frac{\partial\sigma}{\partial t} + \frac{8}{15}\frac{\partial q}{\partial x} + \frac{4}{3}\frac{\partial v}{\partial x} - \epsilon\frac{6}{5}\frac{\partial^2\sigma}{\partial x^2} = -\sigma \ ,$$

$$\frac{\partial q}{\partial t} + \frac{5}{2}\frac{\partial\theta}{\partial x} + \frac{\partial\sigma}{\partial x} - \epsilon\frac{18}{5}\frac{\partial^2 q}{\partial x^2} = -\frac{2}{3}q \ ,$$

where $\epsilon = 0$ for the Grad 13 equations, and $\epsilon = 1$ for the regularization.

The computation of the dispersion relation (10.2) for all of these cases, and the evaluation of these, is a rather straightforward problem, of which no details are given (see Problem 10.3). The characteristic polynomial $\det\mathcal{G}_{AB} = 0$ has several solutions, denoted as modes. All modes have to be considered for stability tests.

Figure 10.1 shows $\Omega = \Omega_r(k) + i\Omega_i(k)$ in the complex plane with k as parameter. The results show that the NSF, augmented Burnett, Grad 13, and R13 equations are stable for all frequencies, since the solutions of the dispersion relation always have positive imaginary parts, i.e. positive damping. Burnett and super-Burnett equations, however, are unstable, since they have contributions with negative damping [56].

The difference between super-Burnett and augmented Burnett equations is small, and lies essentially in the different sign between the coefficients

---

[3] Here Kn $= 1$, and the formal smallness parameter ε is used to indicate the order of the equations. Recall that $\varepsilon = 1$ for actual calculations.

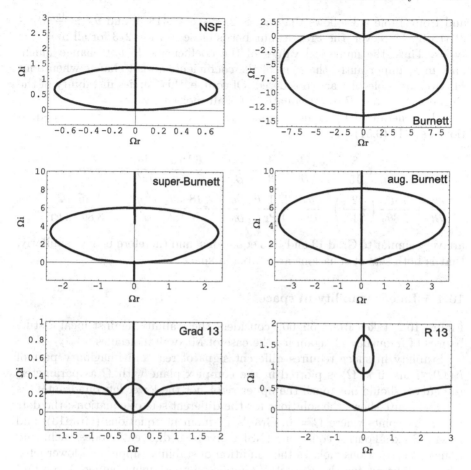

Fig. 10.1. The solutions $\Omega(k)$ of the dispersion relation in the complex plane with k as parameter for Navier-Stokes-Fourier, Burnett, Super-Burnett, augmented Burnett, Grad 13, and R13 equations.

$(-157/16)$ in the super-Burnett equations, and $(11/16)$ in the augmented Burnett equations. This change in sign changes the character of the equations so that they become stable in time.

Similar figures for non-Maxwellian molecules are not shown. Indeed, the relevant equations are very similar to the equations for Maxwell molecules. The Burnett equations read

$$\sigma_{CE} = -\varepsilon\frac{4}{3}\frac{\partial v}{\partial x} - \varepsilon^2\left[\frac{2\varpi_2}{3}\frac{\partial^2\rho}{\partial x^2} + \frac{2}{3}(\varpi_2 - \varpi_3)\frac{\partial^2\theta}{\partial x^2}\right] + \cdots,$$

$$q_{CE} = -\varepsilon\frac{5}{2\,\mathrm{Pr}}\frac{\partial\theta}{\partial x} - \varepsilon^2\frac{2}{3}(\theta_4 - \theta_2)\frac{\partial^2 v}{\partial x^2} - \cdots.$$

and from Table 9.3 follows that $\varpi_2 \simeq 2$, $\varpi_3 \in [2.418, 3]$ and $\theta_2 \simeq \frac{45}{8}$, $\theta_4 \in [2.418, 3]$. Moreover, the Prandtl number is close to $\text{Pr} = 2/3$ for all molecule types. Thus, the numerical values of the coefficients do not change much, and, most importantly, the signs of the coefficients do not change when non-Maxwellian molecules are considered. Of course, this implies instability of the Burnett and super-Burnett equations for all molecule types.

The generalized 13 moment equations ($\epsilon = 0$), and the Jin-Slemrod equations ($\epsilon = 1$) (9.51),

$$\frac{\partial \sigma}{\partial t} + \frac{8}{15} \frac{\text{Pr} \, \varpi_3}{\varpi_2} \frac{\partial q}{\partial x} + \frac{4}{3} \frac{2}{\varpi_2} \frac{\partial v}{\partial x} - \epsilon \frac{6}{5} \frac{\text{Pr} \, \hat{\omega}_4}{2\varpi_2} \frac{\partial^2 \sigma}{\partial x^2} = -\frac{2}{\varpi_2} \sigma \,,$$

$$\frac{\partial q}{\partial t} + \frac{5}{2} \frac{45}{8\theta_2} \left(\frac{2}{3\text{Pr}} \right)^2 \frac{\partial \theta}{\partial x} + \frac{5}{4} \frac{1}{\text{Pr}} \frac{\theta_4}{\theta_2} \frac{\partial \sigma}{\partial x} - \epsilon \frac{18}{5} \frac{5\hat{\theta}_4}{18\theta_2} \frac{\partial^2 q}{\partial x^2} = -\frac{2}{3} \frac{45}{8\theta_2} \frac{2}{3\text{Pr}} q,$$

are very similar to Grad 13 and R13 equations, and therefore behave similarly, that is both sets of equations are stable in space.

10.1.5 Linear stability in space

Figure 10.2, taken from [59][60], considers the stability against local disturbances of frequency Ω, again for the case of Maxwell molecules.

Stability in space requires different signs of real and imaginary part of $k(\Omega)$. Thus, if $k(\Omega)$ is plotted in the complex plane with Ω as parameter, the curves should not touch the upper right nor the lower left quadrant.

The figure shows the solutions for the different sets of equations; the dots mark the points where $\Omega = 0$. Grad's 13 moment equations (Grad13), and Navier-Stokes-Fourier equations (NSF) give two different modes each, and none of the solutions violates the condition of stability (upper and lower left). This is different for the Burnett (3 modes, upper right), super-Burnett (4 modes, middle left), and augmented Burnett (4 modes, middle right) equations: the Burnett equations have one unstable mode, and super-Burnett and augmented Burnett have two unstable modes. The R13 equations, shown in the lower left, have 3 modes, all of them are stable.

10.1.6 Discussion

From the two tests for stability in space and time follows that all Burnett type equations, including the augmented Burnett equations, fail the test for stability. The augmented Burnett equations were constructed to have stability in time [62][63], but it was overlooked that they are unstable in space [60].

The instability of the Burnett and super-Burnett equations is one of the major drawbacks of these theories, since it makes accurate numerical computations impossible. Instabilities occur for large values of the wave number k, that is for wavelengths below a critical wavelength $\lambda_c = 2\pi/k$. When the grid spacing Δx in a numerical computation is so small that $\Delta x < \lambda_c$, numerical

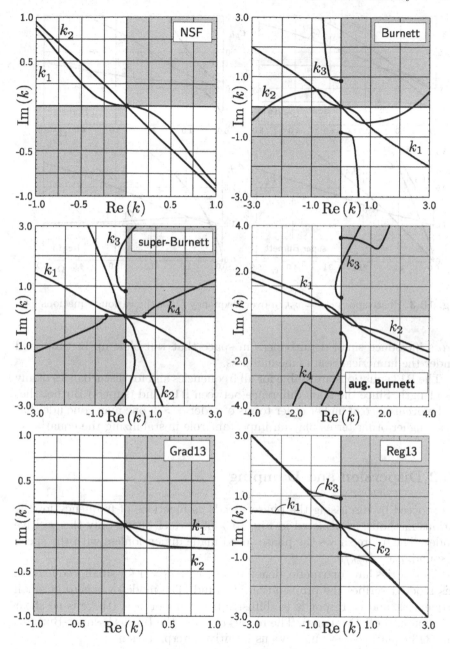

Fig. 10.2. The solutions $k(\Omega)$ of the dispersion relation in the complex plane with Ω as parameter for Navier-Stokes-Fourier, Burnett, Super-Burnett, augmented Burnett, Grad 13, and R13 equations [59][60]. The dots denote the points where $\Omega = 0$.

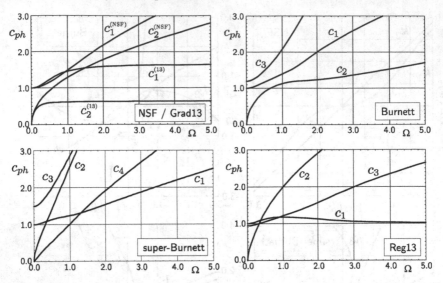

Fig. 10.3. Phase speed $c_{ph} = v_{ph}/a$ over frequency Ω for the various equations [59].

errors introduce small disturbances in space that will blow up in time, and render the numerical results meaningless.

The R13 equations are stable for all frequencies and for disturbances of any wavelength. Since the main difference between R13, and (super-) Burnett lies in the occurrence of higher order terms of order $\alpha > 3$, one may conclude that these higher order terms play an important role in stabilizing the equations.

10.2 Dispersion and Damping

We proceed by discussing the phase speeds as functions of frequency for the various methods, as depicted in Fig. 10.3 [59], which shows only the positive modes. The figure shows the phase speed made dimensionless with the speed of sound, $c_{ph} = v_{ph}/a$.

All theories have one mode, denoted as c_1 with $c_1 = 1$ for small frequencies, this mode describes the propagation of sound. The mode c_2 is zero at small frequencies, and corresponds to diffusive transport of heat (there is no shear diffusion in the 1-D setting). The modes c_3, c_4 in the Burnett, super-Burnett, and R13 equations have no obvious intuitive interpretation.

At large Ω (note that $\Omega = \mathrm{Kn}_\Omega = 1$ is already large in the present dimensionless description), all curves behave quite differently. Most notably, the phase speeds for Grad's 13 moment equations approach constant values of 1.65 and 0.63, this reflects the hyperbolicity of the equations. The fact that the Grad 13 equations imply finite wave speeds is related to the occurrence of

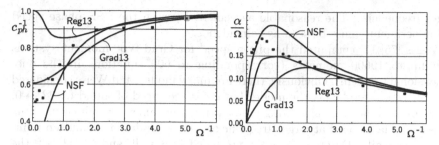

Fig. 10.4. Inverse phase velocity and damping, theoretical results from Navier-Stokes-Fourier, Grad13 and R13 equations, and measurements by Meyer and Sessler [114] (dots).

sub-shocks, which occur if the inflow velocity lies above the maximum wave speed.

In all theories obtained with the Chapman-Enskog expansion, the phase speeds grow monotonously with the frequency, so that signals of infinite frequency travel at infinite speeds. The corresponding damping (not shown) is infinite as well.

The R13 equations show a mixed behavior, with two modes giving monotonously increasing phase speeds, and one mode—the sound mode—remaining finite. The R13 equations include infinite wave speeds and this is related to the fact that they give smooth shock structures for all Mach numbers.

Next, phase speed and damping are compared to high frequency experiments performed by Meyer and Sessler [114]. Figure 10.4 shows the inverse phase speed and the damping (as α/Ω) as functions of the dimensionless inverse frequency $1/\Omega$, computed with the NSF, Grad 13, and R13 equations, and experimental data [114]. For comparison with the experimental data only those modes are considered which give the speed of sound in the low frequency limit ($\Omega \to 0$).

As can be seen, the R13 equations reproduce the measured values of the damping coefficient α for all dimensionless frequencies less than unity, while the NSF and Grad13 equations already fail at $\Omega = 1/4$ and $\Omega = 1/2$, respectively. The agreement of the R13 prediction for the phase velocity is less striking, but also the other theories do not match well. One reason for this might be insufficient accuracy of the measurement. Altogether, the R13 equations give a remarkably good agreement with the measurements for values of $\Omega < 1$.

Equations from expansions in the Knudsen number can be expected to be good only for Kn < 1. As was said above, the proper definition for the Knudsen number for an oscillation with dimensionless frequency Ω is the frequency itself, $\text{Kn}_\Omega = \Omega$. Thus the R13 equations allow a proper description of processes quite close to the natural limit of their validity of $\text{Kn}_\Omega = 1$. It is not surprising that all theories show discrepancies to the experiments for larger frequencies. The reasonable agreement between the NSF phase speed and experiments must be seen as coincidence.

Weiss [75][17] computed the phase speed for Grad-type moment systems with up to 19600 moments (corresponding to 600 equations in the one-dimensional problem). Obviously, there are many modes, and Weiss considered only the sound mode, i.e. the mode that gives the speed of sound for $\Omega \to 0$. His analysis shows that increasing moment numbers allow reproduction of the measured phase speed at larger frequencies, corresponding to Knudsen numbers above unity. Damping is not reproduced that well, and for $\Omega > 2$ the simulation cannot reproduce the experimental data.

To fully understand the high frequency experiments, one would have to simulate the experiment, i.e. solve a boundary value problem. Only that can show, whether other modes besides the sound mode play a role in the experiment. Weiss points out that modes with low damping will be important, and that the sound mode has the lowest damping of all modes at low frequency, but not at high frequency. Boundary conditions for large moment systems are not available, and therefore nothing conclusive can be said about this problem.

10.3 Stability analysis for the ES-BGK Burnett equations

The stable NSF equations result from the first order Chapman-Enskog expansion, while the unstable Burnett and super-Burnett equations result from the CE expansion to second and third order. Thus, one might think that higher order CE expansions lead to unstable equations. In order to test this conjecture, Zheng and Struchtrup considered the stability in time of the Burnett equations for the ES-BGK model [57]. This section repeats their analysis which follows closely the analysis of Bobylev [56].

The linear Burnett equations (9.44) with the ES-BGK Burnett coefficients (9.6) read

$$\sigma_{ES} = -\frac{4}{3}\frac{\partial v}{\partial x} - \left[\frac{4}{3}\frac{\partial^2 \rho}{\partial x^2} + \frac{4}{3}\left(1 - \frac{1}{\Pr}\right)\frac{\partial^2 \theta}{\partial x^2}\right] + \cdots,$$

$$q_{ES} = -\frac{5}{2\Pr}\frac{\partial \theta}{\partial x} + \frac{4}{3\Pr}\left(1 - \frac{5}{4}\frac{1}{\Pr}\right)\frac{\partial^2 v}{\partial x^2} + \cdots.$$

Note that the value of the viscosity exponent ω plays no role in the linear equations. The corresponding matrix $\mathcal{G}_{AB}$ assumes the form

$$\mathcal{G}_{AB} = \begin{bmatrix} \lambda & -ik & 0 \\ -ik\left(1 + \frac{4}{3}k^2\right) & \lambda + \frac{4}{3}k^2 & -ik\left(1 + \frac{4}{3}\left(1 - \frac{1}{\Pr}\right)k^2\right) \\ 0 & -ik + ik\frac{4}{3\Pr}\left(1 - \frac{5}{4}\frac{1}{\Pr}\right)k^2 & \frac{3}{2}\lambda + \frac{5}{2\Pr}k^2 \end{bmatrix}_{AB}$$

where $\lambda = i\Omega = i\Omega_r - \Omega_i$. Therefore, stability in time requires that the real part of λ is negative. λ is obtained as the solution of the characteristic equation $\det \mathcal{G}_{AB} = 0$,

$$\lambda^3 + A\lambda^2 + B\lambda + C = F(\lambda, k) = 0 \qquad (10.4)$$

with the abbreviations

$$A = \frac{4}{3}\left(1 + \frac{5}{4\,\mathrm{Pr}}\right)k^2 ,$$

$$B = \left[\frac{5}{3} + \frac{4}{9}\left(5 + \frac{1}{\mathrm{Pr}} + \frac{5}{2\,\mathrm{Pr}^2}\right)k^2 - \frac{32}{27\,\mathrm{Pr}}\left(\frac{1}{\mathrm{Pr}} - 1\right)\left(\frac{5}{4\,\mathrm{Pr}} - 1\right)k^4\right]k^2 ,$$

$$C = \frac{5}{3\,\mathrm{Pr}}\left(1 + \frac{4}{3}k^2\right)k^4 . \qquad (10.5)$$

The ES-BGK model requires that $-1/2 \leq b \leq 1$, or, since $b = 1 - \frac{1}{\mathrm{Pr}}$,

$$\frac{2}{3} \leq \mathrm{Pr} \leq \infty .$$

To discuss the roots of the above cubic equation for λ, and the linear stability of one-dimensional processes, the range of Pr is now divided into three regions.

Region I: $\mathrm{Pr} \to \infty$

In this case the cubic equation reduces to

$$\lambda^3 + \frac{4}{3}k^2\lambda^2 + \left(\frac{5}{3} + \frac{20}{9}k^2\right)k^2\lambda = 0$$

which has the roots

$$\lambda_1 = 0 , \quad \lambda_{2,3} = -\frac{2}{3}k^2 \pm ik\sqrt{\frac{5}{3} + \frac{16}{9}k^2} .$$

Therefore $\mathrm{Re}\,[\lambda_{1,2,3}] \leq 0$ which means that the Burnett equations for the ES-BGK model are linearly stable for $\mathrm{Pr} \to \infty$. This case corresponds to a gas with infinite viscosity, or zero heat conductivity.

Region II: $\frac{2}{3} \leq \mathrm{Pr} < 1$ and $\frac{5}{4} < \mathrm{Pr} < \infty$

It is easy to see that $F(0, k) > 0$ and $F(\lambda \to \infty, k) \to +\infty$. Moreover, for large values of k one finds that

$$F\left(\lambda = k^2, k\right) = -\frac{32}{27\,\mathrm{Pr}}\left(\frac{1}{\mathrm{Pr}} - 1\right)\left(\frac{5}{4\,\mathrm{Pr}} - 1\right)k^8$$

which is negative for the values of Pr considered. Therefore, the equation $F(\lambda, k) = 0$ has at least two positive real roots. It follows that the Burnett equations for the ES-BGK model are linearly unstable for values of Pr in $\frac{2}{3} \leq \mathrm{Pr} < 1$ and $\frac{5}{4} \leq \mathrm{Pr} < \infty$.

Region III: $1 \leq \mathrm{Pr} \leq \frac{5}{4}$

A cubic equation has either three real roots, or one real root and two conjugate complex roots, and both cases are discussed separately.

(a) Three real roots: First it is assumed that all three roots of (10.4) are real. For Prandtl numbers in $1 \leq \mathrm{Pr} \leq \frac{5}{4}$ all coefficients in (10.4) are positive, and $F(\lambda = 0, k) > 0$. Thus $F(\lambda, k) > 0$ for all positive λ, which means that F cannot have positive roots, and that all three real roots (so they exist) will be negative.

(b) One real root and two conjugate complex roots: Now it is assumed that (10.4) has one real root λ_1, and two conjugate complex roots $\lambda_{1,2} = \lambda_R \pm i\lambda_I$. Since $F(\lambda, k)$ increases with λ for $\lambda > 0$, and is positive, λ_1 must be negative. Furthermore, by inserting $\lambda = \lambda_R + i\lambda_I$, follows

$$\lambda_R^3 - 3\lambda_R\lambda_I^2 + A\left(\lambda_R^2 - \lambda_I^2\right) + B\lambda_R + C = 0 ,$$
$$3\lambda_R^2 - \lambda_I^2 + 2A\lambda_R + B = 0 .$$

Elimination of λ_I between these two equations yields

$$\lambda_R^3 + A\lambda_R^2 + \frac{A^2 + B}{4}\lambda_R + \frac{AB - C}{8} = 0 .$$

By (10.5) all coefficients are positive for $1 \leq \mathrm{Pr} \leq \frac{5}{4}$ and it follows that all roots of the above equation must be negative.

Therefore the Burnett equations for the ES-BGK model are linearly stable for $1 \leq \mathrm{Pr} \leq \frac{5}{4}$. For all other values of Pr, except $\mathrm{Pr} = \infty$, however, the Burnett equations for the ES-BGK model are linearly unstable—this includes the physically relevant case of $\mathrm{Pr} = 2/3$.

The stability of the BGK-Burnett equations—where $\mathrm{Pr} = 1$—can be found in the literature [65]–[67]. The analysis above shows that the stability can be obtained for a wider range of values for the Prandtl number. It follows that higher order Chapman-Enskog expansions do not necessarily lead to unstable equations.

Problems

10.1. Stability in time and space
Show that the equation $\frac{\partial y}{\partial t} - \frac{\partial^2 y}{\partial x^2} + \frac{\partial^3 y}{\partial t \partial x^2} = 0$ is stable in space but unstable in time.

10.2. Euler equations and wave equation
Show that the one-dimensional Euler equations can be brought into the form of the wave equation, $\frac{\partial^2 v}{\partial t^2} - a^2 \frac{\partial^2 v}{\partial x^2} = 0$, where a denotes the speed of sound.

10.3. Dispersion relations
(a) Set up the matrices $\mathcal{G}_{AB}$ for the NSF, Burnett, super-Burnett, Grad 13, generalized 13 moment, and R13 equations.
(b) Compute the phase speeds in the limits $\Omega \to 0$ and $\Omega \to \infty$.
(c) Compute the solutions for phase speed and damping for these systems, and plot the results.

10.4. Damping and dispersion of sound waves
(a) A typical sound wave in air at normal conditions has a frequency of 1000 Hz. Compute the distance of the source, at which damping has halved the amplitude of the original signal for a plane wave. Discuss the result.
(b) Two sound waves of 1000 Hz and 2000 Hz are emitted in phase. Compute the sound speeds of both waves, and find at which distance their phase has shifted by half a wave length. Discuss the result.

10.5. Radial sound waves
The previous example showed that damping does not play an important role in a plane wave (which is why sound travels far in a pipe). That we cannot hear well what a person says 10 or 20 m away is due to the radial distribution of sound. To see this, consider the Euler equations in radial geometry, and compute the amplitude of a wave with frequency Ω. Show that the amplitude is inversely proportional to the radius (i.e. the distance from the source).

11

Shock structures

The computation of shock structures is a standard test for macroscopic equations designed to describe rarefied gas flows. This chapter presents results for the equations discussed in the preceding chapters. Most of the material presented stems from a collaboration with Manuel Torrilhon [60].

11.1 The 1-D shock structure problem

11.1.1 Basic definitions

A one-dimensional steady shock structure connects two equilibrium states, defined by the values of density ρ_0, ρ_1, velocity v_0, v_1, and temperature θ_0, θ_1. Values with subscript 0 refer to the state before the shock at $x \to -\infty$ (upstream), while values with subscript 1 refer to states behind the shock at $x \to \infty$ (downstream).

It is convenient to use a dimensionless formulation of the corresponding equations, where the upstream values are used to introduce dimensionless quantities as

$$\hat{\rho} = \frac{\rho}{\rho_0}, \quad \hat{v} = \frac{v}{\sqrt{\theta_0}}, \quad \hat{\theta} = \frac{T}{T_0}, \quad \hat{\sigma} = \frac{\sigma}{\rho_0 \theta_0}, \quad \hat{q} = \frac{q}{\rho_0 \sqrt{\theta_0}^3}. \tag{11.1}$$

Here, $\hat{\sigma} = \sigma_{11}$ represents the non-trivial component of the stress, and $\hat{q}$ denotes the normal heat flux.

The dimensionless space variable is introduced as

$$\hat{x} = \frac{\rho_0 x}{\mu_0 \sqrt{\theta_0}}$$

where μ_0 is the upstream viscosity. For comparison with DSMC calculations the mean free path from [1][2] is used, calculated upstream,

$$\lambda_0 = \frac{4}{5} \frac{\mu_0}{\rho_0 \sqrt{\frac{\pi}{8}\theta_0}} \,. \tag{11.2}$$

Thus, the relation

$$\frac{x}{\lambda_0} = \frac{5}{4}\sqrt{\frac{\pi}{8}}\hat{x} \approx 0.783\,\hat{x}$$

holds for the dimensionless space variable; results will be plotted over x/λ_0.

The shock profile is best considered in the frame of reference moving with the shock, where it is described by the one-dimensional equations of Sec. 9.3 in steady state. The relevant parameter for the shock is the inflow Mach number, defined as the ratio between the inflow velocity relative to the shock and the speed of sound (10.3),

$$M_0 = \frac{v_0}{a} = \frac{v_0}{\sqrt{\frac{5}{3}\theta_0}} = \sqrt{\frac{3}{5}}\hat{v}_0 \,. \tag{11.3}$$

To simplify notation, the bars on the non-dimensional variables will be dropped in the sequel.

11.1.2 Conservation laws and Rankine-Hugoniot relations

For the shock structure problem it is convenient to consider the balance laws in divergence form (see Sec. 3.3.2). With $\frac{D}{Dt} = v\frac{d}{dx}$ in steady state, the conservation laws (9.21) can be written as

$$\frac{d}{dx}[\rho v] = 0 \,,$$

$$\frac{d}{dx}[\rho v^2 + \rho\theta + \sigma] = 0 \,, \tag{11.4}$$

$$\frac{d}{dx}[5\rho v\theta + \rho v^3 + 2\sigma v + 2q] = 0 \,.$$

Far before and behind the shock the gas is in equilibrium states with $\sigma_0 = \sigma_1 = 0$ and $q_0 = q_1 = 0$. Integration of (11.4) between the equilibrium states gives

$$\rho_0 v_0 = \rho_1 v_1 \,,$$

$$\rho_0 v_0^2 + \rho_0 \theta_0 = \rho_1 v_1^2 + \rho_1 \theta_1 \,,$$

$$5\rho_0 v_0 \theta_0 + \rho_0 v_0^3 = 5\rho_1 v_1 \theta_1 + \rho_1 v_1^3 \,.$$

Since dimensionless velocity and temperature before the shock are given by

$$v_0 = v\,(x \to -\infty) = \sqrt{\frac{5}{3}}M \,, \quad \theta_0 = \theta\,(x \to -\infty) = 1 \,, \tag{11.5}$$

the downstream values are

$$\rho_1 = \rho\,(x \to \infty) = \rho_0 \frac{4M_0^2}{M_0^2 + 3} \,,$$

$$v_1 = v\,(x \to \infty) = \sqrt{\frac{5}{3}\frac{M_0^2 + 3}{4M_0}} \,, \qquad (11.6)$$

$$\theta_1 = \theta\,(x \to \infty) = \frac{\left(5M_0^2 - 1\right)\left(M_0^2 + 3\right)}{16M_0^2} \,.$$

These are the well-known Rankine-Hugoniot relations of gas dynamics which give the boundary conditions under which a steady shock wave can be observed.

11.1.3 Entropy production over the shock

The second law (3.42) for the shock reads

$$\frac{d}{dx}\left[\rho v s + \phi\right] \geq 0 \,,$$

where $\phi = \phi_1$ denotes the non-convective entropy flux. Integration between the two equilibrium states—where $\phi = 0$—yields, after division by $\rho_1 v_1 = \rho_0 v_0$, that the entropy must grow over the shock,

$$s_1 - s_0 \geq 0 \,.$$

With (3.43) and the Rankine-Hugoniot relations follows that this happens only for Mach numbers that fulfill[1]

$$\left(\frac{5M_0^2 - 1}{4}\right)^3 \left(\frac{M_0^2 + 3}{4M_0^2}\right)^5 \geq 1 \ \text{or} \ M_0 \geq 1 \,.$$

Thus, a shock can only be observed for Mach numbers $M_0 \geq 1$, i.e. when the inflow velocity is supersonic.

The Mach number behind the shock is

$$M_1 = \frac{v_1}{a_1} = \frac{v_1}{\sqrt{\frac{5}{3}\theta_1}} = \sqrt{\frac{M_0^2 + 3}{5M_0^2 - 1}} \leq 1 \,. \qquad (11.7)$$

Equations (11.7) and (11.6) imply that in a shock the velocity changes from supersonic to subsonic, while density and temperature grow. For a strong shock ($M_0 \to \infty$) density and post-shock Mach number remain finite ($\rho_1 \to 4\rho_0$, $M_1 \to 1/\sqrt{5}$), while the temperature grows quadratically ($\theta_1 \to 5M_0^2/16$).

---

[1] We are only interested in positive velocities, to have a flow from "0" to "1".

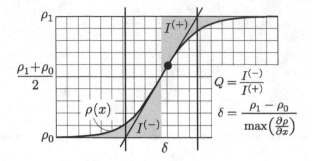

Fig. 11.1. Definition of shock thickness and shock asymmetry for the density profile of a steady shock wave: The dot marks the position of the steepest gradient which is linearly extrapolated to give the shock thickness δ. The shock asymmetry Q follows from the ratio of the two integrals (shaded regions) of the upper and lower half of the profile.

11.1.4 Shock thickness and asymmetry

Figure 11.1 introduces two often used profile characteristics, thickness and asymmetry.

The shock thickness is defined as [88][115][117]

$$\delta = \frac{\rho_1 - \rho_0}{\max\left(\frac{\partial \rho}{\partial x}\right)} , \qquad (11.8)$$

and refers to a linear density curve with a slope corresponding to the steepest density gradient. The shock thickness is infinite for Mach numbers M_0 approaching unity, approaches a minimum of several mean free paths at Mach numbers around 3 through 5, and finally tends slowly to infinity again as $M_0 \to \infty$. This non-monotone behavior represents an interplay between steepening non-linearity and smoothing dissipation. Usually the shock thickness is related to the mean free path in front of the shock λ_0, and the ratio λ_0/δ is plotted.

The shock asymmetry gives additional information about the actual shape of the profile. It is defined as [88]

$$Q = \frac{\int_{-\infty}^{x^*} (\rho(x) - \rho_0)\, dx}{\int_{x^*}^{\infty} (\rho_1 - \rho(x))\, dx} \qquad (11.9)$$

where x^* is defined by $\rho(x^*) = \frac{1}{2}(\rho_0 + \rho_1)$. For realistic shock waves, the asymmetry is close to unity, i.e. the measured profiles are rather symmetric. Typical measured values lie between 0.8 and 1.2, see [117]. An asymmetry value $Q < 1$ indicates a relatively slower relaxation behind the shock.

Shock thickness and asymmetry are usually defined by means of the density field for reasons of experimental feasibility.

11.1.5 Shocks and Knudsen number

The various sets of macroscopic equations were derived in preceding chapters by assuming sufficiently small values of the Knudsen number, which is defined as the ratio between the mean free path and a relevant macroscopic length of the process.

The shock thickness is a meaningful length to characterize a shock, and its use gives the shock Knudsen number as $\mathrm{Kn}_S = \frac{\mu_0 \sqrt{\theta_0}}{p_0 \delta} = \frac{5}{8}\sqrt{\frac{\pi}{2}}\frac{\lambda_0}{\delta}$. Measured values of λ_0/δ reach values of not more than 0.3, corresponding to $\mathrm{Kn}_S \simeq 0.235$. This value is not too large, and thus one might expect that a third order theory, e.g. R13, or super-Burnett, can well describe shock structure experiments.

An alternative choice for the relevant length scale is the desired resolution Δx which gives the shock Knudsen number as $\mathrm{Kn}_{\Delta x} = \frac{\mu_0 \sqrt{\theta_0}}{p_0 \Delta x}$. Depending on the value of Δx, $\mathrm{Kn}_{\Delta x}$ can be very large, which leads to the expectation that a highly accurate description of the shock cannot be delivered by models which are based on an expansion in the Knudsen number to only second or third order.

Indeed, if a low resolution is sufficient, one might chose Δx several orders of magnitude above the mean free path, so that $\mathrm{Kn}_{\Delta x} \ll 1$. Then the Euler equations are sufficient to describe the shock, which appears as a discontinuity between two equilibrium states, connected by the Rankine-Hugoniot relations.

As one starts to resolve the finer features of the shock, e.g. local values of gradients, one reduces the value of Δx, therefore increases the Knudsen number, and cannot expect perfect results from equations derived by assuming small Knudsen numbers.

11.1.6 Solution method

The numerical solution can be found with a strategy given in [89]. The first step is to reduce the number of variables by integrating the conservation laws (11.4) from the upstream equilibrium into the shock to obtain relations for density, stress and heat flux,

$$\rho(v) = \sqrt{\frac{5}{3}\frac{M_0}{v}}\,,$$

$$\sigma(v,\theta) = 1 + \frac{5}{3}M_0^2 - M_0\sqrt{\frac{5}{3}}\left(\frac{\theta}{v} + v\right)\,, \tag{11.10}$$

$$q(v,\theta) = \sqrt{\frac{5}{12}}M_0\left(\frac{5}{3}M_0^2 + 5v^2 - 3\theta\right) - v\left(1 + \frac{5}{3}M_0^2\right)\,.$$

With the relations (11.10), the equations for σ and q provided by the various methods (NSF, Burnett, R13, etc.) form two equations to determine velocity and temperature.

The next material is based on a study of the Grad13/R13 equations. The one-dimensional equations for stress and heat flux (9.28, 9.29) can be recast as

$$\frac{d}{dx}\left[\rho v^3 + 3\rho\theta v + 3\sigma v + \tfrac{6}{5}q + m\right] = -\rho\theta^{1-\omega}\,\sigma\,, \tag{11.11}$$

$$\frac{d}{dx}\left[\frac{1}{2}\rho v^4 + 4\rho\theta v^2 + \frac{5}{2}\sigma v^2 + \frac{16}{5}qv + \frac{7}{2}\theta\sigma\right.$$
$$\left.+\frac{5}{2}\rho\theta^2 + mv + B\right] = -\rho\theta^{1-\omega}\left(\sigma v + \frac{2}{3}q\right)\,, \tag{11.12}$$

where (4.45) was used. ω is the viscosity exponent of the gas, with $\omega = 1$ for Maxwell molecules, and $\omega = 0.5$ for hard spheres.

The quantities $m = m_{111}$ and $B = \frac{1}{2}R_{11} + \frac{1}{6}\Delta$, introduced in (11.11, 11.12), are the dimensionless forms of (9.31),

$$B = -\epsilon\frac{18}{5}\frac{\theta^{\omega-1}}{\rho}\left[\left(\theta - \frac{5}{9}\frac{\sigma}{\rho}\right)\frac{dq}{dx} - \frac{q}{\rho}\frac{d\sigma}{dx} + \frac{11}{6}q\frac{d\theta}{dx}\right.$$
$$\left.-\theta q\frac{d\ln\rho}{dx} + \frac{5}{9}\left(\frac{11}{7}\theta\sigma - \frac{\sigma^2}{\rho}\right)\frac{dv}{dx}\right]\,,$$

$$m = -\epsilon 2\frac{\theta^{\omega-1}}{\rho}\left[\frac{3}{5}\theta\frac{d\sigma}{dx} - \frac{3}{5}\theta\sigma\frac{d\ln\rho}{dx} + \frac{4}{5}q_{\langle i}\frac{dv_j}{dx_{k\rangle}} - \frac{3}{5}\frac{\sigma}{\rho}\frac{d\sigma}{dx}\right]\,. \tag{11.13}$$

The artificial parameter ϵ controls the transition between the classical 13 moment equations of Grad, $\epsilon = 0$, and the regularized equations, $\epsilon = 1$. The impact of this parameter is discussed in Sec. 11.3.1 below.

Substituting m and B (11.13) into (11.11, 11.12) gives the final system of equations in the compact form

$$\frac{dF(v,T)}{dx} = P(v,T) + \frac{d}{dx}\left(D(v,T)\cdot\frac{dG(v,T)}{dx}\right)\,. \tag{11.14}$$

Here, F and P are 2-vectors and the expression $D \cdot dG/dx$ represent both equations (11.13).

Formally shock structures for velocity and temperature are solutions of the system of ordinary differential equations (11.14) with the boundary conditions (11.5, 11.6). Since the right hand side of (11.14) vanishes for both boundary conditions, shock structures represent, speaking in terms of ordinary differential equations, orbits connecting stationary points in the phase space of (11.14). In [115] shock structures of the NSF theory are calculated using detailed information about the orbit and the stationary points. Fortunately, in the present case the orbits are sufficiently well behaved, so that the straightforward finite difference method of [89] could be used.

The shocks are assumed to relax fast, so that (11.5) and (11.6) may be prescribed at the boundaries of a finite computational domain. The computational domain is uniformly discretized into $N + 2$ positions x_i with

$i = 0, 1, 2...N + 1$ and step size Δx, and the system (11.14) is evaluated at positions x_1 through x_N with help of the discretization rules

$$\left.\frac{du}{dx}\right|_i \rightarrow \frac{u_{i+1} - u_{i-1}}{2\Delta x}, \tag{11.15}$$

$$\left.\frac{d}{dx}\left(a\frac{du}{dx}\right)\right|_i \rightarrow \frac{(a_{i-1} + 3a_{i+1})u_{i+1} - 4(a_{i+1} + a_{i-1})u_i + (3a_{i-1} + a_{i+1})u_{i-1}}{4\Delta x^2}. \tag{11.16}$$

The evaluation at position x_1 and x_N requires field values at x_0 and x_{N+1} which are supplied by (11.5) and (11.6). Thus, the discretization of (11.14) yields $2N$ coupled algebraic equations for the N unknown values of velocity and temperature.

The resulting non-linear system was solved by a quasi-Newton method [116] with appropriate $\tanh(x)$ curves as initial guess for velocity and temperature. Strictly speaking, the shock structure description as in (11.14) with (11.5),(11.6) has no unique solution since the profile may be shifted arbitrarily along the space variable. In fact, the numerical method used shows convergence failures if the computational domain is chosen too large. For sufficiently small domains, however, converged solutions with relative accuracy in the magnitude of 10^{-9} could be achieved for the discretized system. Computational results for shock structures of the R13 equations presented here have been obtained with $N = 1000$ ($\Delta x \approx 0.01\lambda_0$) on a usual PC Pentium-III within a few seconds.

The numerical method described above was also used to calculate the shock structures of the various Burnett-type equations. However, due to the instabilities of these equations the calculation must be conducted very carefully and is restricted to coarse discretizations with $\Delta x \approx \lambda_0$.

11.2 Comparison with DSMC results

In this section we shall compare the shock structures obtained from macroscopic equations for rarefied flows to DSMC results obtained with Bird's code [1]. The actual values for interval length, upstream temperature, etc. are the values of Pham-Van-Diep et al. [118]. Note that the calculation of a single low Mach number shock structure by a standard DSMC program takes several hours which is several orders of magnitude slower than the calculation by a continuum model.

We compare to DSMC using two different particle interaction models: Maxwell molecules and hard spheres. The VSS collision modeling of the DSMC method requires two parameters, the viscosity exponent $\omega^{(VSS)}$ and an additional parameter $\alpha^{(VSS)}$. For hard spheres these parameters are given by $\omega^{(VSS)} = 0.5$ and $\alpha^{(VSS)} = 1.0$, while for Maxwell molecules they are $\omega^{(VSS)} = 1.0$ and $\alpha^{(VSS)} = 2.26$, see [1]. Since the DSMC code uses physical units, the mean free path of the upstream region λ_0 defines the length scale. The values used are

$$\lambda_0^{(HS)} = 0.00162 \, \text{m}, \quad \lambda_0^{(MM)} = 0.0014 \, \text{m}$$

for hard spheres and Maxwell molecules, respectively, which corresponds to the definition (11.2) and also reproduces the shock thickness results of [118].

In the next figures we compare the profiles of density and heat flux. The heat flux in a shock wave follows solely from the temperature and velocity via (11.12). Hence, its profile gives a combined impression of the quality of the profiles of temperature and velocity. The soliton-like shape of the heat flux also helps to give a more significant adjudgement of the quality of the structure. Since it is a higher moment, the heat flux is more difficult to match than the stress. Profiles of velocity, temperature and stress are not presented in the following. The density is normalized to give values between zero and unity for each Mach number. Similarly, the heat flux is normalized such that the DSMC result gives a maximal heat flux of 0.9. Note that the non-normalized heat flux is negative.

11.2.1 Failure of classical methods

First we discuss the failure of the classical theories and the standard Burnett models. Figure 11.2 shows the density and heat flux profiles of a $M_0 = 2$ shock calculated with the NSF and Grad13 equations as well as with the Burnett and super-Burnett equations for Maxwell molecules.

The NSF results simply mismatches the profile, while the Grad13 solution shows a strong subshock at $x/\lambda_0 = -4$. The Burnett and super-Burnett solutions are spoiled by oscillations in the downstream part of the shock.

In the Burnett case the oscillations arise if the length of a grid cell is below half of the mean free path. This stands in correspondence to the result of the linear analysis in Chapter 10 which predicts spatial instabilities. The oscillations appear first in the downstream region, since the mean free path is smaller in that region. The oscillations stick to a wave length corresponding to the length of a grid cell and therefore high resolution calculations are impossible. The super-Burnett result shows the same behavior, but the oscillation wavelength is a multiple of the length of a grid cell. Again, the oscillations increase with grid refinement and convergence cannot be established.

The oscillations of both models, Burnett and super-Burnett, increase for shocks with higher Mach number, and are present for all values of the viscosity exponent. Thus, for the description of shock structures the Burnett equations and super-Burnett-equations have to be rejected.

To test the usefulness of the Burnett equations, Salomons and Mareshall [119] used DSMC results for velocity and temperature to compute stress σ and heat flux q from the Burnett equations (9.23, 9.24). Comparison with the actual DSMC results for σ and q showed considerable improvement over

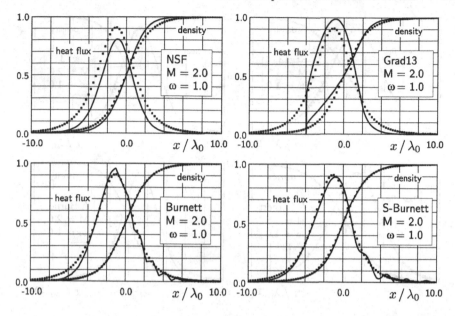

Fig. 11.2. Shock structure solutions for the NSF equations, Grad 13 equations, and Burnett and super-Burnett equations, for Maxwell molecules at Mach number $M_0 = 2$ (solid lines). Both Burnett results exhibit non-physical oscillations in the downstream region. The squares represent the DSMC solution.

the NSF equations (9.22). Thus, the Burnett equations contain the proper physics of the shock, but are useless, since their mathematical structure does not allow to compute a stable solution.

Uribe et al. presented numerical solutions for the shock wave structure of the Burnett equations [120][121]. Due to the instability, they could not solve for the complete shock structure.

Fiscko and Chapman [61] deleted one linear term from the Burnett and super-Burnett equations, and were able to obtain stable shock solutions, in reasonable agreement to DSMC simulations. Obviously, the mathematical properties of the equations are changed by deleting terms ad hoc, and thus it is not surprising that they could obtain stable behavior.

11.2.2 Maxwell molecules

R13 equations and augmented Burnett equations give good results for a wider range of Mach numbers, and results for Maxwell molecules and hard spheres are discussed now.

The augmented Burnett equations are stable in time but not in space. Since the shock structure calculation was formulated as a boundary value problem, the spatial stability is crucial. The results for the augmented Burnett equations were obtained on a coarse grid with $\Delta x \approx 0.5\,\lambda_0$ where only

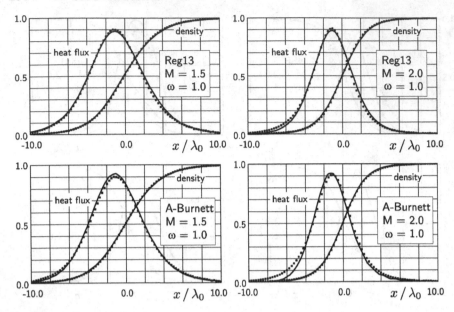

Fig. 11.3. Shock structures in a gas of Maxwell molecules with Mach numbers $M_0 = 1.5$ and $M_0 = 2.0$. The upper row shows the solution of the R13 equations, while the lower row shows the results of the augmented Burnett equations. The squares correspond to the DSMC solution.

small oscillations occur. Note that Zhong et al. solved the augmented Burnett equations by time stepping into steady state [62][63], where no stability problems arise, since the augmented Burnett equations are stable in time.

Figure 11.3 compares shock structures for Maxwell molecules at $M_0 = 1.5$ and $M_0 = 2$ for R13 and augmented Burnett with DSMC results. The results for the density profile exhibit no visible differences and both models match the DSMC results very well. The shape of the heat flux is captured very well by the R13 equations, while the augmented Burnett equations do not reproduce the maximum value and the upstream relaxation. The deviations from the DSMC solutions become more pronounced for higher Mach numbers. Figure 11.4 shows the results for $M_0 = 3$ and $M_0 = 4$. The R13 results begin to deviate from the DSMC solution in the upstream part. In the tail of the augmented Burnett profiles small oscillations are present, indicating the onset of instability.

From the presented figures follows that the results of the R13 system for Maxwell molecules agree better with DSMC results than the solutions of the augmented Burnett equations. For higher Mach numbers both, R13 and augmented Burnett equations, deviate somewhat from DSMC results.

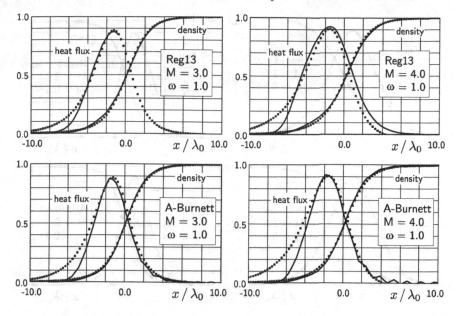

Fig. 11.4. As Fig. 11.3 but with Mach numbers $M_0 = 3$ and $M_0 = 4$. Note the small oscillations in the result of the augmented Burnett equations.

11.2.3 Hard Spheres

Before we discuss results for hard spheres, we consider the limits of the equations used. The derivation of the R13 equations is based on integrals of the Boltzmann collision term for Maxwell molecules, see Chapters 7 and 8. The ad hoc extension to more general particle interactions via the viscosity exponent ω as introduced in (4.45) is only a first approximation. Indeed, in Chapter 8 we saw that the full R13 system for hard spheres must include the hard sphere collision integrals of higher moments, and an analogous statement is valid for the Burnett models. As they are used here, the equations correspond to the first order approximations of the collision integrals in terms of Sonine polynomials, see Chapter 4. The exact computation of the (Burnett) coefficients requires higher order approximations.

To give an impression in the performance of the hard sphere modeling, we present shock structures for a hard sphere gas ($\omega = 0.5$) for Mach numbers $M_0 = 1.5, 2.0, 3.0, 4.0$ in Figs. 11.5 and 11.6. As in the previous section, the results of the continuum theories (solid lines) are compared to the DSMC solution (squares).

The R13 equations have difficulties to describe the upstream relaxation, while the tails of the structures are matched almost perfectly. Note that the upstream part of a shock is more difficult to capture since it is the more rarefied region with larger mean free path. The augmented Burnett equations

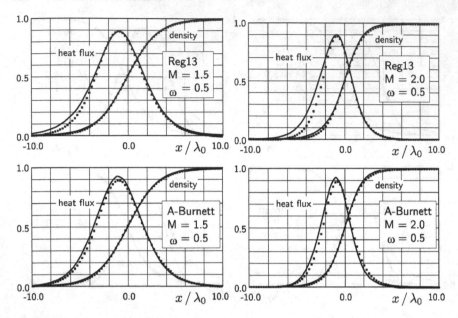

Fig. 11.5. Shock structures in a gas of hard spheres with Mach numbers $M_0 = 1.5$ and $M_0 = 2.0$. The solid lines show the results of the R13 equations (upper row) and the augmented Burnett equations (lower row). The squares correspond to the DSMC solution.

fall short of the tail of the structure ($M_0 \geq 3$). The sharp edged shape of the result of the augmented Burnett equations is due to the coarse resolution.

Altogether the augmented Burnett system seems to give better results for hard spheres than for Maxwell molecules. Note, however, that the coarse resolution and spatial instability of the equations calls their reliability into question. On the other hand, also the R13 equations need improvements for gases of hard spheres.

11.2.4 Jin-Slemrod equations

Shock structure computations with the Jin-Slemrod regularization of the Burnett equations, were shown in [109] where solutions of the one-dimensional NSF, Grad 13, and Jin-Slemrod equations (9.36–9.39) are compared to DSMC calculations for Maxwell molecules.

Figure 11.7 shows density and heat flux for $M_0 = 4$ as presented in [109], the flow direction is from the right to the left. In order to understand the curves, we recall that the Jin-Slemrod equations can be considered as a regularization of the Grad13 equations, just as the R13 equations. For both theories, the regularizing terms are of super-Burnett order, but only the R13 equations are accurate to super-Burnett order in the sense that their CE expansion gives the super-Burnett equations. The density curve in Fig. 11.7

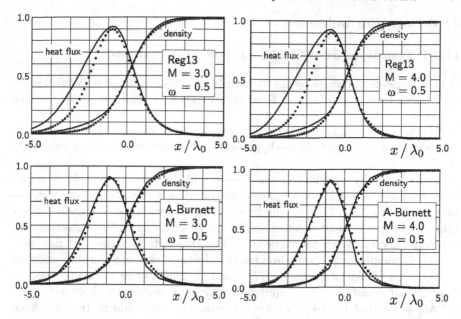

Fig. 11.6. As Fig. 11.5, but for Mach numbers $M_0 = 3$ and $M_0 = 4$.

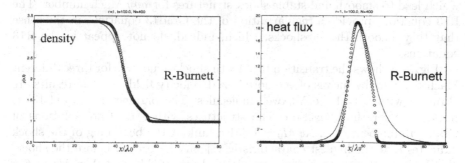

Fig. 11.7. Density and heat flux for the regularized Burnett equations of Jin and Slemrod (9.38, 9.39) for Maxwell molecules at $M_0 = 4$, compared to DSMC calculations [109]. Note that the flow direction is from right to left, and that the dots refer to the solution of the Jin-Slemrod equations while the continuous line is the DSMC result.

displays a very steep region at the beginning of the shock (i.e. on the right side), which shares some similarity to the Grad13 equations which display a discontinuity at the same location. In the Jin-Slemrod equations, this discontinuity is smoothed out somewhat, but it is still visible. The R13 equations, on the other hand, regularize the Grad13 equations to a higher degree, which leads to a complete disappearance of this discontinuity.

11.3 Solution behavior

11.3.1 Transition from Grad's 13 moment equations

Grad's 13 moment equations fail to describe continuous shock structures since they suffer from a subshock in front of the shock beyond the Mach number $M_0 = 1.65$ [8][9][122]. The subshock grows with higher Mach numbers and at $M_0 \approx 3.5$ a second subshock appears in the middle of the shock. Both subshocks are artefacts from the hyperbolic nature of the 13 moment equations [104].

Indeed, any hyperbolic moment theory will yield continuous shock structures only up to the Mach number corresponding to its highest characteristic velocity [89][123]. Further validation of results with measurements shows that Grad-type moment theories succeed to describe shock thickness data accurately only for Mach numbers far below this critical value. For instance, Grad's 13 moment equations describe the shock thickness accurately only up to $M_0 \approx 1.1$. J. Au in his thesis [77] found that up to 680 moments (64 one-dimensional equations) are required to calculate a smooth shock structure for $M_0 = 1.8$ that fits to experimental data. For more information on shock structures in moment theories see [17].

One of the reasons to derive the R13 equations was to obtain field equations which lead to smooth and stable shock structures for *any* Mach number. The R13 equations provide a regularization of the Grad13 equations in the sense that they smooth the subshocks, which, indeed, do not appear in the R13 equations.

Fig. 11.8 shows the transition to smooth shock structures for three different Mach numbers by means of the normalized velocity field v_N. The results are obtained with $\omega = 1$, i.e. Maxwell molecules. The parameter ϵ in (11.13) is responsible for the transition. The structures with $\epsilon = 0$ are solutions of Grad's 13 moment case. At $M_0 = 1.651$ a kink at the beginning of the shock indicates that the highest characteristic velocity is reached before the shock. The kink develops into a pronounced subshock at $M_0 = 3$. For $M_0 = 6$ a second subshock is present towards the end of the structure.

The curves for $\epsilon = 0.1$ follow mainly the results of Grad's 13 moment case. The subshocks are still clearly visible, albeit smoothed out by increased dissipation.

At $\epsilon = 1$, however, the additional terms in the R13 moment equations succeed to annihilate the subshocks completely, and an overall smooth shock structure is obtained. At $M_0 = 6$ the R13 solution ($\epsilon = 1$) exhibits obvious asymmetries which start to appear in the structure with Mach numbers $M_0 > 3$. Since experiments [117] and DSMC simulations [1] predict almost perfect S-shaped profiles, we conclude that the validity of the R13 equations may be lost beyond Mach numbers $M_0 \approx 3.0$.

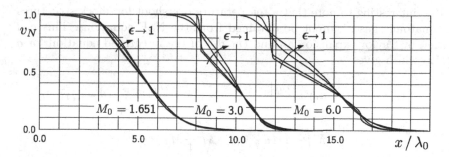

Fig. 11.8. Regularization process of the Grad13 equations. Profiles for three different Mach numbers are shown with different values of $\epsilon = 0.0, 0.1, 0.5, 1.0$. The results of the Grad13 equations ($\epsilon = 0$) exhibt kinks as well as one or two subshocks of increasing strength. These singularities vanish for $\epsilon = 1$.

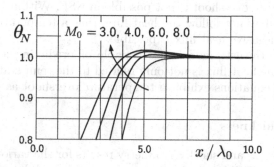

Fig. 11.9. Magnification of the normalized temperature profile for several Mach numbers. The profile shows increasing temperature overshoots starting from Mach number $M_0 = 4$.

11.3.2 Temperature overshoot

Figure 11.9 displays results of the temperature profile for Mach numbers up to $M_0 = 8$. The figure concentrates on the downstream part of the shock structure for a qualitative analysis.

The curves in Fig. 11.9 have been obtained as solutions of the R13 equations with $\omega = 1$, and the temperature field has been normalized by means of the values θ_0 and θ_1 to lie between 0 and 1. As may be read off Fig. 11.9, the R13 equations predict a temperature overshoot for high Mach number shocks. This phenomenon is known from experiments as well as from DSMC simulations [1], especially in the case of polyatomic gases due to the relaxation of internal degrees of freedom. The effect is less pronounced but present in the case of monatomic gases. Here the overshoot is a result of a delayed relaxation of transversal (i.e. inside the shock plane) kinetic energy of the gas particles.

It is interesting to note that a temperature overshoot for shock structures of monatomic gases cannot be obtained from the NSF equations. This can be seen by Taylor expansion around (v_1, θ_1) of the integrated equation for q (11.12) which gives

$$q(v, \theta) \approx q(v_1, \theta_1) + \left.\frac{\partial q}{\partial v}\right|_{(v_1, \theta_1)} (v - v_1) + \left.\frac{\partial q}{\partial \theta}\right|_{(v_1, \theta_1)} (\theta - \theta_1)$$

$$= -\frac{1}{4}(5M_0^2 - 1)(v - v_1) - \sqrt{\frac{15}{4}} M_0 (\theta - \theta_1) .$$

Accordingly, the heat flux q will be negative[2] in the vicinity of the temperature overshoot where $\theta \to \theta_1 + \Delta\theta$ and $v \to v_1 + \Delta v$ with some small $\Delta\theta, \Delta v > 0$. According to Fourier's law this yields a positive gradient of temperature in contradiction to the overshoot where $d\theta/dx$ is must be negative. Thus, a temperature overshoot is not possible in NSF. Within R13, however, the temperature may not follow the heat flux, thus a temperature overshoot together with a negative heat flux is possible.

Similarly, the Burnett models are capable of predicting a temperature overshoot since the heat flux is not only related to the temperature gradient. The Jin-Slemrod equations exhibit a temperature overshoot as well [109].

11.3.3 Shock thickness

Next, shock thickness and shock asymmetry results for the various models are compared, again, to DSMC results [118] and experimental data [88][117]. The oscillatory behavior of the Burnett and super-Burnett results can be ignored, since only the density profile is considered, which is not seriously altered by the oscillations.

In order to evaluate the low Mach number range, shock structures have been calculated for Mach numbers $M_0 = 1.2, 1.25, 1.35, 1.5, 1.651, 1.8, 2.0$ and 2.5 for hard spheres and Maxwell molecules. The shock thickness values obtained are linearly interpolated in order to produce continuous curves.

Figure 11.10 shows the curves for the inverse shock thickness (11.8) of the various continuum models together with the DSMC results as symbols. The left figure shows results for Maxwell molecules and the right hand figure shows those for a gas of hard spheres. The comparison of the curves manifests the clear advantage of the higher order models like R13 or Burnett over NSF and Grad13. The super-Burnett results, however, do not show a better correspondence than the lower order Burnett equations. The curves of R13, Burnett, and augmented Burnett equations follow the symbols quite accurately. Note that the DSMC calculations are subject to statistical errors, so that slight differences between DSMC results and predictions of the continuum models are acceptable.

---

[2] In the figures, it appears as positive only due to normalization.

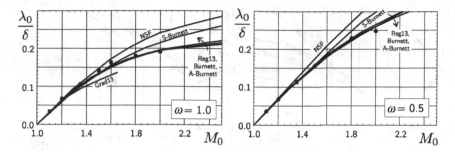

Fig. 11.10. Shock thickness results for Maxwell molecules ($\omega = 1$, left) and hard spheres ($\omega = 0.5$, right) calculated with the R13, Burnett, super-Burnett, augmented Burnett, Navier-Stokes-Fourier, and Grad13 equations. The symbols represent the results of DSMC simulations [118].

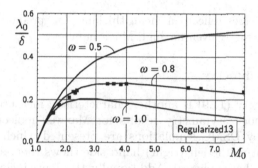

Fig. 11.11. Comparison of the inverse shock thickness for the R13 equations with measurements for Argon (squares, $\omega \approx 0.8$). The curve of the augmented Burnett equations with $\omega = 0.8$ shows a similar agreement.

Additional shock structures have been calculated for Mach numbers $M_0 = 3.0$, 4.0, 6.0 and 8.0. As has been mentioned in Sec. 11.3.1 the R13 profiles begin to loose a trustable appearance for these Mach numbers, because buckles arise. Nevertheless, the profiles stay smooth and the shock thickness is well defined. In Fig. 11.11 the results of the R13 system are compared to measurements in argon [117][88]. The textbook value $\omega = 0.8$ [1][2] for the viscosity exponent in the R13 equations yields a striking agreement with the experimental data. The results of the augmented Burnett equations with $\omega = 0.8$ lead to a similar agreement, while the NSF results lie far off.

To some extent the good agreement of the shock thickness for high Mach number should be seen as a lucky coincidence. The detailed shape of the profiles for high Mach numbers does obviously not match the profiles shown in [117]. It becomes evident that the single parameter δ cannot reflect the complete profile so that the agreement with shock thickness measurements does not imply a reliable description of the complete profile. Nevertheless, the

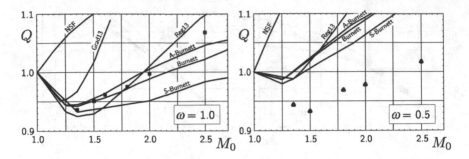

Fig. 11.12. Shock asymmetry result for Maxwell molecules (left) and hard spheres (right) calculated with the NSF, Burnett, super-Burnett, augmented Burnett, Grad13, and R13. The symbols represent the results of DSMC simulations [118].

information that δ does reflect—a mean thickness—is predicted by the R13 equations very accurately even for high Mach numbers.

11.3.4 Shock asymmetry

The shock asymmetry (11.9) is another characteristic measure of the shock profile. In [118] shock asymmetry results for Maxwell molecules and hard spheres obtained by DSMC simulations are presented which are shown in Fig. 11.12. The plot at the left (right) hand side shows the results for gas of Maxwell molecules (hard spheres). Additionally, the figure includes the curves as obtained from NSF, Grad 13, the R13, and the Burnett models. None of these exhibits a reasonable agreement with the DSMC results.

The results of NSF indicate a qualitative failure, since it predicts an asymmetry $Q > 1$ already for small Mach numbers, in clear contradiction to DSMC simulations and measurements [117]. The R13 solution for $\omega = 1$ matches the DSMC results around $M_0 = 2$, while the Burnett and augmented Burnett give a reasonable agreement below $M_0 = 2$. The asymmetry results of the Grad13 and the super-Burnett equations lie far off.

For hard spheres all macroscopic models disagree with the DSMC results. All models predict larger values for asymmetry for hard spheres as compared to Maxwell molecules. In contrast, the DSMC results predict a smaller asymmetry. We recall that the continuum modelling of the hard sphere gas via the viscosity exponent ω is only a first approximation. A more careful investigation with better approximations is needed.

11.3.5 Positivity of the phase density

Close to and in equilibrium the phase density for the microscopic velocity of the gas particles is a Maxwellian

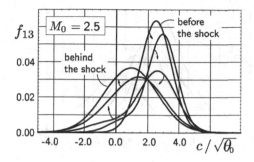

Fig. 11.13. The phase density (11.18) in normal direction ($\vartheta = 0$) before and behind the shock as well as at three positions inside the shock for $M_0 = 2.5$. Inside the shock the phase density becomes a bimodal distribution.

$$f_M = \frac{\rho}{\sqrt{2\pi\theta}^3} \exp\left[-\frac{C^2}{2\theta}\right] , \qquad (11.17)$$

where density ρ, temperature θ and peculiar velocity of the particles C_i are dimensionless according to (11.1). In non-equilibrium flows, especially in shock waves, the phase density deviates markedly from the Maxwellian (11.17). Since the R13 equations still consider Grad's 13 moments as variables, we assume the Grad13 distribution (6.11) as a reasonable approximation to the actual phase density in the flow. In one-dimensional formulation, Grad's distribution reads

$$f_{|13} = f_M \left(1 + \frac{\sigma}{4\rho\theta^2}\left(3C_x^2 - C^2\right) - \frac{q}{\rho\theta^2}C_x\left(1 - \frac{1}{5\theta}C^2\right)\right) , \qquad (11.18)$$

where C_x is the dimensionless peculiar velocity component normal to the shock. Both, C and C_x, are given by the the absolute velocity c, the flow velocity v, and an angle of direction ϑ as

$$C_x = c\cos\vartheta - v , \quad C^2 = c^2 - 2cv\cos\vartheta + v^2 .$$

The angle $\vartheta = 0$ represents the direction normal to the shock. Thus, Grad's distribution $f_{|13}(c, \vartheta \,|\, \rho, v, T, \sigma, q)$ can be considered as a function of the flow quantities and c and ϑ.

With the results from the numerical calculations of the shock profiles the phase density (11.18) can be plotted at any position in the shock. Figure 11.13 shows $f_{|13}$ at three positions inside a Mach 2.5 shock ($\omega = 1$) together with the initial and final Maxwell distributions. The shock transforms a right-positioned and narrow distribution into a more left and flat one, which indicates the transition from high velocity and low temperature to low velocity and high temperature. Inside the shock the phase density exhibits an asymmetric shoulder representing the onset of bimodality which is also observed in DSMC simulations [1]. Indeed, the classical Mott-Smith method solves the

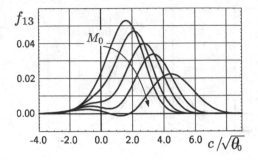

Fig. 11.14. Phase density normal to the shock ($\vartheta = 0$) inside of shock profiles with increasing Mach numbers $M_0 = 1.651$, 2.0, 2.5, 3.0, 4.0. The phase density is taken at x^*, where $\theta(x^*) = \frac{1}{2}(\theta_0 + \theta_1)$.

shock structure problem successfully by assuming a bimodal phase density which is a superposition of the two Maxwellians at both ends of the shock [124].

Grad's distribution is known to admit non-physical negative values, since the polynomial expression in (11.18) may change its sign. Indeed, in Fig. 11.13 negative values for $f_{|13}$ are present but these occur only for large velocities ($c > 6$) where the Maxwellian in (11.18) already guarantees that $f_{|13}$ is almost zero (at $c = 6$, f_{13} is about 10^{-4} and decreases for larger c). Thus, $f_{|13}$ is positive where its absolute value is significantly different from zero, and the results depicted in Fig. 11.13 indeed refer to a valid description of a Mach 2.5 shock.

However, the threat of negative values of the phase density becomes more significant for higher Mach numbers. Figure 11.14 displays distribution functions inside shocks of various Mach numbers, all taken at x^*, where $\theta(x^*) = \frac{1}{2}(\theta_0 + \theta_1)$, at which the deviation from a Maxwellian was found to be most pronounced. For a shock with $M_0 = 4$ the distribution function becomes significantly negative inside a relevant range. This negativity can be considered as an indicator for the loss of validity for the R13 equations.

11.4 Concluding remarks

In this chapter we showed shock structure computations for the most important macroscopic models for rarefied gas flows, i.e. the NSF, Burnett, super-Burnett, augmented Burnett, Grad 13, and R13 equations.

The NSF equations describe shock structures qualitatively well, but fail in a quantitative description.

For the Burnett and super-Burnett equations the instability of the equations leads to unphysical oscillations in the computed shock structure. Numerical methods for their solutions work only when the grid spacing is not too small compared to the mean free path.

The augmented Burnett equations show oscillations as well when a spatial method is used for their solution, and are restricted to coarse grid resolutions. In time dependent methods, they do not show oscillations. This behavior follows from the stability properties of the augmented Burnett equations, which are stable in time, but unstable in space, in the sense of Chapter 10.

The Grad 13 equations, and all other Grad-type equations, introduce unphysical subshocks into the shock structure, which are a reflection of their hyperbolicity.

The best results are obtained from the R13 and the augmented Burnett equations. However, the R13 equations should be preferred, for two reasons: (a) they are stable in space and time, and thus instabilities do not introduce oscillations into numerical solutions. (b) They are accurate to super-Burnett order (for Maxwell molecules), and are derived from the Boltzmann equation, while the augmented Burnett equations contain ad hoc terms, that were introduced to give stability (in time), but cannot be derived from the Boltzmann equation.

The R13 equations yield smooth shock structures over a wide range of Mach numbers, and thus resolve the subshock problem in Grad's original equations. The agreement with DSMC is better for the R13 equations than the augmented Burnett equations in the case of Maxwell molecules and vice versa in the case of hard spheres. The full third order theory for non-Maxwellian molecules is not available at present, but it can be expected to give better results than the approximation for hard spheres used above.

R13 equations and the Burnett models were derived by expansions in the Knudsen number, and we have argued that the appropriate Knudsen number for a shock wave depends on the desired resolution. From the comparison with DSMC results and from an analysis of the positivity of Grad's distribution function inside the shock we concluded that quantitative features of the shock will only be captured up to a Mach number $M_0 \approx 3.0$. Still, the equations describe the shock thickness—which is a low resolution measure—well even at larger Mach numbers. Thus, when quantitative features—i.e. a high resolution—are less important, they can be useful at larger Mach numbers nevertheless.

Boundary value problems

In order to solve extended macroscopic equations for engineering flow problems, boundary conditions are required for higher moments or for gradients of temperature and velocity. Since the macroscopic equations were derived from the Boltzmann equation by averaging, the boundary conditions should follow from the boundary conditions for the Boltzmann equation through averaging as well.

To find proper boundary conditions for the extended equations is a difficult topic that is widely discussed in the literature, but, unfortunately, no conclusive answers are available. Nevertheless, the discussion in this chapter will show that extended macroscopic models are able to describe processes that are affected by the boundaries.

Before we discuss boundary value problems for the extended sets of equations, the well-known Maxwell-Smoluchowski jump and slip boundary conditions for the Navier-Stokes-Fourier equations are derived.

Higher order equations are then considered for the problem of plane Couette flow. Their solution is split into a bulk part which is obtained from a Chapman-Enskog expansion, and linear Knudsen boundary layers which are solutions of the linearized equations. Superpositions of bulk and Knudsen layer solutions are compared with DSMC simulations.

12.1 Boundary conditions for moments

12.1.1 Basic considerations

Maxwell's boundary condition for the phase density was introduced already in Sec. 2.4, where we wrote the phase density directly at the wall as (2.23)

$$\bar{f} = \begin{cases} \chi f_W + (1 - \chi) f_N \left(-C_k^W n_k \right) & , C_k^W n_k \geq 0 \\ f_N \left(C_k^W n_k \right) & , C_k^W n_k \leq 0 . \end{cases} \qquad (12.1)$$

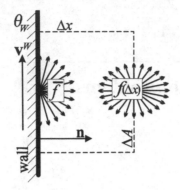

Fig. 12.1. The phase density in front of a wall at temperature θ_W with velocity v_W. $f(\Delta x)$ denotes the phase density in distance Δx from the wall.

Here, χ is the accommodation coefficient, n_k is the wall normal pointing into the gas, and

$$f_W = \frac{\rho_W}{m} \sqrt{\frac{1}{2\pi\theta_W}}^{-3} \exp\left[-\frac{C_W^2}{2\theta_W}\right] . \tag{12.2}$$

is the wall Maxwellian; f_N denotes the phase density of the incoming particles ($C_k^W n_k \leq 0$).

We use Fig. 12.1 to study the boundary conditions on the level of moments. For this, we integrate the moment equation for a moment $u_A = \int \Psi_A f \, dc$ in divergence form (6.1) over the volume $\Delta V = \Delta A \Delta x$, to obtain

$$\frac{d}{dt} \int_{\Delta V} u_A dV + \oint_{\partial \Delta V} F_{Ak} \bar{n}_k dA = \int_{\Delta V} P_A dV ,$$

where $\partial \Delta V$ denotes the surface of ΔV, and $\bar{n}_k$ its outer normal. For the following, we assume that the surface element ΔA is so small that moments and fluxes do not vary on ΔA. The volume integrals vanish in the limit $\Delta x \to 0$ which thus gives

$$\left(\bar{F}_{Ak} - F_{Ak}^N\right) n_k \Delta A = 0 . \tag{12.3}$$

The fluxes are defined in the rest frame of the wall, as

$$\bar{F}_{Ak} = \int \Psi_A C_k^W \bar{f} \, dc \quad \text{and} \quad F_{Ak}^N = \lim_{\Delta x \to 0} \int \Psi_A C_k^W f(\Delta x) \, dc . \tag{12.4}$$

Any accurate solution of the Boltzmann equation with the Maxwell boundary condition will fulfill condition (12.3) for all moments, since by definition $\lim_{\Delta x \to 0} f(\Delta x) = f_N$ (for $C_k^W n_k \leq 0$).

The collisions between particles and wall change the phase density drastically, so that the distributions of incoming and outgoing particles will be very different. Most importantly, the phase density directly at the wall, $\bar{f}$, will be discontinuous in velocity space unless thermal equilibrium prevails. Collisions will relax this discontinuity within a layer of few mean free paths away from the wall, the so-called Knudsen layer.

All sets of macroscopic equations, from NSF, over the Burnett and Grad models, to the R13 equations, are associated with an approximation to the phase density, f_a, that depends on the local values of the moments e.g. $f_a = f_{|13}$ (6.11), or on the local values of the gradients of density, velocity and temperature, e.g. $f_a = f_{|CE} = f_M + f^{(1)}$ (4.25). Typically the phase densities f_a are Maxwellians multiplied with polynomials in the peculiar velocity.

Use of macroscopic equations to describe the flow is tantamount to describing the flow with the phase densities f_a—everywhere in the domain. Thus, in order to have an agreement with solutions of the Boltzmann equation, the approximation f_a should fulfill $\lim_{\Delta x \to 0} f_a(\Delta x) = \bar{f}$, i.e. should reproduce the discontinuous phase density at the wall.

The f_a used—Maxwellian times polynomial—are not well suited to describe discontinuous functions, and one will expect some difficulties for the macroscopic models to describe the gas close to the wall, and in the Knudsen layer. Nevertheless, as will be seen, macroscopic equations can describe boundary influenced rarefied flow problems within certain limits.

For macroscopic theories, one has to replace $f = f_N = f_a$ in the equations above. The evaluation of (12.3) yields relations between the wall properties θ_W, v_i^W and the moments, or their gradients, i.e. boundary conditions. Whether these are meaningful depends on the choice of Ψ_A, and on the quality of the approximative phase density.

We recall useful relations between the particle velocity c_i, the particle velocity relative to the wall C_i^W, and the peculiar velocity C_i:

$$C_i^W = c_i - v_i^W = C_i + V_i \ , \quad C_i^W n_i = c_i n_i - v_i^W n_i = C_i n_i \ .$$

$V_i = v_i - v_i^W$ denotes the slip velocity (2.21), which is parallel to the wall, $V_k n_k = 0$.

The interesting functions Ψ_A are either even or odd in $C_k^W n_k$, and we write $\Psi_A\left(-C_k^W n_k\right) = \pm\Psi_A\left(C_k^W n_k\right)$. Insertion of (12.1) and (12.4) into (12.3) then yields

$$(1-\chi) F_{Ak}^N n_k = \chi \int_{C_k^W n_k \geq 0} \Psi_A C_k^W n_k \left(\pm f_W - f_a\right) d\mathbf{c}$$

$$+ (1 \mp 1) \int_{C_k^W n_k \geq 0} \Psi_A C_k^W n_k f_a d\mathbf{c} . \quad (12.5)$$

The computations for this chapter require half-space integrals of Maxwellians, and Appendix A.3.3 outlines their calculation, and gives specific results. Moreover the half-space moments of f_a are defined as

$$H_{i_1 \cdots i_n}^r = m \int_{C_i n_i \geq 0} C^{2r} C_{i_1} \cdots C_{i_n} f_a d\mathbf{c} . \quad (12.6)$$

The condition for conservation of mass (2.24) follows from (12.5) for $\Psi_A = 1$,

$$\frac{1}{2}\sqrt{\frac{2}{\pi}}\rho_W\sqrt{\theta_W} = H_k^0 n_k \quad \text{or} \quad \rho_W = \sqrt{\frac{2\pi}{\theta_W}}H_k^0 n_k \,. \tag{12.7}$$

12.1.2 Tangential momentum

The momentum parallel to the wall, measured in the restframe of the wall, is given by

$$\Psi_{c_i}^{\parallel} = C_i^W - n_i C_r^W n_r = C_i - n_i C_r n_r + V_i \,,$$

and is even in $C_k^W n_k$, that is $\Psi_{c_i}^{\parallel}\left(-C_k^W n_k\right) = \Psi_{c_i}^{\parallel}\left(C_k^W n_k\right)$. Evaluation of (12.5) gives at first[1]

$$\frac{1-\chi}{\chi}\left(\sigma_{ik}n_k - n_i\sigma_{kr}n_k n_r\right) = m\int_{C_k^W n_k \geq 0}\left(C_i^W - n_i C_r^W n_r\right)C_k^W n_k f_W \,d\mathbf{c}$$

$$- m\int_{C_k^W n_k \geq 0}\left(C_i - n_i C_r n_r + V_i\right)C_k n_k f_N d\mathbf{c} \,. \tag{12.8}$$

The half-space integrals of the Maxwellian can be computed easily (Appendix A.3.3) to give

$$\frac{1-\chi}{\chi}\left(\sigma_{ik}n_k - n_i\sigma_{jk}n_j n_k\right) = -V_i H_k^0 n_k - \left(\delta_{ij} - n_i n_j\right)H_{jk}^0 n_k \,. \tag{12.9}$$

This equation will be used to determine the velocity slip. Note that the tangential stress $(\sigma_{ik}n_k - n_i\sigma_{kr}n_k n_r)$ vanishes for $\chi \to 0$—there are no shear stresses at an elastically reflecting wall.

12.1.3 Energy flux

Next we consider the energy flux, that is chose

$$\Psi_{\frac{1}{2}c^2} = \frac{m}{2}C_W^2 = \frac{m}{2}\left(C^2 + 2V_i C_i + V^2\right)$$

which is even as well, $\Psi_{\frac{1}{2}c^2}\left(-C_k^W n_k\right) = \Psi_{\frac{1}{2}c^2}\left(C_k^W n_k\right)$. Evaluation of (12.5) gives at first

$$\frac{1-\chi}{\chi}\left(q_k n_k + V_i\sigma_{ik}n_k\right) = \frac{m}{2}\int_{C_k^W n_k \geq 0}\left(C_W^2 - V^2\right)C_k^W n_k f_W \,d\mathbf{c}$$

$$- \frac{m}{2}\int_{C_k^W n_k \geq 0}\left(C^2 + 2V_i C_i\right)C_k n_k f_N d\mathbf{c} \,. \tag{12.10}$$

---

[1] Recall the definitions $\int C_i f_a d\mathbf{c} = 0$, $m\int C_i C_j f_a d\mathbf{c} = p\delta_{ij} + \sigma_{ij}$.

V^2 can be eliminated by means of the scalar product of (12.8) with V_i to obtain

$$\frac{1-\chi}{\chi}\left(q_k n_k + \frac{1}{2}V_i\sigma_{ik}n_k\right) = 2\theta_W H^0_k n_k - \frac{1}{2}H^1_k n_k - \frac{1}{2}V_i H^0_{ik}n_k . \quad (12.11)$$

This equation will be used to determine the temperature jump. Note that the normal heat flux $q_k n_k$ vanishes for $\chi \to 0$—an elastically reflecting wall is adiabatic.

12.1.4 Maxwell-Smoluchowski boundary conditions

The boundary relations (12.9, 12.11) become only meaningful after the half-space moments $H^r_{i_1\cdots i_n}$ are computed by means of a particular phase density f_a. We shall be content with some first insight into jump and slip phenomena, and thus we chose Grad's 13 moment phase density (6.11), which can be reduced to the first order Chapman-Enskog phase density (4.25) by inserting the laws of Navier-Stokes and Fourier for stress and heat flux (Sec. 6.3).

Evaluated with (6.11), and after some re-shuffling of terms, (12.9, 12.11) give explicit expressions for velocity slip and temperature jump in terms of the gas properties [125],

$$V_i = \frac{-\frac{2-\chi}{\chi}\sqrt{\frac{\pi}{2}}\sqrt{\theta}\left(\sigma_{ik}n_k - n_i\sigma_{jk}n_jn_k\right) - \frac{1}{5}\left(q_i - n_iq_jn_j\right)}{\rho\theta + \frac{1}{2}\sigma_{jk}n_jn_k} ,$$

$$\theta - \theta_W = -\frac{\frac{2-\chi}{\chi}\frac{1}{2}\sqrt{\frac{\pi}{2}}\sqrt{\theta}q_kn_k + \frac{1}{4}\theta\sigma_{jk}n_jn_k}{\rho\theta + \frac{1}{2}\sigma_{jk}n_jn_k} + \frac{V^2}{4} . \quad (12.12)$$

These relations are the jump and slip boundary conditions for Grad's 13 moment equations.

In order to obtain jump and slip conditions for the Navier-Stokes-Fourier theory, stress and heat flux must be replaced by the NSF laws $\sigma_{ij} = -2\mu\frac{\partial v_{\langle i}}{\partial x_{j\rangle}}$ and $q_k = -\frac{5\mu}{2\Pr}\frac{\partial\theta}{\partial x_k}$. To simplify notation, the derivative in normal direction is introduced as $\frac{\partial}{\partial n} = n_k\frac{\partial}{\partial x_k}$, and the tangential derivative as $\frac{\partial}{\partial\tau_i} = \frac{\partial}{\partial x_i} - n_i\frac{\partial}{\partial n}$. Moreover, the velocity is split into its tangential and normal parts, $v_i = v_i^\tau + n_iv_n$. Note that $v_n = 0$ directly at the wall, which implies $\frac{\partial v_n}{\partial\tau_k} = 0$, but not $\frac{\partial v_n}{\partial n} = 0$.

Since the NSF equations are of first order in the Knudsen number, normally only the leading order terms of (12.12) are considered,[2]

$$V_i = \frac{2-\chi}{\chi}\sqrt{\frac{\pi}{2}}\frac{\mu}{\rho\sqrt{\theta}}\frac{\partial v_i^\tau}{\partial n} + \frac{1}{2\Pr}\frac{\mu}{\rho\theta}\frac{\partial\theta}{\partial\tau_i} ,$$

$$\theta - \theta_W = \frac{2-\chi}{\chi}\frac{5}{4\Pr}\sqrt{\frac{\pi}{2}}\frac{\mu}{\rho\sqrt{\theta}}\frac{\partial\theta}{\partial n} + \frac{1}{6}\frac{\mu}{\rho}\left(2\frac{\partial v_n}{\partial n} - \frac{\partial v_r^\tau}{\partial\tau_r}\right) . \quad (12.13)$$

---

[2] One can use the formal smallness parameter ε to scale the viscosity as $\mu \to \varepsilon\mu$, and then expand in ε. Also, rather flat walls are assumed, where $\frac{\partial n_i}{\partial\tau_j} \approx 0$.

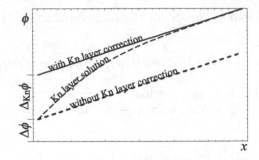

Fig. 12.2. Knudsen layer contribution to jump and slip for an unspecified quantity ϕ (after [3]).

These are the jump and slip boundary conditions for the NSF equations. According to these, a slip velocity occurs due to a velocity gradient normal to the wall (classical slip), or due to a temperature gradient along the wall (thermal creep, thermal transpiration).

A temperature jump results from a temperature gradient perpendicular to the wall, or due to velocity gradients. The velocity effects are usually ignored in the literature, probably because they are very small, and cannot be controlled in experiments.

Lockerby et al. suggest to replace σ_{ij} and q_i in (12.12) by the Burnett equations, even to provide boundary conditions for the NSF system [126].

The measurement of temperature in experiments relies on the assumption that temperature is continuous at the interface between thermometer and gas. The temperature jumps, which can be substantial, therefore compromise the ability to measure temperature.

12.1.5 Knudsen layer correction

The NSF equations cannot resolve Knudsen layers, which, however, give additional contributions to jump and slip. This can be seen from Fig. 12.2, which, following [3], shows the Knudsen layer contribution $\Delta_{\mathrm{Kn}}\phi$ to the jump $\Delta\phi$ of an unspecified quantity ϕ. The simplest way to account for the additional contributions is to introduce correction factors α_α, β_α into the boundary conditions, so that

$$V_i = \frac{2-\chi}{\chi}\alpha_1\sqrt{\frac{\pi}{2}}\frac{\mu}{\rho\sqrt{\theta}}\frac{\partial v_i^\tau}{\partial n} + \frac{\alpha_2}{2\,\mathrm{Pr}}\frac{\mu}{\rho\theta}\frac{\partial\theta}{\partial\tau_i} \,,$$

$$\theta - \theta_W = \frac{2-\chi}{\chi}\frac{5\beta_1}{4\,\mathrm{Pr}}\sqrt{\frac{\pi}{2}}\frac{\mu}{\rho\sqrt{\theta}}\frac{\partial\theta}{\partial n} + \beta_2\frac{1}{6}\frac{\mu}{\rho}\left(2\frac{\partial v_n}{\partial n} - \frac{\partial v_r^\tau}{\partial\tau_r}\right) \,. \qquad (12.14)$$

The above derivation, which ignores Knudsen layer corrections, gives $\alpha_1 = \alpha_2 = \beta_1 = \beta_2 = 1$. Many authors have computed these coefficients, mostly from numerical solutions of the Boltzmann equation, or of related kinetic models, see, e.g. [127]. Sharipov and Seleznev give a detailed overview on the literature [128][129]. They introduce the viscous slip coefficient σ_P, the thermal slip coefficient σ_T, and the temperature jump coefficient ζ_T which are related to our coefficients by (for $\mathrm{Pr} = 2/3$)[3]

$$\alpha_1 = \frac{\chi}{2-\chi}\frac{2}{\sqrt{\pi}}\sigma_p \quad , \quad \alpha_2 = \frac{4}{3}\sigma_T \quad , \quad \beta_1 = \frac{\chi}{2-\chi}\frac{16}{15\sqrt{\pi}}\zeta_T \ .$$

Typical values for the coefficients, computed with diffusive boundary conditions, i.e. $\chi = 1$, are [129][127] $\sigma_P = 0.987$, $\sigma_T = 1.009$, $\zeta_T = 1.873$ which gives

$$\alpha_1 = 1.114 \quad , \quad \alpha_2 = 1.34533 \quad , \quad \beta_1 = 1.127 \ .$$

Thus, our simple approach, which is based on assuming the validity of the first order CE phase density through the whole flow domain towards the wall, introduces an error of 10-35%. Considering the relative simplicity of the argument, it is remarkably successful, since it describes the effects leading to slip and temperature jump quite well.

Other models for jump and slip boundary conditions can be found in the literature, but these shall not be discussed here [22][130][131].

12.2 Plane Couette flow

12.2.1 Couette geometry and conservation laws

Plane Couette flow is a simple flow where rarefaction effects such as velocity slip, temperature jumps, and Knudsen boundary layers can be studied. Figure 12.3 shows the geometry of the flow: Two infinite parallel plates at constant distance L move with the constant velocities v_W^0, v_W^L relative to each other in their respective planes, and are kept at constant temperatures θ_W^0, θ_W^L. The coordinate frame is chosen such that the planes move into the direction $x = x_1$, and $y = x_2/L$ is the direction perpendicular to the plates; note that y is introduced as a dimensionless measure. Body forces, e.g. gravity, are ignored, and only the steady state will be studied.

We have the freedom of choosing an inertial frame for the description of the experiment, and chose a frame that rests with the left wall, so that $v_W^0 = 0$. For most of the following, we assume equal wall temperatures, $\theta_W^0 = \theta_W^L = \theta_W$.

Due to the symmetry of the problem, all variables will depend only on the coordinate y. Since the walls are impermeable the velocity of the gas must point into the x-direction, that is

$$v_i = \{v(y),0,0\}_i \quad \text{and thus} \quad \frac{\partial v_k}{\partial x_k} = 0 \ , \quad \frac{D}{Dt} = 0 \ .$$

---

[3] Information on the coefficient β_2 cannot be found in the literature, we shall use $\beta_2 = 1$.

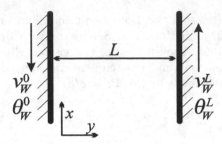

Fig. 12.3. Couette flow between two infinite paralell plates at distance L. Temperatures and velocites of the plates are controlled.

Furthermore, since the set-up is independent of the third space coordinate, $z = x_3$, neither stress nor heat flux should be associated with that direction, so that $\sigma_{13} = \sigma_{23} = q_3 = 0$, and

$$\sigma_{ij} = \begin{bmatrix} \sigma_{11}(y) & \sigma_{12}(y) & 0 \\ \sigma_{12}(y) & \sigma_{22}(y) & 0 \\ 0 & 0 & -\sigma_{11}(y) - \sigma_{22}(y) \end{bmatrix}_{ij} , \quad q_i = \{q_1(y), q_2(y), 0\}_i .$$
(12.15)

The macroscopic description is based on the conservation laws (9.1) which here reduce to

$$0 = 0 , \quad \frac{d\sigma_{12}}{dy} = 0 , \quad \frac{d\rho\theta}{dy} + \frac{d\sigma_{22}}{dy} = 0 , \quad \frac{d\sigma_{32}}{dy} = 0 , \quad \frac{dq_2}{dy} = -\sigma_{12}\frac{dv}{dy} .$$

The first and fourth equation are trivial, and two of the remaining three can be integrated so that the conservation laws assume the form

$$\sigma_{12} = \sigma_{12}^0 = \text{const.} , \quad \rho\theta + \sigma_{22} = P_0 = \text{const.} , \quad \frac{dq_2}{dy} = -\sigma_{12}\frac{dv}{dy} . \quad (12.17)$$

The further evaluation depends on the laws considered for σ_{ij} and q_i.

12.2.2 Navier-Stokes-Fourier equations

In Couette geometry, the NSF laws (9.2) reduce to

$$\sigma_{12} = -\frac{\mu}{L}\frac{dv}{dy} , \quad q_2 = -\frac{5}{2\Pr}\frac{\mu}{L}\frac{d\theta}{dy} , \quad (12.18)$$

all other components of stress σ_{ij} and heat flux q_i vanish. From (12.17) follows that the pressure is constant, $\rho\theta = P_0$, and the two other conservation laws assume the form

$$\frac{dv}{dy} = -\frac{L\sigma_{12}^0}{\mu} , \quad \mu\frac{d}{dy}\left(\mu\frac{d\theta}{dy}\right) = -\frac{2\Pr\left(L\sigma_{12}^0\right)^2}{5} . \quad (12.19)$$

These must be solved under the boundary conditions (12.14) which reduce to

$$v - v_W^\alpha = \frac{2-\chi}{\chi}\alpha_1\sqrt{\frac{\pi}{2}\frac{\mu\sqrt{\theta}}{LP_0}}\frac{dv}{dy}n^\alpha \quad , \quad \theta - \theta_W = \frac{2-\chi}{\chi}\frac{5\beta_1}{4\,\mathrm{Pr}}\sqrt{\frac{\pi}{2}\frac{\mu\sqrt{\theta}}{LP_0}}\frac{d\theta}{dy}n^\alpha ,$$

$$(12.20)$$

where $\alpha = 0, L$ and $n^0 = -n^L = 1$. Recall that the viscosity is a function of temperature alone, $\mu = \mu(\theta)$.

In order to give a flavor of the result, we consider a linearized version of the equations by assuming only small temperature variations around θ_W. Then the viscosity can be assumed to be constant, $\mu = \mu(\theta_W)$, and the equations read

$$\frac{dv}{dy} = a = \text{const.} \quad , \quad \frac{d^2\theta}{dy^2} = -\frac{2\,\mathrm{Pr}}{5}a^2 , \tag{12.21}$$

with the linearized boundary conditions

$$v - v_W^\alpha = \frac{2-\chi}{\chi}\alpha_1\sqrt{\frac{\pi}{2}}\mathrm{Kn}\,a\,n^\alpha \quad , \quad \theta - \theta_W = \frac{2-\chi}{\chi}\frac{5\beta_1}{4\,\mathrm{Pr}}\sqrt{\frac{\pi}{2}}\mathrm{Kn}\frac{d\theta}{dy}n^\alpha , \tag{12.22}$$

where $\mathrm{Kn} = \mu(\theta_W)\sqrt{\theta_W}/(P_0 L)$ is the appropriate Knudsen number.

The equations (12.21) with the boundary conditions (12.22) are easy to integrate, and the result can be written as

$$v = \frac{v_w^L}{2} + a\left(y - \frac{1}{2}\right) \quad \text{with} \quad a = \frac{v_w^L}{1 + \frac{2-\chi}{\chi}\sqrt{2\pi}\alpha_1\mathrm{Kn}} ,$$

$$\theta = \theta_W + \frac{2-\chi}{\chi}\frac{\beta_1}{4}\sqrt{\frac{\pi}{2}}\mathrm{Kn}\,a^2 + \frac{\mathrm{Pr}}{5}a^2 y(1-y) , \tag{12.23}$$

$$\sigma_{12} = -\mathrm{Kn}\frac{P_0 a}{\sqrt{\theta_W}} \quad , \quad q_2 = \mathrm{Kn}\frac{P_0 a^2}{\sqrt{\theta_W}}\left(y - \frac{1}{2}\right) .$$

Solution curves for non-linear Couette flow will be shown below, and thus solutions of the linear equations are not printed.

The features of the linear solution can be read off easily from the equations: The solutions are symmetric with respect to the point $y = \frac{1}{2}$, velocity v and heat flux q_2 are straight lines, while the temperature θ is a parabola, and, of course, σ_{12} is a constant. The temperature rises due to heating by friction, and is largest at $y = \frac{1}{2}$.

It is instructive to study temperature jump and slip in the linear case. Due to the symmetry, it suffices to consider both at $y = 0$, where

$$v^0 - v_W^0 = \frac{v_w^L}{2}\frac{\frac{2-\chi}{\chi}\alpha_1\sqrt{2\pi}\mathrm{Kn}}{1 + \frac{2-\chi}{\chi}\sqrt{2\pi}\alpha_1\mathrm{Kn}} \quad , \quad \theta^0 - \theta_W^0 = \frac{\frac{2-\chi}{\chi}\frac{5\beta_1}{4\,\mathrm{Pr}}\sqrt{\frac{\pi}{2}}\mathrm{Kn}}{\left(1 + \frac{2-\chi}{\chi}\sqrt{2\pi}\alpha_1\mathrm{Kn}\right)^2}\frac{(v_w^L)^2}{5\,\mathrm{Pr}} .$$

Obviously, the Knudsen number Kn strongly affects jump and slip. In the limit $\mathrm{Kn} \to 0$, jump and slip vanish, $v^0 = v_W^0$, $\theta^0 = \theta_W^0$. As the Knudsen number grows, the slip grows and approaches $\frac{v_w^L}{2}$ as Kn $\rightarrow \infty$. In the latter case, the velocity is constant throughout the domain, $a = 0$, which reflects that particles to not interact anymore. Thus, there is no frictional heating ($q_2 = 0$), and the temperature is constant at θ_W.

At small Knudsen numbers, jump and slip can be ignored, but they play a considerable role as the Knudsen number grows. Note also their dependence on the shear rate a.

12.2.3 Grad 13 equations

Next we consider the Grad 13 equations in Couette geometry. It needs careful consideration of the definitions of trace-free tensors to reduce the balance laws for stress (9.9) and heat flux (9.10) for Couette geometry, where they read

$$\frac{2}{5}\frac{dq_1}{dy} + P_0\frac{dv}{dy} = -\frac{\rho\theta}{\mu}\sigma_{12} ,$$

$$-\frac{6}{5}\sigma_{12}\frac{dv}{dy} = -\frac{\rho\theta}{\mu}\sigma_{22} ,$$

$$\frac{7}{5}q_2\frac{dv}{dy} + \frac{7}{2}\sigma_{12}\frac{d\theta}{dy} = -\frac{2}{3}\frac{\rho\theta}{\mu}q_1 , \qquad (12.24)$$

$$\frac{5}{2}P_0\frac{d\theta}{dy} + \frac{d\sigma_{22}\theta}{dy} + \frac{2}{5}q_1\frac{dv}{dy} = -\frac{2}{3}\frac{\rho\theta}{\mu}q_2 .$$

Moreover, one finds $\sigma_{11} = -\frac{4}{3}\sigma_{22}$. The jump and slip boundary conditions (12.12) are not sufficient to solve the problem, and three additional boundary conditions are required. One condition follows by prescribing the mass of gas between the walls as, see (12.17)$_2$,

$$M = \int_0^1 \rho dy = \int_0^1 \frac{P_0 - \sigma_{22}}{\theta}dy . \qquad (12.25)$$

Two additional conditions could be, e.g., boundary conditions for σ_{22} at $y = 0$ and $y = 1$. Until now, no reliable boundary conditions for σ_{22} are available, see Sec. 12.6 for a discussion.

Marques and Kremer presented an elegant analytical solution of the Grad 13 equations with jump and slip, under the additional condition that the pressure $p = \rho\theta$ is a constant [132]. Their results show good agreement with DSMC simulations for Knudsen numbers up to Kn = 0.25, which, however, ignores Knudsen layers. Liu and Rincon, using an iterative approximation method, found a similar solution numerically [133]; they were not interested in incorporating jump and slip. Similar methods are presented in [134][135].

Outside the Knudsen layer, the pressure is constant indeed, see [7] for a comprehensive discussion, so that these solutions have some interest to describe the bulk flow.

12.2.4 R13 equations

The R13 equations, (9.9, 9.10) with (9.13) read for Couette geometry

$$\frac{8}{5}\sigma_{12}\frac{dv}{dy} + \frac{dm_{112}}{dy} = -\frac{pL}{\mu}\sigma_{11} ,$$

$$\frac{2}{5}\frac{dq_1}{dy} + P_0\frac{dv}{dy} + \frac{dm_{122}}{dy} = -\frac{pL}{\mu}\sigma_{12} ,$$

$$-\frac{6}{5}\sigma_{12}\frac{dv}{dy} + \frac{dm_{222}}{dy} = -\frac{pL}{\mu}\sigma_{22} ,$$

$$\frac{7}{5}q_2\frac{dv}{dy} + \frac{7}{2}\sigma_{12}\frac{d\theta}{dy} + \frac{1}{2}\frac{dR_{12}}{dy} + m_{112}\frac{dv}{dy} = -\frac{2}{3}\frac{pL}{\mu}q_1 ,$$

$$\frac{5}{2}P_0\frac{d\theta}{dy} + \sigma_{22}\frac{d\theta}{dy} + \theta\frac{d\sigma_{22}}{dy} + \frac{2}{5}q_1\frac{dv}{dy} + \frac{dB}{dy} + m_{122}\frac{dv}{dy} = -\frac{2}{3}\frac{pL}{\mu}q_2 ,$$

$$R_{12} = -\frac{4}{7}\frac{1}{\rho}\left(\sigma_{11} + \sigma_{22}\right)\sigma_{12}$$

$$-\frac{12}{5}\frac{\mu}{pL}\left[\theta\frac{dq_1}{dy} + q_1\frac{d\theta}{dy} - \theta q_1\frac{d\ln\rho}{dy} + \frac{5}{7}\theta\left(\sigma_{11} + \sigma_{22}\right)\frac{dv}{dy}\right] , \quad (12.26)$$

$$B = -\frac{1}{7}\frac{1}{\rho}\left(\sigma_{11}^2 + 3\sigma_{12}^2 + 3\sigma_{22}^2 + \sigma_{11}\sigma_{22}\right)$$

$$-\frac{18}{5}\frac{\mu}{pL}\left(\theta\frac{dq_2}{dy} + \frac{5}{7}\theta\sigma_{12}\frac{dv}{dy} + \frac{11}{6}q_2\frac{d\theta}{dy} - \theta q_2\frac{d\ln\rho}{dy}\right) ,$$

$$m_{112} = -\frac{2}{3}\frac{\mu}{pL}\left[\theta\left(\frac{d\sigma_{11}}{dy} - \frac{2}{5}\frac{d\sigma_{22}}{dy}\right)\right.$$

$$\left. -\theta\left(\sigma_{11}\frac{d\ln\rho}{dy} - \frac{2}{5}\sigma_{22}\frac{d\ln\rho}{dy}\right) + \frac{16}{25}q_1\frac{dv}{dy}\right] ,$$

$$m_{122} = -\frac{\mu}{pL}\frac{16}{15}\left[\theta\frac{d\sigma_{12}}{dy} - \theta\sigma_{12}\frac{d\ln\rho}{dy} + \frac{2}{5}q_2\frac{dv}{dy}\right] ,$$

$$m_{222} = -\frac{\mu}{pL}\frac{6}{5}\left[\theta\frac{d\sigma_{22}}{dy} - \theta\sigma_{22}\frac{d\ln\rho}{dy} - \frac{4}{15}q_1\frac{dv}{dy}\right] .$$

Obviously, the R13 equations require even more additional boundary conditions, and at present no reliable set of boundary conditions is available. Nevertheless, it is worthwhile to study whether the R13 equations, or the Burnett equations, are capable of describing Couette flow problems in rarefied gases, i.e. at larger Knudsen numbers.

12.3 Bulk equations

In order to gain insight into the solution behavior for extended models in Couette flow, we assume that the non-equilibrium quantities can be split into bulk (B) and boundary layer (L) contributions, $\phi = \phi_B + \phi_L$. It is assumed that the boundary layer contributions vanish in some distance from the wall; they will be discussed in Sec. 12.4.

The bulk equations follow from a Chapman-Enskog expansion of the steady state equations in Couette geometry. Mostly, we are interested in equations up to second order, which can be derived from the Grad 13 equations or the Burnett equations.[4] Indeed, since the Burnett equations can be extracted by CE expansion from the Grad 13 equations, both sets of equations should give the same results when the CE expansion is applied to their steady state form in Couette geometry.

In order to have some insight into the third order, we shall give results of the expansion of the R13 equations, up to third order. For the expansion, we introduce the formal smallness parameter ε as usual, i.e. to scale viscosity, so that $\mu \to \varepsilon\mu$ in the R13 equations (12.26). Then, all non-equilibrium moments appearing in these (σ_{11}, σ_{12}, σ_{22}, q_1, q_2, R_{12}, B, m_{112}, m_{122}, m_{222}) are expanded in a CE series as

$$\phi = \varepsilon\phi^{(1)} + \varepsilon^2\phi^{(3)} + \varepsilon^3\phi^{(3)} + \cdots .$$

The evaluation proceeds similarly to the evaluation of the CE expansion in Sec. 6.3, but is simpler, since no time derivatives occur. The resulting equations are highly non-linear, and we show the full results at second order, but give only the linear terms of the third order contributions. With the non-linear terms of third order removed, the result reads

$$\sigma_{12} = -\left[\frac{\mu\sqrt{\theta}}{pL}\right]\frac{P_0}{\sqrt{\theta}}\frac{dv}{dy} - \frac{16}{15}\left[\frac{\mu\sqrt{\theta}}{pL}\right]^3\frac{P_0}{\sqrt{\theta}}\frac{d^3v}{dy^3} ,$$

$$\sigma_{22} = -\frac{3}{4}\sigma_{11} = -\frac{6}{5}\left[\frac{\mu\sqrt{\theta}}{pL}\right]^2\frac{P_0}{\theta}\left(\frac{dv}{dy}\right)^2 ,$$

$$q_1 = \frac{105}{8}\left[\frac{\mu\sqrt{\theta}}{pL}\right]^2\frac{P_0}{\theta}\frac{d\theta}{dy}\frac{dv}{dy} , \tag{12.27}$$

$$q_2 = -\frac{15}{4}\left[\frac{\mu\sqrt{\theta}}{pL}\right]\frac{P_0}{\sqrt{\theta}}\frac{d\theta}{dy} - \frac{81}{4}\left[\frac{\mu\sqrt{\theta}}{pL}\right]^3\frac{P_0}{\sqrt{\theta}}\frac{d^3\theta}{dy^3} .$$

Note that $\left[\frac{\mu\sqrt{\theta}}{pL}\right]$ is a (local) Knudsen number. The (linear) third order terms are underlined. These equations must be used together with the conserva-

---

[4] Of course, the molecular interaction model must be the same. All results shown here are for Maxwell molecules.

tion laws (12.17) and appropriate boundary conditions. Since the expressions
$(12.27)_{1,4}$ contain third order derivatives, the jump and slip conditions do not
suffice, and additional boundary conditions would be required to solve the
equations.

However, when only terms up to second order are considered, (12.27) reduce to

$$\sigma_{12} = -\frac{\mu P_0}{\rho\theta L}\frac{dv}{dy} \quad , \quad q_2 = -\frac{15}{4}\frac{\mu P_0}{\rho\theta L}\frac{d\theta}{dy} \quad ,$$

$$\sigma_{22} = -\frac{6}{5}\frac{\sigma_{12}\sigma_{12}}{P_0} \quad , \quad q_1 = \frac{7}{2}\frac{\sigma_{12}q_2}{P_0} \quad . \tag{12.28}$$

The first two equations, for σ_{12} and q_2, are simply the laws of Navier-Stokes
and Fourier (12.18) multiplied with the factor $P_0/\rho\theta = P_0/p$. When these are
used with the conservation laws (12.17), it suffices to prescribe the jump and
slip boundary conditions (12.12), which now read

$$v^\alpha - v_W^\alpha = \frac{-\frac{2-\chi}{\chi}\alpha_1\sqrt{\frac{\pi}{2}}\sqrt{\theta}\sigma_{12}n^\alpha - \frac{1}{5}\alpha_2 q_1}{\rho\theta + \frac{1}{2}\sigma_{22}} \quad ,$$

$$\theta^\alpha - \theta_W^\alpha = -\frac{\frac{2-\chi}{2\chi}\beta_1\sqrt{\frac{\pi}{2}}\sqrt{\theta}q_2 n^\alpha + \frac{1}{4}\theta\sigma_{22}}{\rho\theta + \frac{1}{2}\sigma_{22}} + \frac{V^2}{4} \quad . \tag{12.29}$$

Here, the correction factors $\alpha_1, \alpha_2, \beta_1$ were introduced as in (12.14) . The
constant P_0 follows from the prescribed mass between the plates (12.25) as

$$M = \int_0^1 \rho\, dy = \int_0^1 \frac{P_0 - \sigma_{22}}{\theta}\, dy \quad . \tag{12.30}$$

For DSMC simulations M can be computed from the initial conditions, and
thus is available data.

While the factor $P_0/\rho\theta$ has some influence, it can be expected that the
solution of the second order bulk equations (12.28) gives velocity and temperature curves very similar to the NSF equations (12.18).

More interesting are the equations for the normal stress, σ_{22}, and the heat
flux parallel to the wall, q_1. Both vanish in the NSF theory, and thus their
non-zero values describe pure rarefaction effects. In particular it must be noted
that there is no temperature gradient in the x-direction: q_1 is a heat flux that
is not driven by a temperature gradient.

From $(12.28)_3$ follows that σ_{22} is a constant, while q_1 is proportional to the
heat flux q_2. In order to see that both, in particular q_1, can have substantial
values, the linear NSF results (12.23) are substituted to estimate their values
as

$$\sigma_{22} = -\frac{6}{5}\mathrm{Kn}^2\frac{P_0}{\theta_W}a^2 \quad , \quad q_1 = -\frac{7}{2}\mathrm{Kn}^2\frac{P_0}{\theta_W}a^3\left(y - \frac{1}{2}\right) \quad .$$

Numerical results will be shown in Sec. 12.5.

12.4 Linear Knudsen boundary layers

12.4.1 Scaling and Knudsen layers

The second order bulk solution computed in the last section gives only non-linear contributions to σ_{22} and q_1. Depending on the set of equations used, both quantities can have linear Knudsen layer contributions as well. The CE expansion, that gave the bulk solution, discards these linear parts, for reasons that have to do with the scaling used. A boundary layer contribution will have an exponential decay over a distance of few mean free paths, and thus should be described with a rescaled space variable $\xi = y/\varepsilon$ as, e.g., $\phi(\xi)$. Then the derivative of ϕ is scaled as $\frac{\partial \phi}{\partial y} = \frac{1}{\varepsilon} \frac{\partial \phi_L}{\partial \xi}$. When the viscosity is scaled appropriately, by writing $\varepsilon \mu$, the smallness parameter ε cancels from all sets of macroscopic equations, e.g. (12.26). It follows that the full equations must be considered to study the Knudsen layer.

We assume that the non-linear contributions are well described through the bulk solution, while the linear contributions are most important for the description of the Knudsen layers. For this reason, we now study the linear equations, as given in Sec. 9.4. We will not discuss boundary conditions, but only the ability of the equations to properly describe Knudsen layers. The subsequent discussion follows [136].

In Couette geometry the linearized conservation laws (9.42) read

$$\frac{d\rho}{dy} + \frac{d\theta}{dy} + \frac{d\sigma_{22}}{dy} = 0 \;, \quad \frac{d\sigma_{12}}{dy} = 0 \;, \quad \frac{dq_2}{dy} = 0 \;. \tag{12.31}$$

The last two equations yield that σ_{12} and q_1 are constants of integration,

$$\sigma_{12} = \text{const.} \;, \quad q_2 = \text{const.}$$

The Knudsen layers computed in the sequel are hyperbolic sine and cosine functions of the form $\cosh\left[\lambda \frac{y-\frac{1}{2}}{\text{Kn}}\right]$, $\sinh\left[\lambda \frac{y-\frac{1}{2}}{\text{Kn}}\right]$, where λ is a constant of the order of unity. Figure 12.4 shows these functions for a variety of Knudsen numbers, in order to emphasize the boundary layer structure of the solution (12.38). These functions have the typical shape of a boundary layer, their largest values are found at the walls, and the curves decrease to zero within several mean free paths away from the walls. As Kn grows, the width of the boundary layers is growing as well. For Knudsen numbers above ~ 0.1 one cannot speak of boundary layers any longer, since the functions $\cosh\left[\lambda \frac{y-\frac{1}{2}}{\text{Kn}}\right]$, $\sinh\left[\lambda \frac{y-\frac{1}{2}}{\text{Kn}}\right]$ are non-zero in the region between the plates. In this case boundary effects can have an important influence on the flow pattern. Note that we consider linear boundary layers, as solutions of linearized equations. These depend only on the Knudsen number, but, as a consequence of linearization, not on the Mach number.

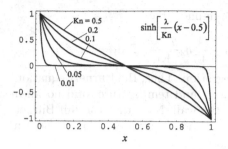 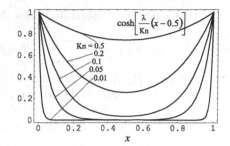

Fig. 12.4. Boundary layer functions $\sinh[\lambda x/\mathrm{Kn}]$ and $\cosh[\lambda x/\mathrm{Kn}]$ for various Knudsen numbers.

12.4.2 Navier-Stokes-Fourier and Grad 13 equations

Interestingly, the linear NSF equations (9.43) and the linear Grad 13 equations (9.48) reduce to the same set of equations, viz.

$$\sigma_{12} = -\mathrm{Kn}\frac{dv}{\partial y} \ , \ q_2 = -\frac{15}{4}\mathrm{Kn}\frac{d\theta}{dy} \ , \ \sigma_{22} = 0 \ , \ q_1 = 0 \ .$$

Insertion into the balance laws and integration gives

$$v = v_0 - \sigma_{12}\frac{y - \frac{1}{2}}{\mathrm{Kn}} \ , \ \theta = \theta_0 - \frac{4}{15}q_2\frac{y - \frac{1}{2}}{\mathrm{Kn}} \ , \ \rho = \rho_0 - \frac{4}{15}q_2\frac{y - \frac{1}{2}}{\mathrm{Kn}}$$

where v_0, θ_0, ρ_0, σ_{12}, q_1 are constants of integration. Accordingly, all profiles are linear, and the variation in density is due solely to heat transfer.[5]

Note that $\sigma_{22} = q_1 = 0$ for the NSF and Grad 13 equations. Non-zero values for the anisotropic contribution to pressure, σ_{22}, and the heat flux parallel to the wall, q_1, appear only in higher order theories. DSMC calculations show non-zero values for these quantities at larger Knudsen numbers (above ~ 0.05), and this agrees with the expectation that the NSF equations are limited to small Knudsen numbers.

The Grad 13 equations yield the second order bulk equations for σ_{22} and q_1, (12.28), and thus can describe some of the rarefaction effects. However, they cannot describe linear Knudsen layers.

12.4.3 Burnett equations

The linear Burnett expressions (9.44) in Couette geometry are

$$\sigma_{12} = -\mathrm{Kn}\frac{dv}{dy} \ , \ \sigma_{22} = -\frac{4}{3}\mathrm{Kn}^2\frac{d^2\rho}{dy^2} + \frac{2}{3}\mathrm{Kn}^2\frac{d^2\theta}{dy^2} \ ,$$

$$q_1 = \frac{3}{2}\mathrm{Kn}^2\frac{d^2v}{dy^2} \ , \ q_2 = -\frac{15}{4}\mathrm{Kn}\frac{d\theta}{dy}$$

---

[5] For symmetry reasons q_2 will vanish when both walls have the same temperature.

Together with the conservation laws this implies that

$$v = v_0 - \sigma_{12}\frac{y - \frac{1}{2}}{\mathrm{Kn}} \ , \quad \theta = \theta_0 - \frac{4}{15}q_2\frac{y - \frac{1}{2}}{\mathrm{Kn}} \ , \quad q_1 = 0 \ ,$$

where $v_0, \sigma_{12}, T_0, q_2$ are constants of integration. Thus, the Burnett equations do not predict boundary layers for velocity and temperature, and no linear contribution to the heat flux parallel to the wall. Note that the full Burnett equations yield the bulk equations (12.28), and thus imply a non-linear contribution to q_1.

The equations for ρ and σ_{22} lead to Knudsen layer solutions for density and stress σ_{22}, viz.

$$\rho = \rho_0 + \frac{4}{15}\frac{q_2}{\mathrm{Kn}}\left(y - \frac{1}{2}\right) + A\sinh\left[\frac{\sqrt{3}}{2}\frac{y - \frac{1}{2}}{\mathrm{Kn}}\right] + B\cosh\left[\frac{\sqrt{3}}{2}\frac{y - \frac{1}{2}}{\mathrm{Kn}}\right]$$

$$\sigma_{22} = -A\sinh\left[\frac{\sqrt{3}}{2}\frac{y - \frac{1}{2}}{\mathrm{Kn}}\right] - B\cosh\left[\frac{\sqrt{3}}{2}\frac{y - \frac{1}{2}}{\mathrm{Kn}}\right]. \tag{12.34}$$

However, the Burnett equations fail to predict a heat flux parallel to the wall ($q_1 = 0$), and to predict boundary layers for temperature and velocity.[6]

The solution of the Burnett equations requires two additional boundary conditions, e.g. a condition for σ_{22} at each wall [137].

12.4.4 Super-Burnett and augmented Burnett equations

For linear Couette flow the super-Burnett equations (9.45) reduce to

$$\sigma_{12} = -\mathrm{Kn}\frac{dv}{dy} - \frac{2}{3}\mathrm{Kn}^3\frac{d^3v}{dy^3} \ , \quad \sigma_{22} = -\frac{4}{3}\mathrm{Kn}^2\frac{d^2\rho}{dy^2} + \frac{2}{3}\mathrm{Kn}^2\frac{d^2\theta}{dy^2} \ ,$$

$$q_1 = \frac{3}{2}\mathrm{Kn}^2\frac{d^2v}{dy^2} \ , \quad q_2 = -\frac{15}{4}\mathrm{Kn}\frac{d\theta}{dy} - \frac{157}{16}\mathrm{Kn}^3\frac{d^3\theta}{dy^3} - \frac{5}{8}\mathrm{Kn}^3\frac{d^3\rho}{dy^3} \ . \tag{12.35}$$

For the augmented Burnett equations the expressions for σ_{22} and q_1 remain unchanged, while the expressions for σ_{12} and q_2 read

$$\sigma_{12|aB} = -\mathrm{Kn}\frac{dv}{dy} + \frac{1}{6}\mathrm{Kn}^3\frac{d^3v}{dy^3} \ , \quad q_{2|aB} = -\frac{15}{4}\mathrm{Kn}\frac{d\theta}{dy} + \frac{11}{16}\mathrm{Kn}^3\frac{d^3\theta}{dy^3} - \frac{5}{8}\mathrm{Kn}^3\frac{d^3\rho}{dy^3} \ . \tag{12.36}$$

The velocity can be computed as

$$v_{|SB} = v_0 - \sigma_{12}\frac{y - \frac{1}{2}}{\mathrm{Kn}} + A\cos\left[\sqrt{\frac{3}{2}}\frac{y - \frac{1}{2}}{\mathrm{Kn}}\right] + B\sin\left[\sqrt{\frac{3}{2}}\frac{y - \frac{1}{2}}{\mathrm{Kn}}\right] \ , \tag{12.37}$$

$$v_{|aB} = v_0 - \sigma_{12}\frac{y - \frac{1}{2}}{\mathrm{Kn}} + A\cosh\left[\sqrt{6}\frac{y - \frac{1}{2}}{\mathrm{Kn}}\right] + B\sinh\left[\sqrt{6}\frac{y - \frac{1}{2}}{\mathrm{Kn}}\right] \ , \tag{12.38}$$

---

[6] It is surprising that the Grad 13 equations do not give linear Knudsen layers, while the Burnett equations—which can be derived form the Grad 13 equations by CE expansion—imply a linear Knudsen layer.

for super-Burnett and augmented Burnett equations, respectively. v_0, σ_{12}, A, B are constants of integration.

Both sets of equations yield the linear Navier-Stokes solution, $v_0 - \sigma_{12}\frac{y-1/2}{\text{Kn}}$, plus a correction. The correction is periodic in space for the super-Burnett equations (12.37), and a Knudsen boundary layer for the augmented Burnett equations (12.38). Thus the change in sign at the third order derivative in σ_{12} between (12.35) and (12.36) has severe impact on the solution behavior—the super-Burnett equations yield results that are unphysical, but the solution of the augmented Burnett equations have support in physics.

For temperature the results are similar: From the super-Burnett equations follows

$$\theta_{|SB} = \theta_0 - \frac{4}{15}q_2 \frac{y - \frac{1}{2}}{\text{Kn}} + C \cos\left[\lambda_1 \frac{y-\frac{1}{2}}{\text{Kn}}\right] + D \sin\left[\lambda_1 \frac{y-\frac{1}{2}}{\text{Kn}}\right]$$

$$+ F \cosh\left[\lambda_2 \frac{y - \frac{1}{2}}{\text{Kn}}\right] + G \sinh\left[\lambda_2 \frac{y - \frac{1}{2}}{\text{Kn}}\right]$$

$$\text{with}\quad \lambda_1 = \sqrt{\frac{3\sqrt{53129}-201}{1216}} = 0.635, \quad \lambda_2 = \sqrt{\frac{3\sqrt{53129}+201}{1216}} = 0.857$$

where the constants $\theta_0, q_2, C, D, F, G$ must be obtained from boundary conditions. This solution corresponds to the linear Fourier solution, $\theta_0 - \frac{4}{15}q_2\frac{y-1/2}{\text{Kn}}$, plus corrections, of which one is again periodic in space, while the other is of boundary layer type.

From the augmented Burnett equations one finds

$$\theta = \theta_0 - \frac{4}{15}q_2 \frac{y - \frac{1}{2}}{\text{Kn}} + C \cosh\left[\gamma_1 \frac{y-\frac{1}{2}}{\text{Kn}}\right] + D \sinh\left[\gamma_1 \frac{y-\frac{1}{2}}{\text{Kn}}\right]$$

$$+ F \cosh\left[\gamma_2 \frac{y - \frac{1}{2}}{\text{Kn}}\right] + G \sinh\left[\gamma_2 \frac{y - \frac{1}{2}}{\text{Kn}}\right]$$

$$\text{with}\quad \gamma_1 = \sqrt{\frac{3\sqrt{5081}+303}{128}} = 2.009, \quad \gamma_2 = \sqrt{\frac{303-3\sqrt{5081}}{1216}} = 0.835,$$

so that both corrections are of boundary layer type.

At this point we can draw the following conclusions:

(a) The super-Burnett equations lead to unphysical periodic solutions in steady state flows, and should not be considered for steady state problems. We encountered oscillations in the steady state Burnett equations earlier when we discussed shock profiles, and, of course, found them to be unstable in transient problems. The periodic solutions can be avoided by choosing boundary conditions such that $A = B = C = D = 0$ [138], but it seems better to avoid unphysical equations right away.

(b) The augmented Burnett equations seem to describe Knudsen boundary layers for velocity and temperature, and might be useful. The main point against them is that the terms of super-Burnett order that were used to augment the Burnett equations are not based on rational arguments as discussed in section 4.3.4. Polemically one could say that in the augmented Burnett equations the signs of those coefficients "that cause the trouble" are inverted, which results in good behavior of the equations, but is difficult, if not impossible, to justify by an argument based in physics. Another point against the augmented Burnett equations is that they are unstable with respect to local fluctuations at high frequencies, see Sec. 10.1 and [60].

12.4.5 Regularized 13 moment equations

The linear R13 equations (9.47, 9.49) reduce in Couette geometry to

$$\frac{dv}{dy} + \frac{2}{5}\frac{dq_1}{dy} = -\frac{\sigma_{12}}{\mathrm{Kn}} = \text{const.} \ , \quad q_1 = \frac{9}{5}\mathrm{Kn}^2\frac{d^2 q_1}{dy^2} \ ,$$

$$\frac{5}{2}\frac{d\theta}{dy} + \frac{d\sigma_{22}}{dy} = -\frac{2}{3}\frac{q_2}{\mathrm{Kn}} = \text{const.} \ , \quad \sigma_{22} = \frac{6}{5}\mathrm{Kn}^2\frac{d^2\sigma_{22}}{dy^2} \ ,$$

so that velocity and temperature are obtained as

$$v = v_0 - \sigma_{12}\frac{y-\frac{1}{2}}{\mathrm{Kn}} - \frac{2}{5}q_1 \text{ with } q_1 = A\sinh\left[\sqrt{\frac{5}{9}}\frac{y-\frac{1}{2}}{\mathrm{Kn}}\right] + B\cosh\left[\sqrt{\frac{5}{9}}\frac{y-\frac{1}{2}}{\mathrm{Kn}}\right] ,$$
(12.39)

$$\theta = \theta_0 - \frac{4}{15}q_2\frac{y-\frac{1}{2}}{\mathrm{Kn}} - \frac{2}{5}\sigma_{22} \text{ with } \sigma_{22} = C\sinh\left[\sqrt{\frac{5}{6}}\frac{y-\frac{1}{2}}{\mathrm{Kn}}\right] + D\cosh\left[\sqrt{\frac{5}{6}}\frac{y-\frac{1}{2}}{\mathrm{Kn}}\right] .$$
(12.40)

where v_0, σ_{12}, A, B and θ_0, q_1, C, D are constants of integration. We can identify $\left(-\frac{2}{5}q_1\right)$ and $\left(-\frac{2}{5}\sigma_{22}\right)$ as the Knudsen boundary layers for velocity and temperature according to the R13 equations, and they are of the same form as those discussed above, and depicted in Fig. 12.4.

12.4.6 The heat flux parallel to the flow

A closer inspection of the last equations shows that the R13 equations predict the heat flux parallel to the flow as

$$q_{1|R13} = -\frac{9}{2}\mathrm{Kn}^2\frac{\partial^2 v}{\partial y^2} \ ,$$
(12.41)

while the corresponding expression from the Burnett equations reads

$$q_{1|B} = \frac{3}{2}\mathrm{Kn}^2\frac{\partial^2 v}{\partial y^2} \ .$$
(12.42)

Note that the augmented Burnett equations and the super-Burnett equations agree with the Burnett equations up to the order of Kn^2, so that they also predict the result (12.42). Thus, the R13 equations and the Burnett equations predict different signs for q_1.

The question which sign relates heat flux and the second gradient of velocities in a rarefied gas can be answered by means of a DSMC simulation for Couette flow in a gas of Maxwell molecules at Kn = 0.1 with $v(0) = 0$, $v(L) = 200\text{m/s}$, and $\theta_W = \frac{k}{m}273\text{K}$, performed with the free code supplied by Bird [1].[7]

The strong scatter in the DSMC results does not allow us to compute the derivatives of the velocities, and therefore the argument is based on $(12.39)_1$. Solving for q_1 yields, after reinserting of dimensions,

$$q_{1|R13,lin} = -\frac{5}{2}p_0\left[v(y) - v\left(\frac{1}{2}\right) + \frac{L}{\mu_0}\sigma_{12}\left(y - \frac{1}{2}\right)\right]. \qquad (12.43)$$

A similar equation could be written for the super-Burnett equations, but this is not done here, since the super-Burnett equations, as was shown, yield periodic solutions, and are unphysical. Integration of the corresponding equation of the augmented Burnett equations $(12.38)_1$ yields with (12.42)

$$q_{1|aB,lin} = 9p_0\left[v(y) - v\left(\frac{1}{2}\right) + \frac{L}{\mu_0}\sigma_{12|aB}\left(y - \frac{1}{2}\right)\right], \qquad (12.44)$$

that is a similar curve, albeit with different sign and amplitude.

In order to compare these results from the linear equations to DSMC simulations it is best to subtract the bulk solution (12.28) from the DSMC result, i.e. compare to

$$q_{1|DSMC,lin} = q_{DSMC} - \frac{7}{2}\frac{\sigma_{12}q_2}{p + \sigma_{22}}.$$

The result is shown in Fig. 12.5, which shows the heat flux $q_{1|DSMC,lin}$ as computed directly in the simulation, and $q_{1|R13,lin}$ as computed from (12.43), where the DSMC data was used for the evaluation of the right hand side. While $q_{1|R13,lin}$ differs from the simulation value $q_{1|DSMC,lin}$, it is clear that the relation (12.43) yields the proper direction of q_1, while (12.44) would yield the opposite—wrong—direction. The difference between $q_{1|R13,lin}$ and $q_{1|DSMC,lin}$ will partly be due to non-linear effects, and partly to the occurrence of additional contributions to the velocity Knudsen layer.

The figure also shows the function $q_1 = A\sinh\left[\sqrt{\frac{5}{9}}\frac{y-\frac{1}{2}}{\text{Kn}}\right]$ with fitted amplitude A. Obviously, the fitted R13 result $(12.39)_1$ agrees well with the DSMC heat flux $q_{1|DSMC,lin}$. This agreement indicates that the R13 equations can describe the physical Knudsen boundary layer for q_1. This observation leads to considering superpositions between the bulk solutions and the linear Knudsen layer corrections that will be discussed in Sec. 12.5.

---

[7] More details on the data for the DSMC simulations can be found in Sec. 12.5.

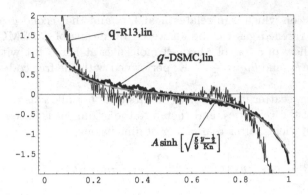

Fig. 12.5. Kn $= 0.1$. q-DSMC,lin: from DSMC simulation, q-R13,lin: from DSMC result for velocity and (12.43). The grey line is the prediction of the R13 equations, $q_1 = A \sinh \left[\sqrt{\frac{5}{9}} \frac{x - \frac{1}{2}}{Kn} \right]$ (12.39).

12.4.7 26 and more moments

The linear Knudsen layers for the 26 moment system (9.50) were computed by Reitebuch and Weiss [139] who used a minimax principle for the entropy production [140] to compute the amplitudes. The results are qualitatively correct, but the authors did not include jump and slip boundary conditions, nor provide comparisons with DSMC calculations. The status of the minimax principle remains unclear [141]. It must be noted that already in case of a low moment number its application is forbiddingly complicated, and that it will be almost impossible to use in the case of large moment numbers.

The author considered pure one-dimensional heat transfer ($v_W^0 = v_W^L = 0$, $\theta_W^0 \neq \theta_W^L$) with a large number of moments [142], and showed that the temperature curve contains terms of the form $A_\alpha \sinh \left[\frac{\lambda_\alpha}{Kn} \left(y - \frac{1}{2} \right) \right] + B_\alpha \cosh \left[\frac{\lambda_\alpha}{Kn} \left(y - \frac{1}{2} \right) \right]$ where the coefficients λ_α are the solutions of an eigenvalue problem. Moment systems with more moments have a wider spectrum of eigenvalues, and thus can resolve more details of the Knudsen layer. The importance of the various contributions depends on the amplitudes A_α, B_α, which are, presently, not known, due to lack of boundary conditions. It is a straightforward conclusion that likewise the velocity curve in Couette flow will have a large number of boundary layer contributions of different widths.

Barbera et al. considered one-dimensional heat transfer in the framework of COET [111], and used fluctuation theory to determine the amplitudes of linear boundary layers [143]. The results show good agreement to numerical solutions of the Boltzmann equation for heat transfer [144]. Barbera et al. did not include temperature jumps into their argument, but it should pose no problems to include these. The extension of the method to non-linear problems seems to be a challenge, however.

12.5 Superpositions

The bulk solution contains the most important non-linear parts of the equations, while the Knudsen layer correction gives their leading linear content. Superpositions of both should therefore give relatively accurate results, which are easier to obtain, and study, than solutions of the full non-linear equations. This approach has been used successfully for the solution of the Burnett equations in steady state 1-D heat transfer [145].

For constructing superpositions of bulk and Knudsen layer solutions to describe Couette flow [146], the solutions of the bulk equations (12.28) under the boundary conditions (12.29, 12.30) are denoted as $v_{|B}$, $\theta_{|B}$, P_0, $\sigma_{12|B}$, $\sigma_{22|B}$, $q_{1|B}$, $q_{2|B}$. The boundary layer results for q_1 and σ_{22} (12.39, 12.40) are denoted as $q_{1|L}$ and $\sigma_{22|L}$.

The superposition of both is done according to

$$v = v_{|B} - \frac{2}{5}q_{1|L} , \quad \theta = \theta_{|B} - \frac{2}{5}\sigma_{22|L} , \quad \sigma_{12} = \sigma_{12|B} , \quad \sigma_{22} = \sigma_{22|B} + \sigma_{22|L} ,$$

$$p = P_0 - \sigma_{22|L} , \quad \rho = \frac{p}{\theta} , \quad q_1 = q_{1|B} + q_{1|L} , \quad q_2 = q_{2|B} . \tag{12.45}$$

This model will be referred to as the SBKn model[8], and its results are compared to DSMC simulations for Couette flow in a gas of Maxwell molecules ($\omega = 1$) with wall temperatures $\theta_W^0 = \theta_W^L = \frac{k}{m}273\mathrm{K}$ and wall velocities $v_W^0 = 0$, $v_W^L = 200\frac{\mathrm{m}}{\mathrm{s}}, 600\frac{\mathrm{m}}{\mathrm{s}}$. Molecular mass and reference viscosity are taken as the values for argon, $\frac{k}{m} = 208\mathrm{J/kgK}$, $\mu = \mu_0\left(\theta/\theta_W^0\right)$ with $\mu_0 = 1.955 \times 10^{-5}\mathrm{Ns/m^2}$. The mass density at rest is fixed as $\rho_0 = 9.288 \times 10^{-6}\mathrm{kg/m^3}$ (corresponding to the rest pressure $p_0 = \rho_0\theta_W = 0.527\mathrm{Pa}$). The Knudsen number (9.41) is controlled by adjusting the length as

$$L = \frac{\mu_0}{\rho_0\sqrt{\theta_W}\mathrm{Kn}} ,$$

so that $L = 0.17666\mathrm{m}$ for $\mathrm{Kn} = 0.05$, $L = 0.08833\mathrm{m}$ for $\mathrm{Kn} = 0.1$, and $L = 0.017666\mathrm{m}$ for $\mathrm{Kn} = 0.5$. The integrated mass (12.30) is fixed at $M = \rho_0 L = \mu_0/\left(\sqrt{\theta_W}\mathrm{Kn}\right)$. Note that the above definition of the Knudsen number differs from that in [1].

Bird's code [1] was used for the DSMC calculations with modifications to obtain output for all moments of interest. The solution for the bulk equations is obtained from a straightforward finite volume scheme [146]. Results for the NSF equations (12.19) with the boundary conditions (12.20) are shown as well.

The boundary layer solutions need additional input to determine their amplitudes, and these are simply fitted to the DSMC results. Due to the geometry of the flow, $q_{1|L}$ is antisymmetric with respect to the center and $\sigma_{22|L}$ is symmetric, so that (12.39, 12.40) reduce to

---

[8] Superposition of **B**ulk solution and **K**nudsen layer.

Table 12.1. Amplitudes in the boundary layer contributions for q_1, σ_{22}.

	Kn = 0.05	Kn = 0.1	Kn = 0.5
$v_W^L = 200\frac{m}{s}$	$A = 0.009$	$A = 0.015$	$A = 0.03$
	$D = 0.0015$	$D = 0.003$	$D = 0.02$
$v_W^L = 600\frac{m}{s}$	$A = 0.018$	$A = 0.001$	
	$D = 0.015$	$D = 0.03$	

$$q_{1|L} = A \sinh\left[\sqrt{\tfrac{5}{9}}\tfrac{y-\frac{1}{2}}{\mathrm{Kn}}\right] \quad , \quad \sigma_{22|L} = D \cosh\left[\sqrt{\tfrac{5}{6}}\tfrac{y-\frac{1}{2}}{\mathrm{Kn}}\right] .$$

Figures 12.6–12.10 show comparisons of DSMC, NSF, and SBKn, (12.45) for a variety of Knudsen numbers and wall velocities v_W^L. The corresponding values for the constants A and D are given in Table 12.1.

For very small Knudsen numbers (Kn ≤ 0.01) all three models agree perfectly (not shown).

At Knudsen number Kn = 0.05, shown in Figs. 12.6 and 12.7, SBKn still agrees very well with DSMC; the most visible difference is in the value for σ_{12} (1.5 − 3%). The temperature curves agree well, with smaller differences close to the boundaries, and excellent agreement in the maximum value. The profiles for velocity v, parallel heat flux q_1, and normal heat flux q_2 show only minor differences. SBKn captures the shape of the normal stress σ_{22} very well, but does not meet the values in the bulk, i.e. in the middle part of the domain. All differences are more prominent at larger velocities (Fig. 12.7).

Moreover, the balance between bulk solution and Knudsen layer differs with velocity. The Knudsen layer is more important at lower velocity, where all variables are smaller. At larger velocities, all other fields grow as well, and the non-linear contributions in the bulk equations gain more importance.

At Kn = 0.05 the NSF results already show more pronounced differences. Most important is that NSF gives $\sigma_{22} = q_1 = 0$, but another important difference is observed in the temperature curves, where the midpoint maximum is not met. It follows that second and third order contributions—in the equations and the boundary conditions—cannot be neglected at Kn = 0.05, in particular at larger flow velocities.

When the Knudsen number is increased to Kn = 0.1, Figs. 12.8 and 12.9, the results are similar: SBKn matches the DSMC simulations quite well; as before the most visible differences lie in the bulk values for σ_{12} and σ_{22}. The temperature maximum is reproduced very well, while some differences can be observed at the boundaries.

NSF, on the other hand, cannot match the temperature maximum at all.

Only one set of results is shown for the larger Knudsen number Kn = 0.5, at $v_W^L = 200\frac{m}{s}$ (Fig. 12.10). Now both, SBKn and NSF, cannot match the velocity curve, nor the normal heat flux q_2. At larger velocities (not shown), this discrepancy becomes stronger. From the velocity curve one might infer that the results could be improved by reducing the coefficients α_1, α_2 in the boundary condition (12.29). This, however, results in a larger absolute value for σ_{12} ($\simeq -0.11$ Pa), and thus worsens the results for other moments. Clearly, this Knudsen number is outside the range of the described method .

In order to understand the results better, it must be considered that the NSF equations, and their boundary equations, are only of first order accuracy in the Knudsen number. Thus, their failure at larger Knudsen numbers is not surprising. Considering their limitations they recover the curves for velocity and normal heat flux q_2 remarkably well. Their most important shortcoming is the failure to reproduce the maximum of temperature around the midpoint. Moreover, they give a value of zero for parallel heat flux q_1 and normal stress σ_{22}.

The bulk solution is of second order in Kn, and the linear Knudsen layer of the R13 equations is of third order. Their superposition, SBKn, thus should be of second order accuracy. It gives a clear improvement over the NSF equations, since it recovers the temperature maximum, and the curves for parallel heat flux and normal stress quite well. An important shortcoming is that boundary conditions for σ_{22} and q_1 are presently not known.

The NSF equations do not describe boundary layers at all. Although the SBKn model (12.45) adds boundary layers to velocity v and temperature θ, it does not describe the boundary layers for v and θ in agreement with DSMC, since these consist of many contributions, as was mentioned in Sec. 12.4.7. Nevertheless, both models describe the overall velocity very well, since the velocity curve is dominated by the linear first order contribution.

The discrepancies at the boundaries are more pronounced in the temperature curves. The overall result for the SBKn model is good nevertheless, since the jump boundary condition $(12.29)_2$ compensates effectively for the discrepancy at the boundary.

While one may criticize the R13 equations for not reproducing the boundary layers in velocity and temperature [138], one should not forget that they are able to describe the second order quantities q_1 and σ_{22} very well.

It is evident that higher Knudsen numbers demand more refined models. The first step would be to consider the fully non-linear R13 equations (12.26). This requires a full set of boundary conditions, which are not available at present. However, the results presented here show that it will be a worthwhile task to develop suitable boundary conditions for the R13 equations.

Zheng [42] used solutions of the ES-BGK equation for testing the ability of macroscopic equations to reproduce higher moments in accordance with exact solutions. For this he wrote the equations for stress and heat flux as $\bar{\sigma}_{ij} = -\frac{\mu}{p}\mathcal{F}_{ij}(\rho, v_i, \theta, \sigma_{ij}, q_i)$, $\bar{q}_i = -\frac{\mu}{p}\mathcal{G}_i(\rho, v_i, \theta, \sigma_{ij}, q_i)$ where $\mathcal{F}_{ij}$ and $\mathcal{G}_i$ are the appropriate functionals to reproduce the proper equations, e.g. the Burnett equations (9.4, 9.5), or the R13 equations (9.9, 9.10) with (9.12). $\mathcal{F}_{ij}$ and $\mathcal{G}_i$ are evaluated with numerical results for the full ES-BGK model, and then the computed values $\bar{\sigma}_{ij}$, $\bar{q}_i$ are compared to the numerical solutions for σ_{ij} and q_i. This analysis, which shows some similarity to the method of [119], supports the results presented above. In particular it indicates that Grad 13 and Burnett equations have severe difficulties to reproduce the gas behavior

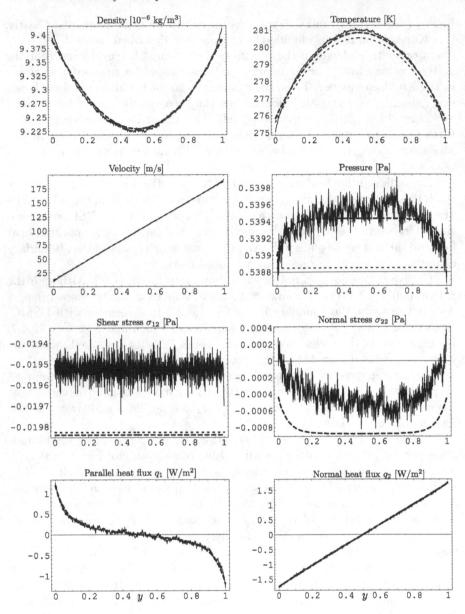

Fig. 12.6. Couette flow at Kn $= 0.05$, with $v_W^L = 200 \frac{\text{m}}{\text{s}}$. Continuous line: DSMC, finely dashed line (- - - -): NSF, dashed line (− − − −): superposition of bulk solution and linear Knudsen layer solution. Recall that NSF implies $q_1 = \sigma_{22} = 0$ (curves not shown).

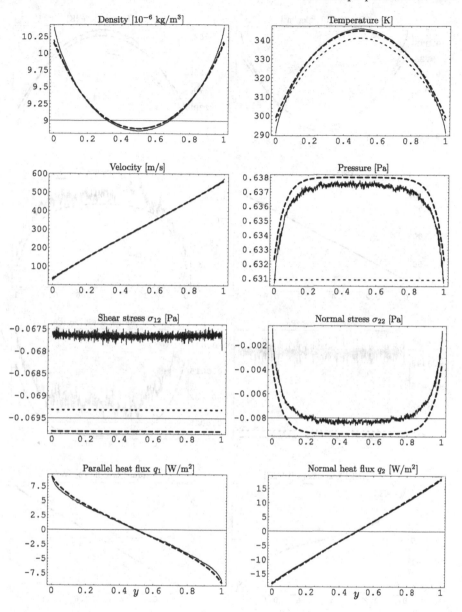

Fig. 12.7. Couette flow at Kn $= 0.05$, with $v_W^L = 600\frac{m}{s}$. Continuous line: DSMC, finely dashed line (- - - -): NSF, dashed line (– – – –): superposition of bulk solution and linear Knudsen layer solution. Recall that NSF implies $q_1 = \sigma_{22} = 0$ (curves not shown).

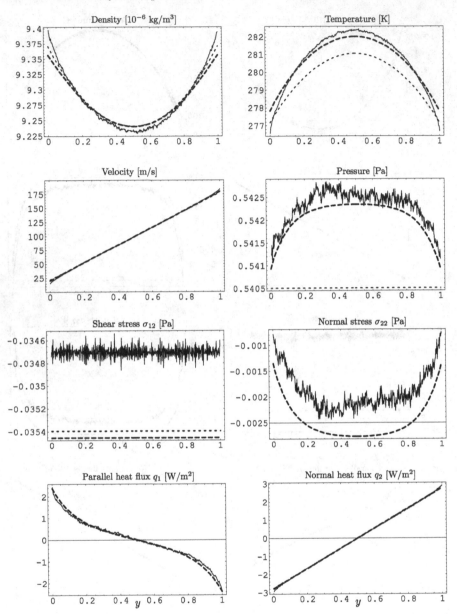

Fig. 12.8. Couette flow at Kn $= 0.1$, with $v_W^L = 200 \frac{m}{s}$. Continuous line: DSMC, finely dashed line (- - - -): NSF, dashed line ($- - - -$): superposition of bulk solution and linear Knudsen layer solution. Recall that NSF implies $q_1 = \sigma_{22} = 0$ (curves not shown).

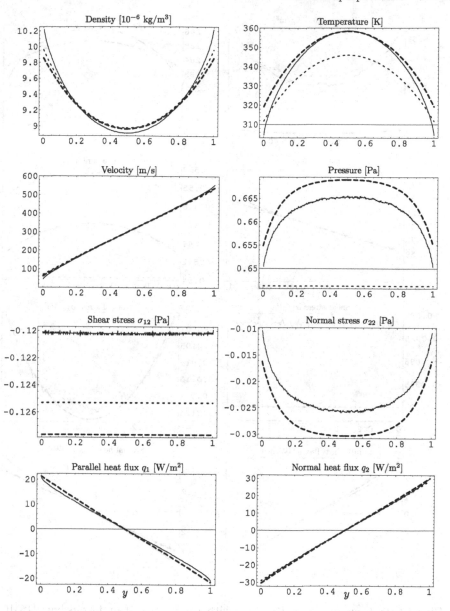

Fig. 12.9. Couette flow at Kn = 0.1,with $v_W^L = 600\frac{m}{s}$. Continuous line: DSMC, finely dashed line (- - - -): NSF, dashed line (– – – –): superposition of bulk solution and linear Knudsen layer solution. Recall that NSF implies $q_1 = \sigma_{22} = 0$ (curves not shown).

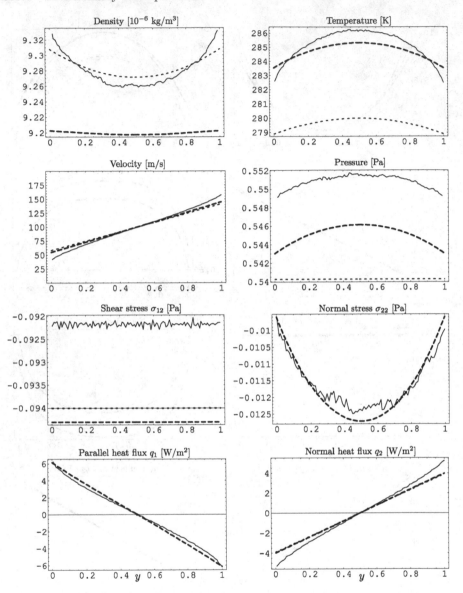

Fig. 12.10. Couette flow at Kn = 0.5,with $v_W^L = 200\frac{m}{s}$. Continuous line: DSMC, finely dashed line (- - - -): NSF, dashed line (— — — —): superposition of bulk solution and linear Knudsen layer solution. Recall that NSF implies $q_1 = \sigma_{22} = 0$ (curves not shown).

in the Knudsen layers, while the R13 equations are far more successful for Knudsen numbers below 0.5.

Superpositions of the second order bulk solutions with the linear Knudsen layers according to the Burnett or augmented Burnett equations are possible as well, but results for these are not shown. The Burnett equations cannot describe the boundary layer for q_1, but are well capable in describing the σ_{22} boundary layer (12.34).

The Burnett equations were solved numerically by Xue, Ji and Shu [147][148], and by Lockerby and Reese [149] with a more stable code. In both cases, the authors set $\sigma_{22} = 0$ at the boundaries, and a superposition of bulk solution and Burnett Knudsen layer (12.34), i.e. the SBKn model, with that boundary condition shows good agreement with the results presented in the cited papers. The boundary condition $\sigma_{22} = 0$ is not supported by DSMC calculations and overestimates the amplitude of the Knudsen layer for σ_{22} considerably. This leads to the inverted density curves reported in [149].

12.6 A boundary condition for normal stress

Above general moment boundary conditions were presented in form of (12.5), which is now considered for providing a boundary condition for σ_{22} [146]. The result should not be seen as a definite answer, but as a base for further discussion of boundary conditions for extended systems.

The momentum perpendicular to the wall, measured in the restframe of the wall, is given by

$$\Psi_{c_i}^\perp = C_r^W n_r = C_r n_r \ ,$$

and is obviously odd in $C_k^W n_k$. Evaluation of (12.5) thus gives

$$\frac{1-\chi}{\chi} \left(p + \sigma_{jk} n_j n_k \right) = \frac{2-\chi}{\chi} H_{jk}^0 n_j n_k - \sqrt{\frac{\pi}{2}} \theta_W H_k^0 n_k \ . \tag{12.46}$$

Evaluation of the half-space integrals H_k^0, H_{jk}^0 with Grad's 13 moment distribution (6.11) results in

$$\sigma_{jk} n_j n_k = \frac{-\frac{2-\chi}{\chi} \gamma \frac{2}{5} \sqrt{\frac{2}{\pi}} q_j n_j + \eta p \left(\sqrt{\theta_W} - \sqrt{\theta} \right)}{\sqrt{\theta} - \frac{1}{2} \sqrt{\theta_W}} \ .$$

Also here, as in (12.14), errors due to insufficient consideration of the boundary layers can be expected. The fitting parameters γ, η have been introduced to compensate for that error.

In Couette geometry, the boundary condition reduces to

$$\sigma_{22} = \frac{-\frac{2-\chi}{\chi} \gamma \frac{2}{5} \sqrt{\frac{2}{\pi}} q_2 n_2 + \eta p \left(\sqrt{\theta_W} - \sqrt{\theta} \right)}{\sqrt{\theta} - \frac{1}{2} \sqrt{\theta_W}} \ . \tag{12.47}$$

From the bulk solution it is known that for Couette flow q_2 and σ_{12} are $\mathcal{O}\left(\text{Kn}\right)$ and q_1 and σ_{22} are $\mathcal{O}\left(\text{Kn}^2\right)$.

For Couette flow with the NSF equations one finds $\sigma_{22} = 0$. The NSF equations are the equations of first order in Kn and the boundary conditions should be in accordance with this. Thus, it seems reasonable to require that the leading order of the boundary condition (12.47) is the second order. This requirement can be used to determine the fitting parameter γ.

The order of magnitude of the moments is made explicit by writing εq_2 instead of q_2 and so on. Then, expansion of the jump condition $(12.29)_2$ in ε gives

$$\frac{\theta}{\theta_W} = 1 - \varepsilon \frac{2-\chi}{2\chi} \beta_1 \sqrt{\frac{\pi}{2}} \frac{q_2 n_2}{\sqrt{\theta_W} p} + \mathcal{O}\left(\varepsilon^2\right) \ .$$

This is used in the equation for σ_{22} to obtain, again by expansion in ε,

$$\sigma_{22} = \frac{2-\chi}{\chi} \varepsilon \frac{4}{5} \sqrt{\frac{2}{\pi}} \left(\frac{5\pi}{16} \eta \beta_1 - \gamma\right) \frac{q_2 n_2}{\sqrt{\theta_W}} + \mathcal{O}\left(\varepsilon^2\right) \ . \tag{12.48}$$

The first order contribution vanishes for

$$\gamma = \frac{5\pi}{16} \eta \beta_1$$

for which then the boundary condition becomes

$$\sigma_{22} = \eta \frac{-\frac{2-\chi}{\chi} \frac{1}{4} \sqrt{\frac{\pi}{2}} \beta_1 q_2 n_2 + p\left(\sqrt{\theta_W} - \sqrt{\theta}\right)}{\sqrt{\theta} - \frac{1}{2}\sqrt{\theta_W}} \ . \tag{12.49}$$

The superposition of bulk equation and linear Knudsen layer reads

$$\sigma_{22} = \sigma_{22|B} + D \cosh\left[\sqrt{\frac{5}{6}} \frac{y - \frac{1}{2}}{\text{Kn}}\right] = -\frac{6}{5} \frac{\sigma_{12}\sigma_{12}}{P_0} + D \cosh\left[\sqrt{\frac{5}{6}} \frac{y - \frac{1}{2}}{\text{Kn}}\right] \ ,$$

and thus the boundary condition allows to relate amplitude D and correction factor η by

$$\eta = \frac{\left(D \cosh\left[\sqrt{\frac{5}{6}} \frac{-\frac{1}{2}}{\text{Kn}}\right] - \frac{6}{5}\frac{\sigma_{12}\sigma_{12}}{P_0}\right)\left(\sqrt{\theta} - \frac{1}{2}\sqrt{\theta_W}\right)}{-\frac{2-\chi}{\chi}\frac{1}{4}\sqrt{\frac{\pi}{2}}\beta_1 q_2 n_2 + \left(P_0 + \frac{6}{5}\frac{\sigma_{12}\sigma_{12}}{P_0}\right)\left(\sqrt{\theta_W} - \sqrt{\theta}\right)} \ .$$

For the presented examples, the new boundary condition (12.49) gives good agreement with the fitted values for the amplitude D as shown in Table 12.1 for $\eta \simeq 0.6$.

While the results are satisfactory, it is rather unsatisfying that the quality of the boundary conditions should depend on the fitting coefficients γ and η, which were introduced as ad hoc corrections. More refined arguments are needed to produce more reliable boundary conditions for the R13 equations, or other extended sets of equations.

12.7 Further discussion

The discussion on scaling of boundary layers in Sec. 12.4 showed that the moment equations cannot be further reduced in the Knudsen layer, since the Knudsen number cancels from the equations. This implies that at first one needs to consider *all* moment equations, i.e. the complete infinite set. As more moment equations are added, the number of boundary layer contributions $(A_\alpha \sinh\left[\frac{\lambda_\alpha}{\text{Kn}}\left(y - \frac{1}{2}\right)\right] + B_\alpha \cosh\left[\frac{\lambda_\alpha}{\text{Kn}}\left(y - \frac{1}{2}\right)\right])$ increases [142], and their relative importance depends on the associated amplitudes A_α, B_α, i.e. on the boundary conditions.

Obviously, these depend on the Knudsen number, i.e. the degree of rarefaction, since for small Knudsen numbers all boundary layer contributions vanish. In other words, the problem can only be tackled with a proper theory of boundary conditions, that gives information about the amplitudes of the various Knudsen layer contributions. This theory is still in development, and the thoughts presented above can at most be seen as a beginning. Of course, Knudsen layers are widely discussed in kinetic theory, see, e.g., [19], but the question of appropriate boundary conditions for extended moment equations was not tackled until recently.

Larger sets of Grad-type moment equations were solved for steady state heat transfer between rigid walls in [76][150] by means of a numerical method, known as kinetic schemes [151][152][100]. This method splits the fluxes F_{Ak} into half-space integrals of the phase density, and allows to directly implement the boundary conditions for the phase density. For given Knudsen number and temperature difference, the number of moments was increased until further increase did not change the resulting temperature curve. It was found that the number of moments required for converged results depends on the Knudsen number, with a larger Knudsen number requiring a larger number of moments. As the Knudsen number grows, the results differ markedly from the Fourier law, with pronounced Knudsen layers.[9]

From these findings for heat transfer one can extrapolate to more complicated cases, e.g. Couette flow, that at moderate Knudsen numbers a lower number of moments is sufficient to describe the behavior of the gas, including the Knudsen boundary layers.

Another possibility to implement solutions of extended macroscopic models are solutions of kinetic equations that force the phase density into a certain form, e.g. the Burnett phase density. Since these methods operate on the level of the distribution function, they allow direct implementation of the boundary conditions for the phase density [153]–[155].

Finally, one could consider hybrid approaches, where a microscopic equation—the Boltzmann equation or a kinetic model equation, e.g. the ES-BGK equation—is solved numerically [156]–[158] in the close vicinity of the wall and

---

[9] Note that at large Knudsen numbers the typical boundary layer expressions $\sinh\left[\frac{\lambda_\alpha}{\text{Kn}}\left(y - \frac{1}{2}\right)\right], \cosh\left[\frac{\lambda_\alpha}{\text{Kn}}\left(y - \frac{1}{2}\right)\right]$ appear as straight lines.

in other flow regions of high Knudsen numbers, and macroscopic equations are solved in those domains where the Knudsen number is sufficiently small. Hybrid approaches require meaningful coupling conditions at the interface between the microscopic and the macroscopic descriptions. Hybrid models are well-known for coupling of the Euler or Navier-Stokes-Fourier equations to numerical Boltzmann solvers or DSMC simulations [159]–[163], and the known methods can be extended to macroscopic equations of higher order. It must be noted that the macroscopic equations require less computer time for their numerical solutions as compared to the microscopic equations. The use of extended macroscopic models, e.g. the R13 equations, in hybrid methods will allow to extend the domain in which macroscopic equations can be used, and thus speed up computations. Of course, the R13 equations can be reduced to equations of lower order, Euler, NSF or Grad13, in regions of smaller Knudsen numbers. At the same time, microscopic equations would be used where the macroscopic equations lead to difficulties, e.g. directly at the walls.

Problems

12.1. Non-linear heat transfer with temperature jump

Consider the steady state heat transfer for a gas at rest between two parallel plates at distance L with temperatures θ_0, θ_L. Assume Knudsen numbers small enough to allow the use of Fourier's law $q_k = -\kappa_0 \left(\frac{\theta}{\theta_0}\right)^\omega \frac{\partial \theta}{\partial x_k}$ where κ_0 is the heat conductivity at θ_0.

(a) Simplify the balances for mass, momentum, energy, and the jump and slip boundary conditions $(12.13)_2$ for this problem. Show that the pressure is constant. Introduce the dimensionless temperature and space coordinate as $\hat{\theta} = \theta/\theta_0$ and $\hat{x} = x/L$, respectively. Identify the Knudsen number.

(b) Solve the problem for $\theta_L = 1.01$, 1.1, 1.5 and Knudsen numbers Kn = 0.001, 0.01, 0.1 for $\omega = 1$ and other meaningful values, e.g. $\omega = 0.5$. Hint: Solve first for the bulk, with boundary conditions ϑ_0, ϑ_L. Then find the values for ϑ_0, ϑ_L from the jump conditions for $\hat{\theta}_0 - \vartheta_0$, $\hat{\theta}_L - \vartheta_L$. The use of a computer is highly recommended.

A

Appendix

A.1 Tensor index notation

Most equations in this text are written in tensor index notation, and here the main elements of this notation are briefly recalled, for a Cartesian three-dimensional frame.

Scalar quantities are written in italics: a, b, etc.

Vectors are written in bold: $\mathbf{a}$, $\mathbf{b}$, etc. A vector can be expressed in terms of its components as $\mathbf{a} = \{a_1, a_2, a_3\}$. Often the vector is expressed through a place holder for the components, e.g. a_i ($i = 1, 2, 3$), and the possible values for the index i are not written, if the dimension of the vector (here $D = 3$) is obvious.

Tensors of higher order are also expressed through their components, A_{ij}, B_{ijk}, C_{ijkl}, etc. and are occasionally written in bold as well, $\mathbf{A}$, $\mathbf{B}$, etc. Then the rank of the tensor (the number of free indices) cannot be made visible easily.

The scalar product of two vectors $\mathbf{a}$ and $\mathbf{b}$ is

$$\mathbf{a} \cdot \mathbf{b} = a_1 b_1 + a_2 b_2 + a_3 b_3 = \sum_{i=1}^{3} a_i b_i \ .$$

Einstein's convention declares it unnecessary to write the summation symbol, and states that a sum has to be taken over an index that appears twice, thus

$$\mathbf{a} \cdot \mathbf{b} = a_i b_i \ .$$

This leads to the following rules: (a) Any term in a tensor equation in index notation can contain the same index only once or twice. (b) If an index appears twice, summation is assumed. (c) Free indices are those indices that appear only once in a term; in a tensor equation, all terms must have the same free indices.

A typical example for a well written tensor equation is the product of a vector and a matrix, which gives a vector,

$$d_i = A_{ij}b_j$$

(summation over j, and $i = 1, 2, 3$ implicitly assumed). It would be wrong, however, to write $d_k = A_{ij}b_j$ (different free indices k, i), or $d_j = A_{jj}b_j$ (index j appears three times on the right).

The unit tensor is written as

$$\delta_{ij} = \begin{cases} 1 \text{ , if } i = j \\ 0 \text{ , if } i \neq j \end{cases} .$$

For example, the unit tensor can be used to change the index of an expression

$$a_i = a_k \delta_{ki} ,$$

or to take the trace of a tensor

$$\text{tr} A = A_{kk} = A_{jk}\delta_{jk} .$$

The nabla vector $\nabla = \left\{ \frac{\partial}{\partial x_1}, \frac{\partial}{\partial x_2}, \frac{\partial}{\partial x_3} \right\}$ can also be expressed through its coordinates,

$$\nabla_i = \frac{\partial}{\partial x_i} .$$

Therefore the gradient of a scalar can be written as

$$[\text{grad}\, a]_i = [\nabla a]_i = \frac{\partial a}{\partial x_i} ,$$

the divergence of a vector is

$$\text{div}\, \mathbf{a} = \nabla \cdot \mathbf{a} = \frac{\partial a_i}{\partial x_i}$$

and its gradient is

$$[\text{grad}\, \mathbf{a}]_{ij} = [\nabla \mathbf{a}]_{ij} = \frac{\partial a_i}{\partial x_j} .$$

Similar equations hold for gradient and divergence of higher order tensors, e.g. for $\mathbf{A} = \{A_{ijk}\}$:

$$[\text{div}\mathbf{A}]_{ij} = [\nabla \cdot \mathbf{A}]_{ij} = \frac{\partial A_{ijk}}{\partial x_k} \quad , \quad [\text{grad}\, \mathbf{A}]_{ijkl} = \frac{\partial A_{ijk}}{\partial x_l} .$$

This example shows that the index notation becomes particularly helpful when tensors of higher orders are considered. Indeed, in taking the divergence, it is not always clear on which index the nabla vector should act. Index notation allows to distinguish easily between $\frac{\partial A_{ijk}}{\partial x_k}$, $\frac{\partial A_{ijk}}{\partial x_j}$, $\frac{\partial A_{ijk}}{\partial x_i}$.

The completely antisymmetric tensor is defined as

$$\varepsilon_{ijk} = \begin{cases} 1 \text{ , if } \{i, j, k\} = \{1, 2, 3\} \text{ or } \{3, 1, 2\} \text{ or } \{2, 3, 1\} \\ -1 \text{ , if } \{i, j, k\} = \{3, 2, 1\} \text{ or } \{1, 3, 2\} \text{ or } \{2, 1, 3\} \\ 0 \text{ , else} \end{cases}$$

so that, e.g., the cross product of two vectors is given as

$$(\mathbf{a} \times \mathbf{b})_i = \varepsilon_{ijk}a_j b_k .$$

A.2 Symmetric and trace-free tensors

A.2.1 Symmetry

A tensor $S_{i_1 i_2 \cdots i_n}$ of rank n is called symmetric, if for any pair of indices i_j, i_k

$$S_{i_1 i_2 \cdots i_j \cdots i_k \cdots i_n} = S_{i_1 i_2 \cdots i_k \cdots i_j \cdots i_n} .$$

A non-symmetric tensor $A_{i_1 i_2 \cdots i_n}$ can be symmetrized by the rule

$$A_{(i_1 i_2 \cdots i_n)} = \frac{1}{n!} \left(A_{i_1 i_2 \cdots i_{n-1} i_n} + A_{i_1 i_2 \cdots i_n i_{n-1}} + \cdots \text{all permutations of indices} \right) .$$

Here, the indices in round brackets denote the symmetric part of the tensor. E.g. for a 2-tensor and a 3-tensor, one finds the symmetric parts as

$$A_{(ij)} = \frac{1}{2} \left(A_{ij} + A_{ji} \right) ,$$

$$A_{(ijk)} = \frac{1}{6} \left(A_{ijk} + A_{ikj} + A_{jik} + A_{jki} + A_{kij} + A_{kji} \right) .$$

A.2.2 Trace-free tensors

A tensor $T_{i_1 i_2 \cdots i_n}$ of rank n is called trace-free, or irreducible, if for each pair of indices i_j, i_k

$$T_{i_1 i_2 \cdots i_j \cdots i_k \cdots i_n} \delta_{i_j i_k} = 0 .$$

The trace-free part of a symmetric tensor $S_{i_1 i_2 \cdots i_n}$ can be obtained according to the rule [80]

$$T_{i_1 i_2 \cdots i_n} = S_{\langle i_1 i_2 \cdots i_n \rangle} = \sum_{k=0}^{\lfloor \frac{n}{2} \rfloor} a_{nk} \delta_{(i_1 i_2} \cdots \delta_{i_{2k-1} i_{2k}} S_{i_{2k+1} \cdots i_n) j_1 \cdots j_k j_1 \cdots j_k} \qquad (A.1)$$

where

$$a_{nk} = (-1)^k \frac{n! \, (2n - 2k - 1)!!}{(n - 2k)! \, (2n - 1)!! \, (2k)!!} ,$$

$$\left\lfloor \frac{n}{2} \right\rfloor = \left\{ \frac{n}{2} \text{ if } n \text{ is even} \right\} \text{ and } \left\{ \frac{n-1}{2} \text{ if } n \text{ is odd} \right\} ,$$

$$n!! = n \, (n - 2) \cdots (2 \text{ or } 1) = \prod_{j=0}^{\lfloor \frac{n-1}{2} \rfloor} (n - 2j) .$$

Note that in the above equations, $S_{i_1 i_2 \cdots i_n}$ is symmetric, which is not made explicit by means of round brackets in order to avoid overly complicated notation. For actual computations, it is more convenient to use

$$S_{\langle i_1 i_2 \ldots i_n \rangle} = S_{i_1 i_2 \cdots i_n} + \alpha_{n_1} \left(\delta_{i_1 i_2} S_{i_3 \cdots i_n kk} + \text{permutations} \right) +$$
$$+ \alpha_{n_2} \left(\delta_{i_1 i_2} \delta_{i_3 i_4} S_{i_5 \cdots i_n kkll} + \text{permutations} \right) + \cdots .$$

where only those permutations must be considered, that are really different, i.e. that cannot be related by means of symmetry properties. Then, the coefficients are given as

$$\alpha_{n_k} = \frac{(-1)^k}{\prod\limits_{j=0}^{k-1} (2n - 2j - 1)} .$$

Simple examples are

$$S_{\langle i \rangle} = S_i ,$$

$$S_{\langle ij \rangle} = S_{(ij)} - \frac{1}{3} S_{kk} \delta_{ij} ,$$

$$S_{\langle ijk \rangle} = S_{(ijk)} - \frac{1}{5} \left(S_{(ill)} \delta_{jk} + S_{(jll)} \delta_{ik} + S_{(kll)} \delta_{ij} \right) .$$

The gradient of a symmetric 2-tensor, $\frac{\partial S_{ij}}{\partial x_k}$, gives an instructive example. Its symmetric part is given by

$$\frac{\partial S_{(ij}}{\partial x_{k)}} = \frac{1}{3} \left(\frac{\partial S_{ij}}{\partial x_k} + \frac{\partial S_{ik}}{\partial x_j} + \frac{\partial S_{jk}}{\partial x_i} \right)$$

and its trace-free and symmetric part is

$$\frac{\partial S_{\langle ij}}{\partial x_{k \rangle}} = \frac{1}{3} \left(\frac{\partial S_{ij}}{\partial x_k} + \frac{\partial S_{ik}}{\partial x_j} + \frac{\partial S_{jk}}{\partial x_i} \right) - \frac{1}{15} \left[\left(\frac{\partial S_{ir}}{\partial x_r} + \frac{\partial S_{ir}}{\partial x_r} + \frac{\partial S_{rr}}{\partial x_i} \right) \delta_{jk} \right.$$
$$\left. + \left(\frac{\partial S_{jr}}{\partial x_r} + \frac{\partial S_{jr}}{\partial x_r} + \frac{\partial S_{rr}}{\partial x_j} \right) \delta_{ik} + \left(\frac{\partial S_{kr}}{\partial x_r} + \frac{\partial S_{kr}}{\partial x_r} + \frac{\partial S_{rr}}{\partial x_k} \right) \delta_{ij} \right] .$$

Obviously, the bracket notation allows for a highly condensed notation.

Note that the moments $u_{i_1 \cdots i_n}^a$ that are used in this text are trace-free by definition, $u_{i_1 \cdots i_n}^a = u_{\langle i_1 \cdots i_n \rangle}^a$, and this is not made explicit with brackets. This avoids confusing notation involving several pairs of brackets. For instance, the gradient of u_{ij} made symmetric and trace-free only in two of its three indices, is given by

$$\frac{\partial u_{i \langle j}}{\partial x_{k \rangle}} = \frac{1}{2} \frac{\partial u_{\langle ij \rangle}}{\partial x_k} + \frac{1}{2} \frac{\partial u_{\langle ik \rangle}}{\partial x_j} - \frac{1}{3} \frac{\partial u_{\langle ir \rangle}}{\partial x_r} \delta_{jk} .$$

For this quantity one would have to write $\frac{\partial u_{\langle i \langle j \rangle}}{\partial x_{k \rangle}}$ if the trace-free properties of the moments were made explicit by angular brackets.

The trace-free part of the unit matrix δ_{ij} vanishes, $\delta_{\langle ij \rangle} = 0$, and this implies $A_{\langle i_1 i_2 \ldots i_n} \delta_{jk \rangle} = 0$.

A useful relation for the scalar product of a symmetric trace-free tensor with an arbitrary tensor $A_{i_1 i_2 \ldots i_n j_1 \cdots j_n}$ is that

$$T_{i_1 i_2 \ldots i_n k_1 \cdots k_r} A_{i_1 i_2 \ldots i_n j_1 \cdots j_n} = T_{i_1 i_2 \ldots i_n k_1 \cdots k_r} A_{\langle i_1 i_2 \ldots i_n \rangle j_1 \cdots j_n} \ .$$

A.2.3 Spherical harmonics

Trace-free linear combinations of unit vectors $\nu_{\langle i_1} \ldots \nu_{i_n \rangle}$ are related to spherical harmonics [21]. Here, the vectors ν_i are unit vectors in spherical coordinates,

$$\nu_i = \{\sin\vartheta \sin\varphi, \sin\vartheta \cos\varphi, \cos\vartheta\}_i \ .$$

Spherical harmonics form an orthogonal set of functions, in the sense that integration over the solid angle $d\Omega = \sin\vartheta d\vartheta d\varphi$ yields

$$\int \nu_{\langle i_1} \ldots \nu_{i_n \rangle} \nu_{\langle j_1} \ldots \nu_{j_m \rangle} d\Omega = \begin{cases} 0 & n \neq m \\ \dfrac{4\pi}{\prod\limits_{j=0}^{n}(2j+1)} \delta_{\langle i_1 \cdots i_n \rangle \langle j_1 \cdots j_n \rangle} & n = m \end{cases} .$$

where $\delta_{k_1 \cdots k_l}$ is a generalized unit tensor defined as

$$\delta_{k_1 \cdots k_l} = \delta_{k_1 k_2} \cdots \delta_{k_{l-1} k_l} + \cdots + \left(\frac{l!}{(\frac{l}{2})! 2^{\frac{l}{2}}} \text{ elements} \right) .$$

An important relation following from this definition is that the product of the above integral with a tensor $A_{i_1 i_2 \ldots i_n}$ yields

$$A_{i_1 i_2 \ldots i_n} \int \nu_{\langle i_1} \ldots \nu_{i_n \rangle} \nu_{\langle j_1} \ldots \nu_{j_m \rangle} d\Omega = \begin{cases} 0 & n \neq m \\ \dfrac{4\pi n!}{\prod\limits_{j=0}^{n}(2j+1)} A_{\langle j_1 \cdots j_n \rangle} & n = m \end{cases} .$$

Another important relation is [80]

$$\nu_{\langle i_1} \ldots \nu_{i_n \rangle} \nu_k = \nu_{\langle i_1} \ldots \nu_{i_n} \nu_{k \rangle} + \frac{n}{2n+1} \nu_{\langle \langle i_1} \ldots \nu_{i_{n-1}} \delta_{i_n \rangle k} \ . \tag{A.2}$$

By introducing the direction vector of the microscopic velocity as $\nu_i = C_i / C$, the moments (2.12) can be written as

$$u^a_{i_1 \ldots i_n} = m \int C^{2a+n+2} \nu_{\langle i_1} \nu_{i_2} \cdots \nu_{i_n \rangle} f dC d\Omega \ .$$

This indicates that our definition of moments refers to an expansion of the phase density in spherical harmonics $\nu_{\langle i_1} \nu_{i_2} \cdots \nu_{i_n \rangle}$ and polynomials in the absolute value of the microscopic velocity C.

Equation (A.2) applied to moments yields the useful relation

$$u^\alpha_{\langle i_1 \cdots i_n \rangle k} = u^\alpha_{i_1 \cdots i_n k} + \frac{n}{2n+1} u^{\alpha+1}_{\langle i_1 \cdots i_n-1} \delta_{i_n \rangle k} \ . \tag{A.3}$$

A.3 Integrals of Gaussians and Maxwellians

A.3.1 Gaussian integrals

Integrals of Gaussian distributions are given by

$$\hat{G}_n = \int_{-\infty}^{\infty} x^n \exp\left[-ax^2\right] dx \quad \text{and} \quad G_n = \int_{0}^{\infty} x^n \exp\left[-ax^2\right] dx .$$

If the exponent n is even ($n = 2k$), then $\hat{G}_n = 2G_n$, and if n is odd ($n = 2k+1$), then $\hat{G}_n = 0$; therefore it is sufficient to compute the G_n. Moreover

$$\frac{dG_n}{da} = \frac{d}{da} \int_{0}^{\infty} x^n \exp\left[-ax^2\right] dx = - \int_{0}^{\infty} x^{n+2} \exp\left[-ax^2\right] dx = -G_{n+2}$$

so that all G_n for $n \geq 2$ can be computed from G_0 and G_1 by taking derivatives.

The integral G_1 follows from the simple substitution $y = x^2$ as

$$G_1 = \int_{0}^{\infty} x \exp\left[-ax^2\right] dx = \frac{1}{2} \int_{0}^{\infty} \exp\left[-ay\right] dy = \frac{1}{2a} .$$

G_0 is computed from its square,

$$G_0^2 = \left(\int_{0}^{\infty} \exp\left[-ax^2\right] dx \right)^2 = \int_{0}^{\infty} \int_{0}^{\infty} \exp\left[-a\left(x^2 + y^2\right)\right] dxdy .$$

The two-dimensional integral can be solved easily by introducing polar coordinates $r = \sqrt{x^2 + y^2}$, $\varphi = \arctan \frac{x}{y}$ whereby

$$G_0^2 = \int_{0}^{\infty} \int_{0}^{\frac{\pi}{2}} \exp\left[-ar^2\right] rdrd\varphi = \frac{\pi}{2}G_1 = \frac{\pi}{4a} .$$

An inductive proof then shows that the following results hold for the integrals

$$G_{2k} = \frac{(2k-1)!! \sqrt{\pi}}{2^{k+1}a^{k+1/2}} \quad \text{and} \quad G_{2k+1} = \frac{1}{2a}\frac{k!}{a^k} , \tag{A.4}$$

where $(2k-1)!! = \prod_{j=0}^{k}(2j-1)$. With the definition of the Gamma function these results can be combined into

$$G_n = \frac{\Gamma\left(\frac{n+1}{2}\right)}{2\sqrt{a}^{n+1}} .$$

A.3.2 Integrals of the Maxwellian

The moments of the Maxwellian are defined as

$$\rho^r_{i_1\cdots i_n|M} = m \int C^{2r} C_{i_1} \cdots C_{i_n} f_M d\mathbf{C} ,$$

where f_M is the Maxwellian phase density (2.18). With the direction vector

$$\nu_i = C_i/C = \{\cos\varphi\sin\vartheta, \sin\varphi\sin\vartheta, \cos\vartheta\}_i ,$$

and $d\mathbf{C} = C^2 dC \sin\vartheta d\vartheta d\varphi$ follows

$$\rho^r_{i_1\cdots i_n|M} = \frac{4\rho}{\sqrt{\pi}\sqrt{2\theta}^3} \int_{C=0}^{\infty} C^{2r+2+n} \exp\left[-\frac{C^2}{2\theta}\right] dC\, I_{i_1\cdots i_n}$$

with the integrals

$$I_{i_1\cdots i_n} = \frac{1}{4\pi} \int_{\varphi=0}^{2\pi} \int_{\vartheta=0}^{\pi} \nu_{i_1} \cdots \nu_{i_n} \sin\vartheta d\vartheta d\varphi .$$

The integral over C is of Gaussian type, so that

$$\rho^r_{i_1\cdots i_n|M} = \rho\sqrt{2\theta}^{2r+n} \frac{2}{\sqrt{\pi}} \Gamma\left(\frac{2r+n+3}{2}\right) I_{i_1\cdots i_n}$$

Since there is no particular direction of symmetry, the integrals $I_{i_1\cdots i_n}$ must be isotropic and symmetric, and it follows immediately that

$$I_{i_1\cdots i_n} = 0 \text{ if } n \text{ is odd.}$$

For $n = 0$ one obtains

$$I = \frac{1}{4\pi} \int_{\varphi=0}^{2\pi} \int_{\vartheta=0}^{\pi} \sin\vartheta d\vartheta d\varphi = 1 ,$$

so that the scalar moments are given by

$$\rho^r_{|M} = \rho\theta^r \frac{2^{r+1}}{\sqrt{\pi}} \Gamma\left(\frac{2r+3}{2}\right) = \rho\theta^r (2r+1)!! .$$

Due to isotropy, the integral for $n = 2$ must have the form

$$I_{ij} = \frac{1}{4\pi} \int_{\varphi=0}^{2\pi} \int_{\vartheta=0}^{\pi} \nu_i\nu_j \sin\vartheta d\vartheta d\varphi = \beta\delta_{ij}$$

where the coefficient β can be computed by multiplying with δ_{ij}, i.e. by taking the trace, and subsequent integration, as $\beta = \frac{1}{3}$. Thus,

$$\rho_{ij|M}^r = \rho\sqrt{2\theta}^{2r+2}\frac{2}{3\sqrt{\pi}}\Gamma\left(\frac{2r+5}{2}\right)\delta_{ij} = \rho\theta^{r+1}\frac{(2r+3)!!}{3}\delta_{ij}$$

Again due to isotropy, the forth order integral has the representation

$$I_{ijkl} = \alpha\left(\delta_{ij}\delta_{kl} + \delta_{ik}\delta_{jl} + \delta_{il}\delta_{jk}\right)$$

Multiplication with $\delta_{ij}\delta_{kl}$, that is taking the trace twice, and subsequent integration yields $\alpha = 1/15$ so that

$$\rho_{ijkl|M}^r = \rho\theta^{r+2}\frac{(2r+5)!!}{15}\left(\delta_{ij}\delta_{kl} + \delta_{ik}\delta_{jl} + \delta_{il}\delta_{jk}\right)\ .$$

Note that the trace-free parts of isotropic tensors vanish, so that

$$u_{i_1\cdots i_n|M}^a = \rho_{\langle i_1 i_2\cdots i_n\rangle|M}^\alpha = 0 \text{ for } n \geq 1\ .$$

A.3.3 Half-space moments of the Maxwellian

Half-space-moments of the Maxwellian are defined as

$$h_{i_1\cdots i_n}^r = m\int_{C_i n_i \geq 0} C^{2r}C_{i_1}\cdots C_{i_n} f_M d\mathbf{C}\ .$$

The same substitutions as in the previous section give

$$h_{i_1\cdots i_n}^r = \rho\sqrt{2\theta}^{2r+n}\frac{\Gamma\left(r + \frac{2r+n+3}{2}\right)}{\sqrt{\pi}}J_{i_1\cdots i_n}$$

with the integrals

$$J_{i_1\cdots i_n} = \frac{1}{2\pi}\int_{\varphi=0}^{2\pi}\int_{\vartheta=0}^{\pi/2}\nu_{i_1}\cdots\nu_{i_n}\sin\vartheta d\vartheta d\varphi\ .$$

For $n = 0$ one finds

$$J = \frac{1}{2\pi}\int_{\varphi=0}^{2\pi}\int_{\vartheta=0}^{\pi/2}\sin\vartheta d\vartheta d\varphi = 1\ ,$$

so that

$$h^r = \rho\theta^r\frac{(2r+1)!!}{2} = \frac{\rho_{|M}^r}{2}\ .$$

The direction of the normal n_i is the only distinguished vector in the calculation of the half-space moments. Thus, the vector integral must be of the form

$$I_i = \frac{1}{2\pi}\int_{\varphi=0}^{2\pi}\int_{\vartheta=0}^{\pi/2}\nu_i\sin\vartheta d\vartheta d\varphi = \gamma n_i\ .$$

Scalar multiplication with n_i gives

$$\gamma = \frac{1}{2\pi} \int_{\varphi=0}^{2\pi} \int_{\vartheta=0}^{\pi/2} \cos\vartheta \sin\vartheta d\vartheta d\varphi = \frac{1}{2}$$

and the vectorial moments are

$$h_i^r = \rho\theta^{r+\frac{1}{2}} \frac{2^{r-\frac{1}{2}}}{\sqrt{\pi}} (r+1)! n_i \ .$$

The tensorial integral must be of the form

$$I_{ij} = \frac{1}{2\pi} \int_{\varphi=0}^{2\pi} \int_{\vartheta=0}^{\pi/2} \nu_i \nu_j \sin\vartheta d\vartheta d\varphi = \gamma n_i n_j + \beta \delta_{ij}$$

where the coefficients follow by scalar multiplication with $n_i n_j$ and δ_{ij} and integration as $\gamma = 0$ and $\beta = 1/3$ so that

$$h_{ij}^r = \rho\theta^{r+1} \frac{(2r+3)!!}{6} \delta_{ij} = \frac{\rho_{ij|M}^r}{2} \ .$$

The integral with three indices has the representation

$$I_{ijk} = \frac{1}{2\pi} \int_{\varphi=0}^{2\pi} \int_{\vartheta=0}^{\pi/2} \nu_i \nu_j \nu_k \sin\vartheta d\vartheta d\varphi = \gamma n_i n_j n_k + \beta \left(n_i \delta_{jk} + n_j \delta_{ik} + n_k \delta_{ij} \right)$$

where the coefficients follow again by multiplication with $n_i n_j n_k$ and $n_i \delta_{jk}$ and subsequent integration, as

$$\gamma + 3\beta = \frac{1}{4} \quad \text{and} \quad \gamma + 5\beta = \frac{1}{2} \ , \quad \text{so that} \quad \beta = -\gamma = \frac{1}{8}$$

and

$$h_{ijk}^r = \rho\theta^{r+\frac{3}{2}} \frac{2^{r-\frac{3}{2}}}{\sqrt{\pi}} (r+2)! \left(n_i \delta_{jk} + n_j \delta_{ik} + n_k \delta_{ij} - n_i n_j n_k \right) \ .$$

The fourth order integral has the representation

$$I_{ijkl} = \alpha \left(\delta_{ij}\delta_{kl} + \delta_{ik}\delta_{jl} + \delta_{il}\delta_{jk} \right) + \gamma n_i n_j n_k n_l +$$
$$+ \beta \left(n_i n_j \delta_{kl} + n_i n_k \delta_{jl} + n_i n_l \delta_{jk} + n_j n_k \delta_{il} + n_j n_l \delta_{ik} + n_k n_l \delta_{ij} \right) \ .$$

Multiplication with $n_i n_j n_k n_l$, $n_i n_j \delta_{kl}$ and $\delta_{ij}\delta_{kl}$ and integration yields $\beta = \gamma = 0$ and $\alpha = 1/15$ so that

$$h_{ijkl}^r = \rho\theta^{r+2} \frac{(2r+5)!!}{30} \left(\delta_{ij}\delta_{kl} + \delta_{ik}\delta_{jl} + \delta_{il}\delta_{jk} \right) = \frac{\rho_{ijkl|M}^r}{2} \ .$$

A.3.4 Integrals of the anisotropic Gaussian

The moments of the anisotropic Gaussian[1]

$$f_{ES} = \frac{\rho}{m} \frac{1}{\sqrt{\det[2\pi\Lambda_{ij}]}} \exp\left[-\frac{1}{2}\Lambda_{ij}^{-1} C_i C_j\right] ,$$

are defined as

$$\rho_{i_1\cdots i_n|ES}^r = m \int C^{2r} C_{i_1} \cdots C_{i_n} f_{ES} d\mathbf{C} ;$$

Λ_{ij}^{-1} is a real symmetric matrix. For the solution it is convenient to introduce the vector ζ_i by $C_i = T_{ik}\zeta_k$, so that

$$\Lambda_{ij}^{-1} C_i C_j = T_{ki}^T \Lambda_{ij}^{-1} T_{jl} \zeta_k \zeta_l .$$

The transformation matrix T_{ij} can be chosen as the (orthogonal) matrix of the eigenvectors, so that $T_{ki}^T \Lambda_{ij}^{-1} T_{jl} = 1/\lambda_{\underline{i}} \delta_{ij}$ and $T_{ik} T_{il} = \delta_{ij}$ (the underlined index indicates that no summation over i is performed). Since Λ_{ij}^{-1} is real and symmetric, the eigenvalues $1/\lambda_i$ are real, and it is assumed that they are positive. Moreover, the eigenvectors can be normalized so that $\det[T_{ij}] = 1$, and $d c = d\zeta_1 d\zeta_2 d\zeta_3$, $C^2 = \zeta^2 = \zeta_1^2 + \zeta_2^2 + \zeta_3^2$. Insertion of all this into the definition of the moments gives

$$\rho_{i_1\cdots i_n|ES}^r = \frac{\rho}{\sqrt{2\pi}^3 \sqrt{\lambda_1\lambda_2\lambda_3}} T_{i_1 k_1} \cdots T_{i_1 k_1} \times$$

$$\int \zeta^{2r} \zeta_{k_1} \cdots \zeta_{k_n} \exp\left[-\frac{1}{2\lambda_1}\zeta_1^2 - \frac{1}{2\lambda_2}\zeta_2^2 - \frac{1}{2\lambda_3}\zeta_3^2\right] d\zeta_1 d\zeta_2 d\zeta_3 .$$

Further evaluation shows that

$$\rho_{|ES}^0 = \rho ,$$

$$\rho_{|ES}^1 = \rho(\lambda_1 + \lambda_2 + \lambda_3) = \rho\Lambda_{kk}$$

$$\rho_{|ES}^2 = \rho\left(3\left(\lambda_1^2 + \lambda_2^2 + \lambda_3^2\right) + 2\left(\lambda_1\lambda_2 + \lambda_1\lambda_3 + \lambda_2\lambda_3\right)\right) = 2\rho\Lambda_{ij}\Lambda_{ij} + \rho\Lambda_{kk}^2$$

$$\rho_{ij|ES}^0 = \rho T_{ik} T_{jl} \begin{pmatrix} \lambda_1 & 0 & 0 \\ 0 & \lambda_2 & 0 \\ 0 & 0 & \lambda_3 \end{pmatrix}_{kl} = \rho\Lambda_{ij}$$

$$\rho_{ij|ES}^1 = \rho T_{ik} T_{jl} \left[2\begin{pmatrix} \lambda_1^2 & 0 & 0 \\ 0 & \lambda_2^2 & 0 \\ 0 & 0 & \lambda_3^2 \end{pmatrix}_{kl} + (\lambda_1 + \lambda_2 + \lambda_3)\begin{pmatrix} \lambda_1 & 0 & 0 \\ 0 & \lambda_2 & 0 \\ 0 & 0 & \lambda_3 \end{pmatrix}_{kl} \right]$$

$$= 2\rho\Lambda_{ik}\Lambda_{kj} + \rho\Lambda_{kk}\Lambda_{ij}$$

and

$$\rho_{i_1\cdots i_n|ES}^r = 0 \text{ for } n = 2k+1 \ (n \text{ uneven}) .$$

---

[1] These integrals appear in evaluations for the ES-BGK model, hence the index "ES".

A.4 The integrals (5.21)

In this appendix the integrals (5.21) are computed, which are defined as

$$I_{i_1 \cdots i_n} = \int_0^{2\pi} \int_0^{\pi/2} k_{i_1} \cdots k_{i_n} \cos^n \Theta \sin \Theta F_M(\Theta) \, d\Theta \, d\varepsilon .$$

In polar coordinates, the collision vector is $\mathbf{k} = \{\cos \varepsilon \sin \Theta, \sin \varepsilon \sin \Theta, \cos \Theta\}$ and we recall that Θ is the angle between $\mathbf{g}$ and $\mathbf{k}$, which implies that $g_i k_i/g = \cos \Theta$. The results contain the coefficients

$$\chi^{(r,s)} = 2\pi \int_0^{\pi/2} \cos^r \Theta \sin^s \Theta F_M(\Theta) \, d\Theta \quad \text{with} \quad \chi^{(n,1)} = \chi^{(n-2,1)} - \chi^{(n-2,3)} .$$

The vector integral must be of the form

$$I_i = \int_0^{2\pi} \int_0^{\pi/2} k_i \cos \Theta \sin \Theta F_M(\Theta) \, d\Theta \, d\varepsilon = a\frac{g_i}{g}$$

and scalar multiplication with g_i/g shows that $a = \chi^{(2,1)}$, and thus

$$I_i = \chi^{(2,1)} \frac{g_i}{g} .$$

The rank-2 tensor integral must be of the form

$$I_{ij} = \int_0^{2\pi} \int_0^{\pi/2} k_i k_j \cos^2 \Theta \sin \Theta F_M(\Theta) \, d\Theta \, d\varepsilon = a\frac{g_i g_j}{g^2} + b\delta_{ij}$$

and by scalar multiplication with δ_{ij} and $\frac{g_i g_j}{g^2}$, respectively, follows that

$$a + 3b = \chi^{(2,1)} \quad \text{and} \quad a + b = \chi^{(4,1)}$$

so that

$$2b = \chi^{(2,1)} - \chi^{(4,1)} = \chi^{(2,3)} \quad \text{and} \quad a = \chi^{(4,1)} - \frac{1}{2}\chi^{(2,3)} = \left(\chi^{(2,1)} - \frac{3}{2}\chi^{(2,3)} \right)$$

and thus

$$I_{ij} = \left(\chi^{(2,1)} - \frac{3}{2}\chi^{(2,3)} \right) \frac{g_i g_j}{g^2} + \frac{1}{2}\chi^{(2,3)} \delta_{ij} .$$

The rank-3 tensor integral must be of the form

$$I_{ijk} = \int_0^{2\pi} \int_0^{\pi/2} k_i k_j k_k \cos^3 \Theta \sin \Theta F_M(\Theta) \, d\Theta \, d\varepsilon = a\frac{g_i g_j g_k}{g^3} + b\delta_{(ij}\frac{g_{k)}}{g}$$

and by scalar multiplication with $\delta_{(ij} g_{k)}/g$ and $\frac{g_i g_j g_k}{g^3}$, respectively, follows that

$$a + \frac{5}{3}b = \chi^{(4,1)} \quad \text{and} \quad a + b = \chi^{(6,1)}$$

so that

$$\frac{2}{3}b = \chi^{(4,1)} - \chi^{(6,1)} = \chi^{(4,3)} \quad \text{and} \quad a = \chi^{(2,1)} - \chi^{(2,3)} - \frac{5}{2}\chi^{(4,3)}$$

and thus

$$I_{ijk} = \left(\chi^{(2,1)} - \chi^{(2,3)} - \frac{5}{2}\chi^{(4,3)}\right)\frac{g_i g_j g_k}{g^3} + \frac{3}{2}\chi^{(4,3)}\delta_{(ij}\frac{g_{k)}}{g} \ .$$

The trace of the rank-4 tensor integral must be of the form

$$I_{ijkk} = \int_0^{2\pi}\int_0^{\pi/2} k_i k_j \cos^4\Theta \sin\Theta F_M(\Theta)\, d\Theta\, d\varepsilon = a\frac{g_i g_j}{g^2} + b\delta_{ij}$$

and by scalar multiplication with δ_{ij}, and $g_i g_j/g^2$, respectively, follows that

$$a + 3b = \chi^{(4,1)} \quad \text{and} \quad a + b = \chi^{(6,1)}$$

so that

$$2b = \chi^{(4,1)} - \chi^{(6,1)} = \chi^{(4,3)} \quad \text{and} \quad a = \chi^{(2,1)} - \chi^{(2,3)} - \frac{3}{2}\chi^{(4,3)}$$

and thus

$$I_{ijkk} = \left(\chi^{(2,1)} - \chi^{(2,3)} - \frac{3}{2}\chi^{(4,3)}\right)\frac{g_i g_j}{g^2} + \frac{1}{2}\chi^{(4,3)}\delta_{ij} \ .$$

A.5 Lagrange multipliers

We ask for the extremum of a function $f(x_i)$ of the variables x_i $(i = 1, \cdots, n)$ under a constraint $g(x_i) = 0$. At the extremum the variation of f vanishes,

$$df = \sum_{i=1}^n \frac{\partial f}{\partial x_i}dx_i = 0\ .$$

Without the constraint, the variables x_i are independent, and one could arbitrarily chose $dx_i = 0$ for $i \neq k$ and $dx_k \neq 0$ so that $\partial f/\partial x_k = 0$ for all k. However, because of the constraint, the variables x_i are not independent, and their variations dx_i are related through

$$\sum_{i=1}^n \frac{\partial g}{\partial x_i}dx_i = 0\ .$$

This implies that the variations $\mathrm{d}x_i$ cannot be chosen independently. To solve the problem, the last equation is multiplied by a Lagrange multiplier λ and added to the other, so that

$$\sum_{i=1}^{n}\left(\frac{\partial f}{\partial x_i} - \lambda \frac{\partial g}{\partial x_i}\right)\mathrm{d}x_i = 0 \,.$$

Still, the variations cannot be chosen independently, but now the Lagrange multiplier can be used to eliminate one term from the sum, by choosing it so that

$$\frac{\partial f}{\partial x_n} - \lambda \frac{\partial g}{\partial x_n} = 0 \,.$$

The remaining condition reads

$$\sum_{i=1}^{n-1}\left(\frac{\partial f}{\partial x_i} - \lambda \frac{\partial g}{\partial x_i}\right)\mathrm{d}x_i = 0$$

where only $(n-1)$ variations $\mathrm{d}x_i$ $(i = 1,\cdots,n-1)$ appear, and these can be chosen independently, so that

$$\frac{\partial f}{\partial x_i} - \lambda \frac{\partial g}{\partial x_i} = 0 \ \text{ for } \ i = 1,\cdots,n-1 \,.$$

In short, finding the extremum of $f - \lambda g$ is equivalent to finding the extremum of f under the constraint $g = 0$.

When several constraints must be considered, $g_\alpha(x_i) = 0$, one just needs to use more Lagrange multipliers, that is one has to find the extremum of $f - \sum_\alpha \lambda_\alpha g_\alpha$.

A.6 Equations for the computation of generalized 13 moment equations

This appendix contains intermediate steps for the computation of the generalized 13 moment equations (8.18, 8.19).

A.6.1 Moment equations for w_i^a

The starting point is the moment equation for u_i^a, (5.15)/(8.1)$_2$. After introducing the w_i^a, w_{ij}^a by means of (8.11), replacing the time derivative of θ by means of the energy balance (8.12)$_2$, cancelling terms of $\mathcal{O}\left(\varepsilon^2\right)$ and higher, and division by $\frac{2\kappa_a}{\kappa_1}\theta^{a-1}$, we see

$$\frac{Dq_i}{Dt} + \frac{a}{2}\frac{\kappa_1}{\kappa_a}\left[\frac{\mu_a - 2\mu_{a-1}}{\mu_0}\right]\sigma_{ik}\frac{\partial\theta}{\partial x_k} - a\frac{\kappa_1\mu_{a-1}}{\kappa_a\mu_0}\theta\sigma_{ik}\frac{\partial\ln\rho}{\partial x_k}$$

$$+\frac{\kappa_1}{2\kappa_a}\left[\frac{\mu_a}{\mu_0} - \frac{(2a+3)!!}{3}\right]\theta\frac{\partial\sigma_{ik}}{\partial x_k} + \left[\frac{25-4a}{15}q_i\frac{\partial v_k}{\partial x_k} + \frac{2a+5}{5}q_k\frac{\partial v_i}{\partial x_k} + \frac{2a}{5}q_k\frac{\partial v_k}{\partial x_i}\right]$$

$$+\frac{1}{2\tau}\sum_b \mathcal{C}_{ab}^{(1)}\frac{\kappa_1}{\kappa_a}\frac{w_i^b}{\theta^{b-1}} + \frac{1}{\tau}\sum_{b,c}\mathcal{Y}_{a,bc}^{1,1,1}\frac{\mu_b\kappa_c}{\mu_0\kappa_a}\frac{\sigma_{ij}q_j}{\rho\theta} = -\frac{a(2a+3)!!}{6}\frac{\kappa_1}{\kappa_a}\rho\theta\left[\frac{\partial\theta}{\partial x_i} + \frac{q_i}{\kappa}\right].$$

Subtracting the balance of the heat flux (8.17) gives

$$\frac{1}{2}\sum_b\left[\mathcal{C}_{1b}^{(1)} - \mathcal{C}_{ab}^{(1)}\frac{\kappa_1}{\kappa_a}\right]\frac{w_i^b}{\tau\theta^{b-1}} = \left[1 + \frac{a}{2}\frac{\kappa_1}{\kappa_a}\frac{\mu_a - 2\mu_{a-1}}{\mu_0} - \frac{1}{2}\frac{\mu_1}{\mu_0}\right]\sigma_{ik}\frac{\partial\theta}{\partial x_k}$$

$$+\left[1 - a\frac{\kappa_1\mu_{a-1}}{\kappa_a\mu_0}\right]\theta\sigma_{ik}\frac{\partial\ln\rho}{\partial x_k} + \left[\frac{5}{2} + \frac{\kappa_1}{2\kappa_a}\left[\frac{\mu_a}{\mu_0} - \frac{(2a+3)!!}{3}\right] - \frac{1}{2}\frac{\mu_1}{\mu_0}\right]\theta\frac{\partial\sigma_{ik}}{\partial x_k}$$

$$+\frac{4-4a}{15}q_i\frac{\partial v_k}{\partial x_k} + \frac{2a-2}{5}q_k\frac{\partial v_i}{\partial x_k} + \frac{2a-2}{5}q_k\frac{\partial v_k}{\partial x_i}$$

$$+\sum_{b,c}\left[\mathcal{Y}_{a,bc}^{1,1,1}\frac{\mu_b\kappa_c}{\mu_0\kappa_a} - \mathcal{Y}_{1,bc}^{1,1,1}\frac{\mu_b\kappa_c}{\mu_0\kappa_1}\right]\frac{\sigma_{ij}q_j}{\tau\rho\theta} - \rho\theta\left[\frac{q_i}{\kappa} + \frac{\partial\theta}{\partial x_i}\right]\left[\frac{5}{2} - \frac{a(2a+3)!!}{6}\frac{\kappa_1}{\kappa_a}\right].$$

The above equation is meaningful for $a \geq 2$ only. The w_i^b follow from inversion of the matrix

$$\mathcal{D}_{ab} = \left[\mathcal{C}_{1b}^{(1)} - \mathcal{C}_{ab}^{(1)}\frac{\kappa_1}{\kappa_a}\right]$$

as

$$\frac{w_i^d}{2\tau\theta^{d-1}} = \sum_a \mathcal{D}_{da}^{-1}\left[1 - \frac{1}{2}\frac{\mu_1}{\mu_0} + \frac{a}{2}\frac{\kappa_1}{\kappa_a}\frac{\mu_a - 2\mu_{a-1}}{\mu_0}\right]\sigma_{ik}\frac{\partial\theta}{\partial x_k}$$

$$+\sum_a\mathcal{D}_{da}^{-1}\left[1 - a\frac{\kappa_1\mu_{a-1}}{\kappa_a\mu_0}\right]\theta\sigma_{ik}\frac{\partial\ln\rho}{\partial x_k}$$

$$+\sum_a\mathcal{D}_{da}^{-1}\left[\frac{5}{2} - \frac{1}{2}\frac{\mu_1}{\mu_0} + \frac{\kappa_1}{2\kappa_a}\left[\frac{\mu_a}{\mu_0} - \frac{(2a+3)!!}{3}\right]\right]\theta\frac{\partial\sigma_{ik}}{\partial x_k}$$

$$+\sum_a\mathcal{D}_{da}^{-1}\left[\frac{4-4a}{15}q_i\frac{\partial v_k}{\partial x_k} + \frac{2a-2}{5}q_k\frac{\partial v_i}{\partial x_k} + \frac{2a-2}{5}q_k\frac{\partial v_k}{\partial x_i}\right]$$

$$+\sum_a\mathcal{D}_{da}^{-1}\sum_{b,c}\left[\mathcal{Y}_{a,bc}^{1,1,1}\frac{\mu_b\kappa_c}{\mu_0\kappa_a} - \mathcal{Y}_{1,bc}^{1,1,1}\frac{\mu_b\kappa_c}{\mu_0\kappa_1}\right]\frac{\sigma_{ij}q_j}{\tau\rho\theta}$$

$$-\sum_a\mathcal{D}_{da}^{-1}\left[\frac{5}{2} - \frac{a(2a+3)!!}{6}\frac{\kappa_1}{\kappa_a}\right]\rho\theta\left[\frac{q_i}{\kappa} + \frac{\partial\theta}{\partial x_i}\right]. \quad (A.5)$$

A.6.2 Moment equations for w_{ij}^a

The equation for the 2-tensors u_{ij}^a, (5.16)/(8.1)$_3$ is manipulated similarly. After introducing the w_i^a, w_{ij}^a by means of (8.11), replacing the time derivative of θ by means of the energy balance $(8.12)_2$, cancelling terms of $\mathcal{O}\left(\varepsilon^2\right)$ and higher, and division by $\frac{\mu_a}{\mu_0}\theta^a$, we see

$$
\frac{D\sigma_{ij}}{Dt} + \frac{4}{5}\frac{\kappa_a + 1 \mu_0}{\kappa_1 \mu_a}\frac{\partial q_{\langle i}}{\partial x_{j\rangle}} + \frac{6a}{7}\sigma_{\langle ij}\frac{\partial v_{k\rangle}}{\partial x_k} + \frac{4a}{5}\sigma_{k\langle i}\frac{\partial v_k}{\partial x_{j\rangle}} - \frac{2}{3}a\sigma_{ij}\frac{\partial v_k}{\partial x_k} + 2\sigma_{k\langle i}\frac{\partial v_{j\rangle}}{\partial x_k}
$$

$$
+\sigma_{ij}\frac{\partial v_k}{\partial x_k} - \frac{4}{5}(2a+5)\frac{\kappa_a \mu_0}{\kappa_1 \mu_a}q_{\langle i}\frac{\partial \ln \rho}{\partial x_{j\rangle}} - \frac{4}{5}\left[(2a+5)\frac{\kappa_a \mu_0}{\kappa_1 \mu_a} - a\frac{\kappa_a + 1 \mu_0}{\kappa_1 \mu_a}\right]q_{\langle i}\frac{\partial \ln \theta}{\partial x_{j\rangle}}
$$

$$
+\frac{\mu_0}{\mu_a \theta^a}\frac{1}{\tau}\sum_b C_{ab}^{(2)}\theta^{a-b}w_{ik}^b + \sum_{b,c}\mathcal{Y}_{a,bc}^{2,0,1}\frac{4\kappa_b \kappa_c \mu_0}{\kappa_1 \kappa_1 \mu_a}\frac{q_{\langle i}q_{k\rangle}}{\tau \rho \theta^2} + \sum_{b,c}\mathcal{Y}_{a,bc}^{2,1,1}\frac{\mu_b \mu_c}{\mu_0 \mu_a}\frac{\sigma_{j\langle i}\sigma_{k\rangle j}}{\tau \rho \theta}
$$

$$
= -\frac{\mu_0}{\mu_a}\frac{(2a+5)!!}{15}\rho\theta\left[\frac{\sigma_{ik}}{\mu} + 2\frac{\partial v_{\langle i}}{\partial x_{j\rangle}}\right].
$$

the result can be further simplified by means of the identities

$$
\sigma_{\langle ij}\frac{\partial v_{k\rangle}}{\partial x_k} = \frac{1}{3}\sigma_{ij}\frac{\partial v_k}{\partial x_k} + \frac{2}{3}\sigma_{k\langle i}\frac{\partial v_{j\rangle}}{\partial x_k} - \frac{4}{15}\sigma_{r\langle i}\frac{\partial v_r}{\partial x_{j\rangle}},
$$

$$
\frac{6a}{7}\sigma_{\langle ij}\frac{\partial v_{k\rangle}}{\partial x_k} + \frac{4a}{5}\sigma_{k\langle i}\frac{\partial v_k}{\partial x_{j\rangle}} - \frac{2}{3}a\sigma_{ij}\frac{\partial v_k}{\partial x_k}
$$

$$
= \frac{4a}{7}\left(\sigma_{k\langle i}\frac{\partial v_{j\rangle}}{\partial x_k} + \sigma_{k\langle i}\frac{\partial v_k}{\partial x_{j\rangle}} - \frac{2}{3}\sigma_{ij}\frac{\partial v_k}{\partial x_k}\right).
$$

Subtraction of the balance for the pressure deviator (8.16) from the main equation yields, after some rearrangements,

$$
\sum_b\left[C_{0b}^{(2)} - \frac{\mu_0}{\mu_a}C_{ab}^{(2)}\right]\frac{w_{ik}^b}{\tau\theta^b} = \frac{4}{5}\left[\frac{\kappa_a + 1 \mu_0}{\kappa_1 \mu_a} - 1\right]\frac{\partial q_{\langle i}}{\partial x_{j\rangle}}
$$

$$
+ \frac{4a}{7}\left(\sigma_{k\langle i}\frac{\partial v_{j\rangle}}{\partial x_k} + \sigma_{k\langle i}\frac{\partial v_k}{\partial x_{j\rangle}} - \frac{2}{3}\sigma_{ij}\frac{\partial v_k}{\partial x_k}\right) - \frac{4}{5}(2a+5)\frac{\kappa_a \mu_0}{\kappa_1 \mu_a}q_{\langle i}\frac{\partial \ln \rho}{\partial x_{j\rangle}}
$$

$$
- \frac{4}{5}\left[(2a+5)\frac{\kappa_a \mu_0}{\kappa_1 \mu_a} - a\frac{\kappa_a + 1 \mu_0}{\kappa_1 \mu_a}\right]q_{\langle i}\frac{\partial \ln \theta}{\partial x_{j\rangle}}
$$

$$
+ \sum_{b,c}\left[\mathcal{Y}_{a,bc}^{2,0,1}\frac{4\kappa_b \kappa_c \mu_0}{\kappa_1 \kappa_1 \mu_a} - \mathcal{Y}_{0,bc}^{2,0,1}\frac{4\kappa_b \kappa_c}{\kappa_1 \kappa_1}\right]\frac{q_{\langle i}q_k\rangle}{\tau \rho \theta^2}
$$

$$
+ \sum_{b,c}\left[\mathcal{Y}_{a,bc}^{2,1,1}\frac{\mu_b \mu_c}{\mu_0 \mu_a} - \mathcal{Y}_{0,bc}^{2,1,1}\frac{\mu_b \mu_c}{\mu_0 \mu_0}\right]\frac{\sigma_{j\langle i}\sigma_{k\rangle j}}{\tau \rho \theta}
$$

$$
- \left[1 - \frac{\mu_0}{\mu_a}\frac{(2a+5)!!}{15}\right]\rho\theta\left[\frac{\sigma_{ij}}{\mu} + 2\frac{\partial v_{\langle i}}{\partial x_{j\rangle}}\right].
$$

The w_{ij}^b follow from inversion of the matrix

$$\mathcal{E}_{ab}=\left[C_{0b}^{(2)}-\frac{\mu_0}{\mu_a}C_{ab}^{(2)}\right]$$

as

$$\frac{w_{ik}^d}{\tau\theta^d}=\sum_a\mathcal{E}_{da}^{-1}\frac{4}{5}\left[\frac{\kappa_{a+1}\mu_0}{\kappa_1\mu_a}-1\right]\frac{\partial q_{\langle i}}{\partial x_{j\rangle}}$$

$$+\sum_a\mathcal{E}_{da}^{-1}\frac{4a}{7}\left(\sigma_{k\langle i}\frac{\partial v_{j\rangle}}{\partial x_k}+\sigma_{k\langle i}\frac{\partial v_k}{\partial x_{j\rangle}}-\frac{2}{3}\sigma_{ij}\frac{\partial v_k}{\partial x_k}\right)$$

$$-\sum_a\mathcal{E}_{da}^{-1}\frac{4}{5}\,(2a+5)\,\frac{\kappa_a\mu_0}{\kappa_1\mu_a}q_{\langle i}\frac{\partial\ln\rho}{\partial x_{j\rangle}}$$

$$-\sum_a\mathcal{E}_{da}^{-1}\frac{4}{5}\left[(2a+5)\,\frac{\kappa_a\mu_0}{\kappa_1\mu_a}-a\frac{\kappa_{a+1}\mu_0}{\kappa_1\mu_a}\right]q_{\langle i}\frac{\partial\ln\theta}{\partial x_{j\rangle}}$$

$$+\sum_a\mathcal{E}_{da}^{-1}\sum_{b,c}\left[\mathcal{Y}_{a,bc}^{2,0,1}\frac{4\kappa_b\kappa_c\mu_0}{\kappa_1\kappa_1\mu_a}-\mathcal{Y}_{0,bc}^{2,0,1}\frac{4\kappa_b\kappa_c}{\kappa_1\kappa_1}\right]\frac{q_{\langle i}q_{k\rangle}}{\tau\rho\theta^2}$$

$$+\sum_a\mathcal{E}_{da}^{-1}\sum_{b,c}\left[\mathcal{Y}_{a,bc}^{2,1,1}\frac{\mu_b\mu_c}{\mu_0\mu_a}-\mathcal{Y}_{0,bc}^{2,1,1}\frac{\mu_b\mu_c}{\mu_0\mu_0}\right]\frac{\sigma_{j\langle i}\sigma_{k\rangle j}}{\tau\rho\theta}$$

$$-\sum_a\mathcal{E}_{da}^{-1}\left[1-\frac{\mu_0}{\mu_a}\frac{(2a+5)!!}{15}\right]\rho\theta\left[\frac{\sigma_{ij}}{\mu}+2\frac{\partial v_{\langle i}}{\partial x_{j\rangle}}\right].\quad\text{(A.6)}$$

A.6.3 Coefficients in (8.18, 8.19)

When the above results (A.5, A.6) are inserted into the equations for stress and heat flux, these assume the form (8.18, 8.19) where the coefficients are defined as

$$\psi_1=\frac{4}{5}\left[1-\sum_{d,a}C_{0d}^{(2)}\mathcal{E}_{da}^{-1}\left[1-\frac{\kappa_{a+1}\mu_0}{\kappa_1\mu_a}\right]\right]$$

$$\psi_2=\sum_{d,a}C_{0d}^{(2)}\mathcal{E}_{da}^{-1}\frac{4a}{7}$$

$$\psi_3=\frac{4}{5}\sum_{d,a}C_{0d}^{(2)}\mathcal{E}_{da}^{-1}\,(2a+5)\,\frac{\kappa_a\mu_0}{\kappa_1\mu_a}$$

$$\psi_4=\frac{4}{5}\sum_{d,a}C_{0d}^{(2)}\mathcal{E}_{da}^{-1}\left[(2a+5)\,\frac{\kappa_a\mu_0}{\kappa_1\mu_a}-a\frac{\kappa_{a+1}\mu_0}{\kappa_1\mu_a}\right]\qquad\text{(A.7)}$$

$$\psi_5=\frac{\mu_0}{2}\left[\sum_{b,c}\frac{4\kappa_b\kappa_c}{\kappa_1\kappa_1}\mathcal{Y}_{0,bc}^{2,0,1}+\sum_{d,a}C_{0d}^{(2)}\mathcal{E}_{da}^{-1}\sum_{b,c}\frac{4\kappa_b\kappa_c}{\kappa_1\kappa_1}\left[\frac{\mu_0}{\mu_a}\mathcal{Y}_{a,bc}^{2,0,1}-\mathcal{Y}_{0,bc}^{2,0,1}\right]\right]$$

$$\psi_6=\frac{\mu_0}{2}\left[\sum_{b,c}\mathcal{Y}_{0,bc}^{2,1,1}\frac{\mu_b\mu_c}{\mu_0\mu_0}+\sum_{d,a}C_{0d}^{(2)}\mathcal{E}_{da}^{-1}\sum_{b,c}\frac{\mu_b\mu_c}{\mu_0\mu_0}\left[\frac{\mu_0}{\mu_a}\mathcal{Y}_{a,bc}^{2,1,1}-\mathcal{Y}_{0,bc}^{2,1,1}\right]\right]$$

$$\psi_7 = 1 - \sum_{d,a} C^{(2)}_{0d} \mathcal{E}^{-1}_{da} \left[1 - \frac{\mu_0}{\mu_a} \frac{(2a+5)!!}{15} \right]$$

and

$$\chi_1 = \frac{1}{2} \frac{\mu_1}{\mu_0} - 1 - \sum_{d,a} C^{(1)}_{1d} \mathcal{D}^{-1}_{da} \left[\frac{1}{2} \frac{\mu_1}{\mu_0} - 1 - a \frac{\kappa_1}{\kappa_a} \left[\frac{1}{2} \frac{\mu_a}{\mu_0} - \frac{\mu_{a-1}}{\mu_0} \right] \right]$$

$$\chi_2 = 1 - \sum_{d,a} C^{(1)}_{1d} \mathcal{D}^{-1}_{da} \left[1 - a \frac{\kappa_1 \mu_{a-1}}{\kappa_a \mu_0} \right]$$

χ_3

χ_4

χ_5

χ_6

The c

with

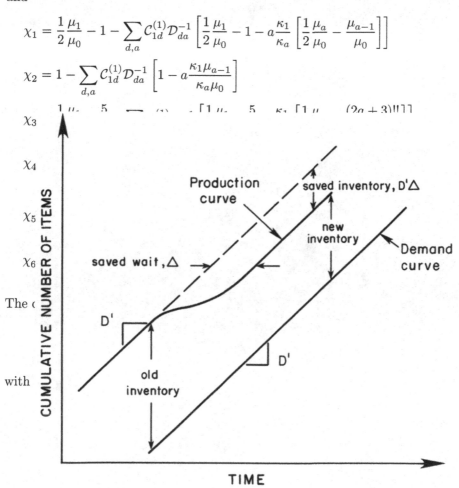

References

1. G. Bird, *Molecular gas dynamics and the direct simulation of gas flows*. Clarendon Press, Oxford 1994
2. S. Chapman and T. G. Cowling, *The Mathematical Theory of Non-Uniform Gases*. Cambridge University Press 1970
3. C. Cercignani, *Theory and Application of the Boltzmann Equation*. Scottish Academic Press, Edinburgh 1975
4. C. Cercignani, *Rarefied Gas Dynamics From Basic Concepts to Actual Calculations*. Cambridge University Press, Cambridge, 2000.
5. C. Cercignani and G.M. Kremer, *The Relativistic Boltzmann Equation: Theory and Applications*. Progress in Mathematical Physics, Vol. 22, Birkhäuser, Basel 2002
6. J.H. Ferziger and H.G. Kaper, *Mathematical theory of transport processes in gases*. North-Holland, Amsterdam 1972
7. V. Garzo and A. Santos, *Kinetic Theory of Gases in Shear Flow*. Kluwer, Dordrecht 2003
8. H. Grad, *On the Kinetic Theory of Rarefied Gases*. Comm. Pure Appl. Math. **2**, 325 (1949)
9. H. Grad, *Principles of the Kinetic Theory of Gases*. in Handbuch der Physik XII: Thermodynamik der Gase, S. Flügge (Ed.), Springer, Berlin 1958
10. S.R. De Groot, W.A. van Leeuwen, and Ch. G. van Weert, *Relativistic Kinetic Theory*. North-Holland, Amsterdam 1980
11. S. Harris, *An Introduction to the Theory of the Boltzmann Equation*. Holt, Rinehart, and Winston, Inc., New York 1971
12. J. O. Hirschfelder, C. F. Curtiss, and R. B. Bird, *Molecular Theory of Gases and Liquids*. Wiley, New York 1954
13. M.N. Kogan, *Rarefied Gas Dynamics*. Plenum Press, New York 1969
14. R.L. Liboff, *The Theory of Kinetic Equations*. Wiley and Sons, New York 1969
15. I. Müller, *Thermodynamics*. Pitman, Boston 1985
16. I. Müller, *Grundzüge der Thermodynamics mit historischen Anmerkungen*. Springer, Berlin 2001
17. I. Müller and T. Ruggeri, *Rational Extended Thermodynamics*. Springer, New York 1998 (Springer Tracts in Natural Philosophy Vol. 37)
18. A. Sommerfeld, *Theoretische Physik Bd. V, Thermodynamik und Statistik*, Verlag Harri Deutsch, Thun 1977 (English translation: *Lectures on theoretical physics, Vol. 5*, Academic Press, New York 1956)

19. Y. Sone, *Kinetic Theory and Fluid dynamics*. Birkhäuser, Boston 2002
20. C. Truesdell and R.G. Muncaster, *Fundamentals of Maxwell's kinetic theory of a simple monatomic gas*. Academic Press, New York 1980
21. L. Waldmann, *Transporterscheinungen in Gasen von mittlerem Druck*. in Handbuch der Physik XII: Thermodynamik der Gase, S. Flügge (Ed.), Springer, Berlin 1958
22. G.M. Karniadakis and A. Beskok, *MicroFlows*. Springer, New York 2001
23. J.A. Fay, *Molecular Thermodynamics*. Addison-Wesley, Reading 1965
24. M. Torrihon, J.D. Au, and H. Struchtrup, *Explicit Fluxes and Productions for Large Systems of the Moment Method based on Extended Thermodynamics*. Cont. Mech. Thermodyn. **15**, 97-111 (2003)
25. L. Boltzmann, *Weitere Studien über das Wärmegleichgewicht unter Gasmolekülen*. Sitzungsberichte der Akademie der Wissenschaften, Wien **66**, 275-370 (1872)
26. Y.A. Cengel and M.A. Boles, *Thermodynamics—An Engineering Approach*. McGraw-Hill, Boston 2002
27. M.J. Moran and H.N. Shapiro, *Fundamentals of Engineering Thermodynamics*. Wiley, New York 1999
28. R.E. Sonntag, C. Borgnakke, and G.J. van Wylen, *Fundamentals of Thermodynamics*. Wiley 2003
29. P.L. Bhatnagar, E.P. Gross, and M. Krook, *A Model for collision processes in gases. I. Small Amplitude Processes in Charged and Neutral One-Component Systems*. Phys.Rev. **94**, 511-525 (1954)
30. E.P. Gross and M. Krook, *A Model for collision processes in gases. I. Small Amplitude Processes in Charged and Neutral One-Component Systems*. Phys.Rev. **102**, 593-604 (1954)
31. M. Krook, *Continuum equations in the dynamics of rarefied gases*. J. Fluid Mech. **6**, 523-541 (1959)
32. L.H. Holway, *New Statistical Models for Kinetic Theory: Methods of Construction*. Physics of Fluids **9**, 1658-1673 (1966)
33. E. M. Shakhov, *Generalization of the Krook kinetic relaxation equation*, Fluid Dynamics (English Translation of Izvestiya Akademii Nauk SSSR, Mekhanika Zhidkosti i Gaza) **3**, 95-96 (1968)
34. E.M. Shakhov, *Kinetic model equations and numerical results*. in 14th International Symposium on Rarefied Gas Dynamics, Tsukuba Japan, 1984
35. F. Sharipov and V. Seleznev, *Data on internal rarefied gas flows*. J. Physical and Chemical Reference Data **27**, 657-706 (1998)
36. G. Liu, *A method for constructing a model form for the Boltzmann equation*. Phys. Fluids **2**, 277-280 (1990)
37. V. Garzo, *Transport equations from the Liu model*. Physics of Fluids A **3**, 1980-1982 (1991)
38. V. Garzo and M. L. d. Haro, *Kinetic model for heat and momentum transport*. Phys. Fluids **6**, 3787-3794 (1994)
39. F. Bouchut and B. Perthame, *A BGK model for Small Prandtl Number in the Navier-Stokes Approximation*. J. Stat. Phys. **71**, 191-207 (1993)
40. H. Struchtrup, *The BGK model with velocity dependent collision frequency*. Continuum Mech. Thermodyn. **9**, 23-32 (1997)
41. L. Mieussens and H. Struchtrup, *Numerical comparison of BGK-models with proper Prandtl number*. Phys. Fluids **16**(8), 2797-2813 (2004)

42. Y. Zheng, *Analysis of kinetic models and macroscopic continuum equations for rarefied gasdynamics.* Ph.D. Thesis, University of Victoria, Canada (2004)

43. P. Andries, P. Le Tallec, J. Perlat, and B. Perthame, *The Gaussian-BGK model of Boltzmann equation with small Prandtl numbers.* European Journal of Mechanics: B Fluids **19**(6), 813-830 (2000)

44. E.S. Oran, C.K. Oh, and B.Z. Cybyk, *Direct Simualtion Monte Carlo: Recent Advances and Applications.* Annu. Rev. Fluid Mech. **30**, 403-441 (1998)

45. A. Ketsdever and P. Muntz (Eds.), *Rarefied Gas Dynamics—23 International Symposium, Whistler, BC 2002,* AIP Conference Proceedings **663** (2003)

46. K. Nanbu, *Direct simulation scheme derived from the Boltzmann equation. I. Multicomponent gases.* J. Phys. Soc. of Japan **45**, 2042-2049 (1980)

47. H. Babovsky and R. Illner, *A convergence proof for Nanbu's simulation method for the full Boltzmann equation.* SIAM J. Numer. Anal. **26**(1), 45–65 (1989)

48. W. Wagner, *A convergence proof for Bird's direct simulation Monte Carlo method for the Boltzmann equation.* J. Statist. Phys. **66**(3-4), 1011–1044 (1992)

49. D. Enskog, Dissertation, Uppsala 1917

50. D. Enskog, *The numerical calculation of phenomena in fairly dense gases,* Arkiv Mat. Astr. Fys. **16**(1) (1921)

51. S. Chapman, *On the law of distribution of Molecular Velocities, and on the Theory of Viscosity and Thermal Conduction, in a Non-uniform Simple Monatomic Gas.* Phil. Trans. R. Soc. A, **216**, 279-348 (1916)

52. S. Chapman, *On the Kinetic Theory of a Gas. Part II: A Composite Monatomic Gas: Diffusion, Viscosity, and Thermal Conduction.* Phil. Trans. R. Soc. A, **217**, 115-197 (1918)

53. D. Burnett, *The distribution of molecular velocities and the mean motion in a non-uniform gas.* Proc. Lond. Math. Soc. **40**, 382-435 (1936)

54. S. Reinecke and G.M. Kremer, *Method of Moments of Grad.* Phys. Rev. A **42**, 815-820 (1990)

55. M.Sh. Shavaliyev, *Super-Burnett Corrections to the Stress Tensor and the Heat Flux in a Gas of Maxwellian Molecules.* J. Appl. Maths. Mechs. **57**(3), 573-576 (1993)

56. A.V. Bobylev, *The Chapman-Enskog and Grad methods for solving the Boltzmann equation.* Sov. Phys. Dokl. **27**, 29-31 (1982)

57. Y. Zheng and H. Struchtrup, *Burnett equations for the ellipsoidal statistical BGK Model.* Cont. Mech. Thermodyn. **16**(1-2), 97-108 (2004)

58. M. Slemrod, *In the Chapman-Enskog Expansion the Burnett Coefficients Satisfy the Universal Relation* $\omega_3 + \omega_4 + \theta_3 = 0$, Arch. Rational Mech. Anal. **161**, 339-344 (2002)

59. H. Struchtrup and M. Torrilhon, *Regularization of Grad's 13 Moment Equations: Derivation and Linear Analysis.* Phys. Fluids **15**(9), 2668-2680 (2003)

60. M. Torrilhon and H. Struchtrup, *Regularized 13-Moment-Equations: Shock Structure Calculations and Comparison to Burnett Models.* J. Fluid Mech. **513**, 171-198 (2004)

61. K.A. Fiscko and D.R. Chapman, *Comparison of Burnett, Super-Burnett and Monte Carlo Solutions for Hypersonic Shock Structure.* in Proceedings of the 16th Symposium on Rarefied Gasdynamics, 374-395, AIAA, Washington 1989

62. X. Zhong, R.W. MacCormack, and D.R. Chapman, *Stabilization of the Burnett Equations and Applications to High-Altitude Hypersonic Flows.* AIAA 91-0770 (1991)

63. X. Zhong, R.W. MacCormack, and D.R. Chapman, *Stabilization of the Burnett Equations and Applications to Hypersonic Flows.* AIAA Journal **31**, 1036 (1993)

64. C.S. Wang Chang, *On the theory of the thickness of weak shock waves.* Studies in Statistical Mechanics V, 27-42, North Holland, Amsterdam (1970)

65. R. Balakrishnan, *Entropy consistent formulation and numerical simulation of the BGK-Burnett equations for hypersonic flows in the continuum-transition regime.* Ph.D. thesis, Wichita State University 1999

66. R. K. Agarwal, K.Y. Yun, and R. Balakrishnan, *Beyond Navier-Stokes: Burnett equations for flows in the continuum-transition regime.* Phys. Fluids **13**, 3061-3085 (2001), Erratum: Phys. Fluids **14**, 1818 (2002)

67. R.K. Agarwal and K.-Y. Yun, *Burnett equations for simulation of transitional flows.* Appl. Mech. Rev. **55**(3), 219-240 (2002)

68. I. Prigogine, *Introduction to Thermodynamics of Irreversible Processes.* Interscience, New York 1967

69. S.R. De Groot and P. Mazur, *Non-equilibrium Thermodynamics.* North-Holland, Amsterdam (1969)

70. H. Struchtrup, *Positive and negative entropy productions and phase densities for approximate solutions of the Boltzmann equation.* J. of Thermophysics and Heat Transfer **15**(3), 372-373 (2001)

71. K.A. Comeaux, D.R. Chapman, and R.W. MacCormack, *An Analysis of the Burnett Equations Based on the Second Law of Thermodynamics.* AIAA Paper 95-0415, 1995.

72. R. Balakrishnan, *An approach to entropy consistency in second-order hydrodynamics equations.* J. Fluid. Mech **503**, 201-245 (2004)

73. C.D. Levermore, W.J. Morokoff, and B.T. Nadiga, *Moment realizability and the validity of the Navier-Stokes equations for rarefied gas dynamics.* Phys. Fluids **10**, 3214-3226 (1998)

74. H. Struchtrup, *Stable transport equations for rarefied gases at high orders in the Knudsen number.* Phys. Fluids **16**(11), 3921-3934 (2004)

75. W. Weiss, *Zur Hierarchie der Erweiterten Thermodynamik.* Ph.D. thesis, Technical University Berlin 1990

76. H. Struchtrup, *Heat Transfer in the Transition Regime: Solution of Boundary Value Problems for Grad's Moment Equations via Kinetic Schemes.* Phys. Rev. E **65**, 041204 (2002)

77. J.D. Au, *Nichtlineare Probleme und Lösungen in der Erweiterten Thermodynamik.* Ph.D. thesis, Technical University Berlin 2000

78. S. Wolfram, *The Mathematica Book, Fifth Edition,* Wolfram Media, 2003

79. K.S. Thorne, *Relativistic Radiative Transfer: Moment Formalisms.* Mon. Not. R. Astr. Soc. **194**, 439-473 (1981)

80. H. Struchtrup, *Zur irreversiblen Thermodynamik der Strahlung.* Ph.D. thesis, Technical University Berlin (1996)

81. H. Struchtrup, *An Extended Moment Method in Radiative Transfer: The Matrices of Mean Absorption and Scattering Coefficients.* Ann. Phys. **257**, 111-135 (1997)

82. H. Struchtrup, *On the number of moments in Radiative transfer problems.* Ann. Phys. **266**, 1-26 (1998)

83. W. Dreyer and H. Struchtrup, *Heat Pulse Experiments Revisited.* Continuum. Mech. Thermodyn. **5**(1), 3-50 (1993)

84. A.M. Anile, V. Romano, and G. Russo, *Extended hydrodynamical model of carrier transport in semiconductors.* SIAM J. Appl. Math **61**(1), 76-101 (2000)

85. A.M. Anile and V. Romano, *Non parabolic band transport in semiconductors: closure of the moment equations.* Continuum Mech. Thermodyn. **11**(5), 307 - 325 (1999)

86. H. Struchtrup, *Extended moment method for electrons in semiconductors.* Physica A **275**, 229-255 (2000)

87. C.S. Wang Chang, *On the propagation of sound in monatomic gases,* Studies in Statistical Mechanics V, 43-98, North Holland, Amsterdam (1970)

88. B. Schmidt, *Electron Beam Density Measurements in Shock Waves in Argon.* J. Fluid Mech. **39**, 361 (1969)

89. W. Weiss, *Continuous shock structure in extended Thermodynamics.* Phys. Rev. E **52**, 5760 (1995)

90. E. Ikenberry and C. Truesdell, *On the pressures and the flux of energy in a gas according to Maxwell's kinetic theory I.* J. of Rat. Mech. Anal. **5**, 1-54 (1956)

91. H. Struchtrup, *Some remarks on the equations of Burnett and Grad.* Transport in Transition Regimes, IMA Vol. Math. Appl. 135, 265-278, Springer, NewYork 2003

92. I. Müller, *On the frame dependence of stress and heat flux.* Arch. Rational Mech. Anal. **45**, 241-250 (1972)

93. S. Reinecke and G.M. Kremer, *Burnett's equations from a (13+9N)-field theory.* Cont Mech. Thermodyn. **8**, 121-130 (1996)

94. S. Reinecke and G.M. Kremer, *A generalization of the Chapman-Enskog and Grad methods.* Continuum Mech. Thermodyn. **3**, 155-167 (1991)

95. W. Dreyer, *Maximization of the Entropy in Non-equilibrium.* J. Phys. A: Math. Gen. **20**, 6505 (1987)

96. C.D. Levermore, *Moment Closure Hierarchies for Kinetic Theories.* J. Stat. Phys. **83**, 1021-1065 (1996)

97. K.O. Friedrichs and P.D. Lax, *Systems of conservartion equations with a convex extension.* Proc.Nat. Acad. Sci. USA **68**, 1686-1688 (1971)

98. T. Ruggeri and A. Strumia, *Main field and convex covariant density for quasilinear hyperbolic systems. Relativistic Fluid Dynamics.* Ann. Inst. H. Poincaré **34A**, 65-84 (1981)

99. A. E. Fischer and J.E. Marsden, *The Einstein evolution equations as a first order quasi-linear symmetric hyperbolic system. I.,* Comm. Math. Phys. **28**, 1–38 (1972)

100. M. Junk, *Kinetic schemes - A new Approach and Applications.* Ph.D. thesis, Shaker, Aachen 1997

101. M. Junk, *Domain of definition of Levermore's five-moment system,* J. Stat. Phys. **93** (5-6): 1143-1167 (1998)

102. W. Dreyer, M. Junk, and M. Kunik, *On the approximation of the Fokker-Planck equation by moment systems,* Nonlinearity **14** (4): 881-906 (2001)

103. M. Junk, *Maximum entropy moment problems and extended Euler equations.* Transport in Transition Regimes, IMA Vol. Math. Appl. 135, 189-198, Springer, NewYork 2003

104. M. Torrilhon, *Characteristic Waves and Dissipation in the 13-Moment-Case.* Cont. Mech. Thermodyn. **12**, 289 (2000)

105. I.V. Karlin, A.N. Gorban, G. Dukek, and T.F. Nonnenmacher, *Dynamic correction to moment approximations.* Phys. Rev. E, **57**(2), 1668-1672 (1998)

106. A.N. Gorban and I.V. Karlin, *Invariant Manifolds for Physical and Chemical Kinetics*. Lecture Notes in Physics Vol. 660, Springer, Berlin 2005

107. I.V. Karlin and A.N. Gorban, *Hydrodynamics from Grad's equations: What can we learn from exact solutions?* Ann. Phys.-Berlin **11**(10-11), 783-833 (2002)

108. S. Jin and M. Slemrod, *Regularization of the Burnett equations via relaxation.* J. Stat. Phys. **103**(5-6), 1009-1033 (2001)

109. S. Jin, L. Pareschi, and M. Slemrod, *A Relaxation Scheme for Solving the Boltzmann Equation Based on the Chapman-Enskog Expansion.* Acta Mathematicas Applicatae Sinica (English Series) **18**, 37-62 (2002)

110. H. Struchtrup, *Derivation of 13 moment equations for rarefied gas flow to second order accuracy for arbitrary interaction potentials.* Multiscale Model. Simul. **3**(1), 211-243 (2004)

111. I. Müller, D. Reitebuch, and W. Weiss, *Extended Thermodynamics - Consistent in Order of Magnitude.* Cont. Mech. Thermodyn. **15**, 113-146 (2003)

112. E.A. Spiegel and J.-L. Thiffeault, *Higher-order continuum approximation for rarefied gases.* Phys. Fluids **15**(11), 3558-3567 (2003)

113. D. Reitebuch, Ph.D. thesis, Technical University Berlin (2004)

114. E. Meyer, and G. Sessler, *Schallausbreitung in Gasen bei hohen Frequenzen und sehr niedrigen Druecken.* Zeitschr. Physik **149**, 15-39 (1957)

115. D. Gilberg and D. Paolucci, *The Structure of Shock Waves in the Continuum Theory of Fluids.* J. Rat. Mech. Anal. **2**, (1953)

116. U. Nowak and L. Weimann, *A Family of Newton Codes for Systems of Highly Nonlinear Equations - Algorithms, Implementation, Applications.* Zuse Institute Berlin, technical report TR 90-10, (1990)

117. H. Alsmeyer, *Density Profiles in Argon and Nitrogen Shock Waves Measured by the Absorbtion of an Electron Beam*, J. Fluid Mech. **74**(3), 497 (1976)

118. G. C. Pham-Van-Diep, D. A. Erwin, and E. P. Muntz, *Testing Continuum Descriptions of Low-Mach-Number Shock Structures.* J. Fluid Mech. **232**, 403 (1991)

119. E. Salomons and M. Mareschal, *Usefulness of the Burnett Description of Strong Shock Waves.* Phy. Rev. Lett. **69**(2), 269-272 (1992)

120. F.J. Uribe, R.M. Velasco, and L.S. García-Colín, *Burnett description of strong shock waves.* Phys. Rev. Lett. 81, 2044-2047 (1998)

121. F. J. Uribe, R. M. Velasco, L. S. García-Colín, and E. Díaz-Herrera, *Shock wave profiles in the Burnett approximation.* Phys. Rev. E **62**(5), 6648–6666 (2000)

122. H. Grad, *The Profile of a Steady Plane Shock Wave.* Comm. Pure Appl. Math **5**, (1952)

123. T. Ruggeri, *Breakdown of Shock Wave Structure Solutions.* Phys. Rev. E **47**, (1993)

124. H.M. Mott-Smith, *The solution of the Boltzmann equation for a shock wave.* Phys. Rev. **82**, 885-892 (1951)

125. H. Struchtrup and W. Weiss, *Temperature jumps and velocity slip in the moment method.* Cont. Mech. Thermodyn. **12**, 1-18 (2000)

126. D.A. Lockerby, J.M. Reese, D.R. Emerson, and R.W. Barber, *Velocity boundary condition at solid walls in rarefied gas calculations.* Phys. Rev. E **70**(1), 017303 (2004)

127. T. Ohwada, Y. Sone, and K. Aoki, *Numerical analysis of the shear and thermal creep flows of a rarefied gas over a plane wall on the basis of the linearized Boltzmann equation for hard-sphere molecules.* Phys. Fluids A **1**(9), 1588-1599 (1989)

128. F. Sharipov and V. Seleznev, *Data on Internal Rarefied Gas Flows*. J. Phys.Chem. Ref. Data **27**(3), 657-706 (1998)

129. F. Sharipov, *Application of the Cercignani–Lampis scattering kernel to calculations of rarefied gas flows. II. Slip and jump coefficients*. European J. of Mechanics - B/Fluids **22**(2), 133-143 (2003)

130. A. Beskok and G.E. Karniadakis, *A model for flows in channels, pipes, and ducts at micro and nano scales*. Microscale Thermophysical Engineering **3**, 43-77 (1999)

131. R.S. Myong, *Gaseous slip models based on the Langmuir adsorption isotherm*. Phys. Fluids **16**(1), 104-117 (2004)

132. W. Marques Jr. and G.M. Kremer, *Couette flow from a thirteen field theory with jump and slip boundary conditions*. Continuum Mech. Thermodyn. **13**, 207-217 (2001)

133. I.-S. Liu and M.A. Rincon, *A boundary value problem in extended thermodynamics: one dimensional steady flows with heat conduction*. Continuum Mech. Thermodyn. **16**, 109-124 (2004)

134. R.E. Street, *Plane Couette Flow by the Methods of Moments*, Proc. Rarefied Gas Dynamics, Ed. L Talbot, Academic Press, New York 1961

135. S. Kubota and Y. Yoshizawa, *Couette flow problems by thirteen moment equations*. Proc. Rarefied Gas Dynamics, M. Becker and M. Fiebig (Eds.), DFVLR Press, Porz-Wahn 1974

136. H. Struchtrup, *Failures of the Burnett and Super-Burnett equations in steady state processes*. Cont. Mech. Thermodyn. (2005, in press)

137. C.J. Lee, *Unique Determination of Solutions to the Burnett Equations*. AIAA Journal **32**(5), 985-990 (1994)

138. D. A. Lockerby, J. M. Reese, and M.A. Gallis, *The usefulness of higher-order constitutive relations for describing the Knudsen layer*, Proceedings of the Conference on Transport Phenomena in Micro and Nanodevices (Hawaii, USA) 2004

139. D. Reitebuch and W. Weiss, *Application of High Moment Theory to the Plane Couette Flow*. Cont. Mech. Thermodyn. **11**(4), 217-225 (1999)

140. H. Struchtrup and W. Weiss, *Maximum of the local entropy production becomes minimal in stationary processes*. Phys. Rev. Lett. 80, 5048-5051 (1998)

141. M. Grmela, I.V. Karlin, and V.B. Zmievski, *Boundary layer variational principles: A case study*. Phys. Rv. E **66**(1), 011201 (2002)

142. H. Struchtrup, *Grad's Moment Equations for Microscale Flows*. Symposium on Rarefied Gasdynamics 23, AIP Conference Proceedings 663, 792-799 (2003)

143. E. Barbera, I. Müller, D. Reitebuch, and N.-R. Zhao, *Determination of boundary conditions in extended thermodynamics via fluctuation theory*, Cont. Mech. Thermodyn. **16**(5), 411 - 425 (2004)

144. T. Ohwada *Heat flow and temperature and density distributions in a rarefied gas between parallel plates with different temperatures. Finite difference analysis of the nonlinear Boltzmann equation for hard sphere molecules*. Phys. Fluids **8**, 2153-2160 (1996)

145. D.W. Mackowski, D.H. Papadopoulos, and D.E. Rosner, *Comparison of Burnett and DSMC predictions of pressure distributions and normal stress in one-dimensional, strongly nonisothermal gas*. Phys. Fluids **11**, 2108-2116 (1999)

146. T. Thatcher, M.A.Sc. thesis, University of Victoria, 2005

147. H. Xue, H.M. Ji, and C. Shu, *Analysis of micro-Couette flow using the Burnett equations*, Int. J. Heat Mass Transfer **44**(21), 4139-4146 (2001)

148. H. Xue, H.M. Ji, and C. Shu, *Prediction of Flow and Heat Transfer Characteristics in micro-Couette Flow*, Microscale Thermophysical Engineering **7**, 51-68 (2003)

149. D.A. Lockerby and J.M. Reese, *High resolution Burnett simulation of micro Couette flow and heat transfer.* J. Comp. Phys. **188**, 333-347 (2003)

150. H. Struchtrup, *Kinetic schemes and boundary conditions for moment equations.* ZAMP **51**, 346-365 (2000)

151. B. Perthame, *Boltzmann type schemes for gas-dynamics and the entropy property.* SIAM J. Num. Anal. **27**, 1405-1421 (1990)

152. P. LeTallec and J.P. Perlat, *Numerical Analysis of Levermore's moment system.* INRIA preprint N° 3124 (1997)

153. K. Xu, *A Gas-kinetic BGK Scheme for the Navier-Stokes Equations, and its Connection with Artificial Dissipation and Godunov Method.* J. Comput. Phys., **171**, 289-335 (2001)

154. K. Xu, *Super-Burnett solutions for Poiseuille flow.* Phys. Fluids **15**(7), 2077-2080 (2003)

155. K. Xu and Z. Li, *Microchannel flow in the slip regime: gas-kinetic BGK-Burnett solution*, J. Fluid Mech. **513**, 87-110 (2004)

156. P. Charrier, B. Dubroca, J.-L. Feugeas, and L. Mieussens, *Modèles à vitesses discrètes pour le calcul d'écoulements hors équilibre cinétique.* C.R. Acad. Sci. Paris **326**, 1347-1352 (1998)

157. L. Mieussens, *Discrete-Velocity Models and Numerical Schemes for the Boltzmann-BGK Equation in Plane and Axisymmetric Geometries.* J. Comp. Physics **162**(2), 429-466 (2000)

158. L. Mieussens *Discrete velocity model and implicit scheme for the BGK equation of rarefied gas dynamics.* M3AS **10**(8), 1121-1149 (2000)

159. A. Klar, *Asymptotic analysis and coupling conditions for kinetic and hydrodynamic equations.* Comp. Math. Appl. **35** (1/2), 127-137, (1998)

160. J.F. Bourgat, P.L. Tallec, and M.D. Tidrir, *Coupling Boltzmann and Navier-Stokes equations by friction.* J. Comp. Phys. **127**, 227-245 (1996)

161. S.P. Popov and F.G. Cheremisin, *Example of simultaneous numerical solution of the Boltzmann and Navier-Stokes equations.* Comp. Math. Math. Phys. **41**, 457-468 (2001)

162. P.L. Tallec and F. Mallinger, *Coupling Boltzmann and Navier-Stokes equations by half fluxes,* J. Comp. Phys. **136**, 51-67 (1997)

163. R. Roveda, D.B. Goldstein, and P.L. Varghese, *A Hybrid Euler/DSMC Approach to Unsteady Flows.* in Rarefied Gas Dynamics, Vol. 2, Proceedings of the 21st Rarefied Gas Dynamics Conference, Edited by R. Brun, R. Campargue, R. Gatignol, and J.-C. Lengrand, 117-124 (1998)

Index

13 moment equations, 3, 89, 150
  Couette flow, 206
  discontinuous shocks, 188
  Knudsen layers, 211
  linearized, 159
  one-dimensional, 155
  shock waves, 182
26 moment equations, 91, 151
  linearized, 159
  one-dimensional, 156

accommodation coefficient, 23, 197
anisotropic Gaussian, 47
  integrals of, 238
augmented Burnett equations, 69
  instability of, 166
  Knudsen layers, 212
  linearized, 159
  one-dimensional, 155
  shock waves, 183
average particle distance, 7

BGK model, 46
BGK-Burnett equations, 70
binary collisions, 29
Boltzmann equation, 33
  linear, 62
Boltzmann's constant, 7
boundary conditions, 23
  for moments, 197
  for normal stress, 225
  for phase density, 22
  jump and slip, 201
  Maxwell's, 23, 197

boundary value problems, 12
bulk equations, 208
Burnett coefficients, 61, 67, 148
Burnett equations, 3, 60, 67, 98, 148
  augmented, 69
  for BGK model, 70
  for ES-BGK model, 61
  for power potentials, 67
  instability of, 164, 170
  Knudsen layers, 211
  linearized, 158
  one-dimensional, 154
  shock waves, 182

Chapman-Enskog expansion, 53, 56,
  147
  and order of magnitude approach,
    139
  for ES-BGK model, 57
  for power potentials, 62
  of moment equations, 96
closure problem, 38, 53
collision angle, 29
collision frequency, 1, 43
collision parameter, 30
collision term, 32
  properties of, 45
collisional invariants, 34
collison frequency, 8
conservation laws, 36, 78, 147
  linearized, 158
  microscopic, 34
  one-dimensional, 154

consistently ordered extended thermo-
    dynamics, 123, 142
continuity equation, 36
convective time derivative, 37
Couette flow, 203
  bulk equations, 208

damping, 162, 168
diameter of gas particles, 7
differential cross section, 31
dispersion, 168
dispersion relation, 162
distribution function, *see* phase density
divergence form, 37
DSMC simulation, 48, 181

energy balance, 36, 50, 78
energy density, 17
energy units, 19
entropy, 39
  and disorder, 40
  maximization of, 41, 104
  of fermions and bosons, 50
entropy flux, 39
entropy production, 39
  over shock wave, 177
equation of transfer, 35
equilibrium, 21, 34
  and maximum disorder, 41
  entropy, 39
ES-BGK model, 47
Euler equations, 3, 59, 129, 148
extended thermodynamics, 95, 104
external forces, 28

Fourier law, 10, 59, 132
Fourier's law, 10
free molecular flow, 2

Gaussian integrals, 234
generalized 13 moment equations, 134,
    152
  linearized, 160
  one-dimensional, 157
Gibbs paradox, 40, 51
Grad distribution, 88
  as non-equilibrium manifold, 110
  for 13 moments, 90
  for 26 moments, 91

Grad method, 87, 140
Grad's 13 moment equations, *see* 13
    moment equations
Grad's 26 moment equations, *see* 26
    moment equations
gravity, 28, 49

H-theorem, 38
hard sphere molecules, 31, 60
heat conductivity
  estimate for, 10
  for ES-BGK model, 59
  for power potentials, 66, 148
heat flux, 10, 19
high altitude flight, 12
hybrid methods, 227
hydrodynamic regime, 2

ideal gas, 6
ideal gas law, 7
interaction potential, 6, 31
internal energy, 18
  balance of, 37
irreversibility, 42

Jin-Slemrod equations, 118, 143, 153
  linearized, 160
  one-dimensional, 157
  shock waves, 186
jump and slip boundary conditions, 4,
    201

kinetic energy, 18
kinetic models, 45
kinetic schemes, 227
Knudsen heat transfer, 25
Knudsen layers, 12, 202, 210
Knudsen number, 1, 11
  and Boltzmann equation, 55
  for shock waves, 179
  for waves, 163

Lagrange multipliers, 240
length scale, 1, 11
linear stability, 162
  in space, 166
  in time, 164
local Maxwellian, 35

Mach number, 176

mass balance, 36, 78
mass density, 17
mass of a particle, 9
maximization of entropy, 104
Maxwell molecules, 32, 60
Maxwell-Smoluchowski boundary
        conditions, 201
Maxwellian distribution, 22, 35
    half-space integrals of, 236
    integrals of, 235
Maxwellian iteration, 99
mean free path, 1, 8
mean free time, 8
microscale processes, 12
microscopic flows, 12
minimal number of moments of order 1,
        127
molecular chaos, 32, 43
molecular weight, 9
moment equations, 75, 150
    infinite hierarchy, 76
moments, 19
    in equilibrium, 22
    relevance of, 21
    trace-free, 20
momentum balance, 36, 78
momentum density, 17

Navier-Stokes law, 11, 59, 132
Navier-Stokes-Fourier equations, 3, 59,
        132, 148
    Couette flow, 204
    Knudsen layers, 211
    limit of validity, 1
    linearized, 158
    one-dimensional, 154
    shock waves, 182
non-equilibrium manifolds, 109
number density, 16

order of magnitude approach, 123

particle number, 16
peculiar velocity, 17
phase density, 15
    as probability, 17
    Maxwellian, 22
    positivity in shocks, 192
phase space, 15

phase velocity, 162
plane wave solutions, 161
power potentials, 31
Prandtl number, 11, 60, 66, 70, 132
pressure, 18
pressure tensor, 19
production term, 35
production terms, 80
    for ES-BGK model, 82
    for Maxwell molecules, 83
    for non-Maxellian molecules, 102
    linear, 84

R13 equations, 4, 113, 138, 151
    Couette flow, 207
    Knudsen layers, 214
    linearized, 159
    one-dimensional, 156
    shock waves, 183
Rankine-Hugoniot relations, 176
rarefaction effects, 12
rarefied gases, 2
Rational Extended Thermodynamics,
        see extended thermodynamics
recurrence of microscopic state, 43
regularization, 109
regularized 13 moment equations, see
        R13 equations
Reinecke-Kremer-Grad method, 101,
        128, 133

SBKn model, 217
second law, 39
    for higher order equations, 71
    for Navier-Stokes-Fourier, 70
shock asymmetry, 178, 192
shock structures, 175
    subshocks, 188
    temperature overshoot, 189
shock thickness, 178, 190
shock waves, 12
slip flow regime, 2
slip velocity, 22, 199
smallness parameter, 56
specific heat, 19
specular reflection, 23
speed of sound, 25, 163
spherical harmonics, 233
stosszahlansatz, 32

stress, 11, 20
super-Burnett equations, 3, 68, 99, 149
    instability of, 164
    Knudsen layers, 212
    linearized, 159
    one-dimensional, 155
    shock waves, 182
symmetric tensors, 231

temperature, 19
temperature jump, 12, 19, 23, 201
    and thermometers, 202
temperature overshoot, 189
tensor index notation, 229
thermal energy, 18
thermal velocity, 25
thermalizing wall, 23
thermodynamics of irreversible
        processes, 71

thermophoresis
    in Chapman-Enskog gas, 74
    in Knudsen gas, 26
trace-free tensors, 231
transition regime, 2

ultrasound, 12, 169

velocity
    macroscopic, 17
    microscopic, 15
    peculiar, 17
    post/pre-collision, 29
viscosity
    estimate for, 11
    for ES-BGK model, 59
    for power potentials, 66, 148
viscosity exponent, 60